Language and New Media

Linguistic, Cultural, and Technological Evolutions

NEW MEDIA: POLICY AND SOCIAL RESEARCH ISSUES

Ronald E. Rice, *Series Editor*

Double Click: Romance and Commitment Among Couples Online
Andrea Baker

Action Reseach and New Media: Concepts, Methods and Cases
Greg Hearn, Jo Tacchi, Marcus Foth, and June Lennie

The Role of Culture and Political Institutions in Media Policy: The Case of TV Privatization in Greece
Theomary Karamanis

Media Policy for the 21st Century in the United States and Western Europe
Yaron Katz

Telecommunications and Development in China
Paul S.N. Lee (ed.)

Impact and Issues in New Media: Toward Intelligent Societies
Paul S.N. Lee, Louis Leung and Clement Y.K. So (eds.)

Global Trends in Communication Research and Education
Kenneth W.Y. Leung, James Kenny, and Paul S.N. Lee (eds.)

Communication Technology and Society: Audience Adoption and Uses
Carolyn A. Lin and David J. Atkin (eds.)

Foundations of Communication Policy: Principles and Process in the Regulation of Electronic Media
Philip M. Napoli

Media Ownership: Research and Regulation
Ronald E. Rice (ed.)

Virtual Politicking: Playing Politics in Electronically Linked Organizations
Celia T. Romm

Language and New Media: Linguistic, Cultural, and Technological Evolutions
Charley Rowe and Eva L. Wyss (eds.)

Automating Interaction: Economic Reason and Social Capital in Addressable Networks
Myles A. Ruggles

The Playful Audience: From Talk Show Viewers to Internet Users
Tony Wilson

Language and New Media

Linguistic, Cultural, and Technological Evolutions

Edited by

Charley Rowe and Eva L. Wyss

With a foreword by

Naomi S. Baron

HAMPTON PRESS, INC.
CRESSKILL, NJ 07626

Printed in the United States of America

Library of Congress Cataloging-in-Publication Data

Language and new media : linguistic, cultural, and technological evolutions / edited by Charley Rowe and Eva L. Wyss ; with a foreword by Naomi S. Baron.
p. cm. — (New media : policy and social research issues)
Includes bibliographical references and index.
ISBN 978-1-57273-929-1 (hardbound) — ISBN 978-1-57273-930-7 (paperbound)
1. Mass media and language. 2. Language and the Internet. 3. Communication and technology. 4. Online etiquette. I. Rowe, Charley. II. Wyss, Eva Lia.
P96.L34L35 2009
306.44—dc22 2009019087

Cover photo by Tanya Hart

Hampton Press, Inc.
23 Broadway
Cresskill, NJ 07626

Contents

Acknowledgments

The editors would like to gratefully acknowledge the hard work and patience that the contributors dedicated to the writing of their chapters. We also acknowledge Susan Herring for her conception and organization of the Language and Media panel at the XVIth International Conference on Historical Linguistics (ICHL) in Copenhagen in 2003, which provided the initial inspiration for the editors to mould the narrower focus of this volume. We are grateful to Naomi Baron for writing the foreword; her thorough research and well-conceived philosophies on new media, always presented in her straightforward writing style, are an inspiration in the field. We appreciate Tanya Hart generously allowing us to use her photography for the cover. A warm thanks also goes to copyeditor Sue Morreale of Partners Composition for her diligence, competence, and good humor throughout the final stages of publication. We thank Barbara Bernstein for her effort in the finalization of the volume. We are particularly indebted to series editor Ron Rice; his prompt and efficient responses to our various queries were invaluable to us, and the special insight, experience, and wide-ranging expertise on all things media that he lent to the volume have benefited the work tremendously.

Finally we wish to dedicate this book to Jonathan and Karam for their moral support, not only during the work of this volume, but throughout the course of our academic careers on a daily basis. They are our faithful sounding boards, without whom many of our ideas may well not have taken root.

Foreword

Twenty-two years ago, I organized a symposium called "Pandora's Box: Computers in Everyday Life." My goal was to create a sense of wonder among the undergraduates at a small university in central Texas about what computers of the day could do. There were speakers on artificial intelligence and on computer graphics, computer-generated music, and computers in education. We had lawn-moving robots and computers that measured Olympians' strength. Even Isaac Asimov was beamed down from New York via satellite.

That world of computing was still ruled by large universities and national governments. In the United States, the Department of Defense—along with a handful of research institutions—constituted the center of computing activity, with mid-sized businesses gradually "computerizing" their operations. Thanks to Apple Computer—and to Seymour Papert, the inventor of the child-friendly computer language known as Logo—personal computers slowly gained recognition as educational technologies. Among adults, word processing was soon to become the killer application for the machine once designed to do mathematical calculations.

What a difference twenty-two years have made. Thanks to the internet, and then the World Wide Web, the growth of broadband and then wireless computing, mobile phones (and a host of software programs designed for personal computers), technology that was originally intended to crunch numbers and then blossomed into a glorified typewriter has now emerged as a ubiquitous tool for interpersonal communication (Baron 2008).

Interest in using computers for communication traces back to the origins of e-mail and of computer conferencing in the early 1970s. Over the next two decades, a steady stream of new communication platforms emerged: MUDs (aka multi-user dungeons) in 1979, newsgroups in 1980, and listservs in 1986. Instant messaging made its debut in the 1980s, but only achieved national prominence (at least in the US) in the early 1990s.

A handful of researchers were there from the beginning—particularly Starr Roxanne Hiltz and Murray Turoff, whose pioneering work on computer conferencing helped redefine the possibilities for communicating at a distance (Hiltz and Turoff 1978). In the 1980s, interest in language online grew, as academics across a range of disciplines—communication, sociology, linguistics, information systems, psychology—started probing the content and context of e-mail, Internet Relay Chat, newsgroups, listservs, and the like. I took an early crack at some of the problems myself (Baron 1984), though 20/20 hindsight demonstrates how little we knew at the time about what would prove to be the relevant variables.

An important landmark was Susan Herring's *Computer-Mediated Communication: Linguistic, Social, and Cross-Cultural Perspectives*, which appeared in 1996. A year earlier, the *Journal of Computer-Mediated Communication* made its debut. By the turn of the millennium, research on the confluence of "new media" and "language" was cropping up in conferences and publications, such as the meetings of the Association of Internet Researchers (first held in 2000) and books with titles like *New Media Language* (Aitchison and Lewis 2003).

These days, speaking in the same breath about language and new media is becoming increasingly commonplace. Happily, the present volume will find a congenial home among scholars and students of communication, linguistics, the internet, and new media more generally. However, Charley Rowe and Eva L. Wyss's *Language and New Media: Linguistic, Cultural, and Technological Evolutions* adds a valuable perspective to the issues at hand by emphasizing questions of language change. It is here in the diachronic arena that a number of particularly interesting real-world challenges lie, and that make this volume especially welcome.

We are all familiar with the truism that languages change over time. Yet making sense of change in process is a challenging undertaking—much like writing a history of the present. It is difficult to know which phenomena are passing blips on the radar screen and which portend significant shifts in the status quo. In the case of language, there is an extra variable at work, namely that of passing judgment on the kind of language (be it slang appearing on the street or acronyms in instant messaging) that people are using.

The public worries, Is language going to hell in a hand basket? Linguists (at least since the early twentieth-century) have steadfastly embraced neutrality on this issue, arguing that no language (or form of language) is superior to any other. Saying "He and I went home" rather than "Him and me went home" is taken to be a matter of social status, not of linguistic efficacy. By the same token, contemporary linguists tend to argue that new media are enriching the written language (given all those IMs, blogs, and entries on social networking sites) and that using an acronym such as "BTW" ('by the way') in a written composition is no more inherently heinous than writing "RSVP." Perhaps it is just a matter of time and attendant shifts in usage conventions before knee-jerk condemnation passes.

Yet the matter is actually more complex. Consider a recent discussion of "thumb novels" in Japan (that is, novels written as concatenations of text messages) that raises the very real question of whether the medium is reshaping the message—and, in this case, the very notion of literature. A woman named Rin is the twenty-one-year-old author of a "cellphone novel" that was the number five best-selling novel in Japan for 2007. Rin argues that "[Members of my generation] don't read works by professional writers because their sentences are too difficult to understand" and "their expressions are intentionally wordy" (Onishi 2008). Sentences "too difficult"? "Wordy" expressions? Maybe there is a new Ernest Hemingway in the making—or perhaps the whole notion of literature is up for redefinition, at least by young people in Japan.

The more we know about the interplay between language and media—in contemporary times, in historical perspective—the better prepared we will be to evaluate whether the medium is really reshaping the message, and, if so, whether the results are for good or for ill. Charley Rowe and Eva L. Wyss are to be commended for assembling an eloquent and thought-provoking collection of essays that help us approach these questions both knowledgeably and objectively.

REFERENCES

Aitchison, J., & Lewis, D. M. (Eds.). (2003). *New Media Language.* London: Routledge.

Baron, N. S. (1984), "Computer Mediated Communication as a Force in Language Change." *Visible Language,* 18: 118–141.

Baron, N. S. (2008). *Always On: Language in an Online and Mobile World.* New York: Oxford.

Herring, S. (Ed.). (1996). *Computer-Mediated Communication: Linguistic, Social, and Cross-Cultural Perspectives.* Amsterdam: John Benjamins.

Hiltz, S. R., & Turoff, M. (1978). *The Network Nation: Human Communication via Computer.* Reading, MA: Addison-Wesley.

Onishi, N. (2008), "Thumbs Race as Japan's Best Sellers Go Cellular." *New York Times,* January 20, A1.

Naomi S. Baron
Professor of Linguistics
American University
Washington, DC USA
January 2009

Introduction

The recent expansion in Internet communication has generated a new paradigm in core linguistic fields, along with a new interest in the effects of technological mediation on language. It can now be said with confidence that the particular characteristics, features, and uses of the medium strongly affect the nature and, usually, the resultant form of the communication. This is nowhere more evident than at points in history when newer media emerge. Of particular interest here is the notion of linguistic change as it occurs synchronically or diachronically in media communication—that is, change at the moment of its occurrence (change in progress) versus changes observable over a longer period of time (e.g. 30 years). Furthermore, the historical context in which emergent media act as catalysts for unique communication acts cannot be understated. It is precisely these facets that this volume addresses.

The editors conceived the idea for this volume following their participation in a panel called Language and Media, organized by Susan Herring for the XVIth International Conference on Historical Linguistics (ICHL) in Copenhagen in 2003. While the concentration of the Language and Media panel was broadly conceived and historically oriented (spanning the range from the earliest media to the most modern), the current volume's focus lies in newer media and the developments leading to their current state (including projections for the future), as well as the internal changes that occur in and as a result of the nature of the new media context. This volume uses much of the Language and Media panel's work as a springboard, and incorporates additional current media research to complete the picture for the newest media. To this end, we have brought together researchers analyzing a wide variety of communication media, with the goal of furthering knowledge about the relationship between technological mediation and language structure and use.

Clearly the current volume is not exhaustive in its coverage, and the editors would like to have included, for example, current research on weblog progeny; however, new media are rapidly emerging, almost faster than they can be adequately described, much less linguistically analyzed. Even as this volume goes to press, still newer web media have emerged and claimed a prominent place in society, notably: social networking sites such as Facebook and Myspace; photo-sharing sites such as Flickr; and difficult to classify microblog interfaces like Twitter. Research on these media is underway, and we must regrettably put a coda on the work of the current volume and leave a compendium on these exciting subfields of new media to another time.

Although this *Language and New Media* volume addresses the topic of change in two quite different dimensions (synchronic and diachronic), the editors have resisted the temptation to order the volume chronologically in order of media emergence. A much more useful approach it is here, it seems, to organize the volume conceptually, toward discussions of the relationship between older and newer media, changes in traditional mass media, and the evolution of new media norms.

The main sections follow that strategy, with chapters treating each of three primary conceptual issues. Part I, Tensions between Newer and Older Media, tackles topics such as visual communication, multimodality, linguistic change, knowledge construction, and the relationship between new media and literacy. Part II, Changes in Traditional Mass Media Language and Culture, discusses the expansive development of radio, as well as the changing nature of television language and presentation. Part III, Evolving New Media Genre Norms, discusses the concepts of stylistic norms and genre text-type challenges in new media. The following describes each chapter within these three main sections.

Part I: Tensions Between Newer and Older Media

Chapter 1 (*Inter-Activity: How New Media Can Help Us to Understand Old Media*) by Rodney H. Jones embraces the current theoretical literature on multimodality, and in so doing, compares new media multitasking with older media multitasking. This chapter proposes a model for the analysis of new media communication which facilitates the tracing of moves and strategies across multiple simultaneous interactions. Anabel Quan-Haase (Chap. 2, *Text-Based Conversations Over Instant Messaging: Linguistic Changes and Young People's Sociability*), taking on the social debate on new literacies, reviews the impact of internet messaging (IM) on standard written language. She concludes that within IM, distinct communication practices and linguistic expressions have developed as a partial result of technological factors, and that these facets reflect creative "reinvention" of language use. Quan-Haase outlines the implications of these changes for discussion of the concept of digital literacy.

In a similar vein, Alex Bergs (Chap. 3, *Just the Same Old Story? The Linguistics of Text Messaging and Its Cultural Repercussions*) investigates the sociolinguistic effects of short message service (SMS, or mobile phone texting). Bergs finds strong parallels not only between this medium and users' offline behavior, but also between older media and emerging newer media. In this regard, he re-exposes the old "complaint tradition" that laments the effects of emergent media on traditional literacy at various points in the evolution of technology.

In Chapter 4 (*E-mail Play and Accelerated Change*), Charley Rowe examines the private language interactions of two sisters and other "intimate" pairs in e-mail correspondence over time. She finds that in all cases, intimacy works in tandem with technology to foster the emergence and development of certain forms. In some cases, she concludes, dramatic changes occur, largely as a result of specific technical and sociopsychological features inherent in the e-mail medium.

Also examining intimate communication, Eva L. Wyss (Chap. 5, *From the Bridal Letter to Written Flirting: Love Letter Text Types From the 19th Century to the Era of Internet and Mobile Phone*) compares traditional love letters and courtship with online dating and flirting rituals. Basing her research on a rich and varied corpus that she collected entitled the Zürich Loveletter Archive (ZLA), Wyss finds dramatic changes in pragmatic, textual, and stylistic differences in amorous expression among the various media.

In Chapter 6 *(From Postcard to E-Card: Tradition and Change in the Smallest Spaces)* Hajo Diekmannshenke focuses on the development of e-cards, beginning with their history as postcard communication since the 19th century. The author finds a growing tendency toward more "creative" compositions. More recently, Diekmannshenke finds, it has been primarily the e-card that—although having the traditional postcard as a "role model"—has accelerated change in the "communication form" of the postcard through innovative features in both layout and composition.

In Chapter 7 (*History Is Now: Media History and Textual History of Wikis*), Christian Kohl, Thomas Metten, and Wolf-Andreas Liebert analyze the Wikipedia virtual encyclopedia as a culture of knowledge that is fundamentally based on linguistic interaction, formed by the process of text genesis, which is in stark contrast to the composition of the traditional encyclopedia. The authors argue that with Wikis, writing becomes a new practice of knowledge construction that shows to what extent linguistically constructed knowledge develops during an unfinished process of dialogic writing in the new media.

Part II: Changes in Traditional Mass Media Language and Culture

Juerg Häusermann (Chap. 8, *Radio Broadcasting: From Secret to Public to Private*) discusses in rich detail the history of radio and the linguistic changes as communication over this medium gradually evolved. In particular, his chapter also raises questions about the future of radio communication as we know it. In this regard, he considers the effect of the more "democratizing" radio-related media, such as podcasting, on journalism and public discourse.

Stefania Spina (Chap. 9, *Is Television Language A Spoken Variety? A Corpus-Based Study of Italian Television*), challenges the notion of television as a purely spoken modality. In fact, Spina argues, a close analysis of grammatical constructions in a corpus of various program types shows that television language occupies an intermediate space between canonical spoken and written discourses. This argument has been made convincingly for e-mail and texting in particular, and for computer-mediated communication in general, but rarely for television subgenres.

Chapter 10 (*Disclosing and Announcing, Interpreting, and Entertaining: A Comparative Study of the History of TV News Presentation*) by Martin Luginbühl examines spoken text as well as other semiotic resources across two culturally distinct television news shows over time. Luginbuehl concludes that presentational norms in TV news are ultimately determined by journalistic culture as well as by the nature of the medium itself.

Part III: Evolving New Media Genre Norms

Sharon Millar (Chap. 11, *Technologies and Techniques: The (n)etiquette of Telephone and E-mail Communication*) provides an interesting look at the evolving codification of etiquette norms in telephone and e-mail communication. Millar demonstrates, in particular, the extent to which new media draw on etiquette norms prevalent and established in older media, in order to deal with informality issues that continuously arise with emergent media.

In a philosophical perspective on new media, Laurie McNeill (Chap. 12, *Diary 2.0? A Genre Moves From Page to Screen*) ponders whether there have been significant pragmatic and technological changes resulting from the migration from paper diary to diary blog. She concludes that Internet diarists appear to be transferring their writing practices more or less directly from their traditional written modality, the primary change, according to McNeill, being the interface itself.

In Chapter 13 (*Wikis, Wikipedia, and Collaborative Technology: New Challenges and Emergent Language Styles*), Antonella Elia examines the redefinition of authorship. In her chapter, Elia demonstrates the discussion-based nature of collaborative authoring of Wikipedia articles, and places it in its own discourse community with its own set of norms, conventions, and unique Wiki-speak jargon. Although the overall topic of Wikipedia and collaborative technology is also treated in Chapter 7, this chapter's focus lies on the Wikipedia community's stylistic norms which feed the process of text creation.

Chapter 14 (*Hypertext and the Evolution of Web Site Genres*) by Eva-Maria Jakobs begins with a review of hypertext and Web site history, and proceeds with an analysis of Web site patterns versus Web site genres. Patterns, she determines, relate a problem to be solved, a social purpose, a situational context, and an action type and its participants. These are amalgamated in part using patterns of structuring, formulation, visualization, and other elements, all of which are very often, Jakobs concludes, culturally based.

Table 1 summarizes the main concepts in each of the chapters.

Table 1. Main Concepts in Each Chapter

Part I: Tensions Between Newer and Older Media

Chapter 1 (Inter-Activity: How New Media Can Help Us to Understand Old Media) by Rodney H. Jones

- Communication practices
- E-mail
- Genre
- Hybridity
- Instant messaging (IM)
- Interactivity
- Literacy
- Mobile phone/cell phone
- Multimodality
- Private communication
- Style
    - Individual
    - Institutional/media
    - Interface
- Spoken language
- Technology
- Telephone
- Text messaging (SMS)
- Text type
- Theory
- Visual communication
- Web site
- Written language

Chapter 2 (Text-Based Conversations Over Instant Messaging: Linguistic Changes and Young People's Sociability) by Anabel Quan-Haase

- Communication practices
- Complaint tradition
- Face-to-face communication
- Hybridity
- Informality
- Instant messaging
- Literacy
- Print/digital culture
- Norms and conventions
- Private communication
- Social change
- Style
    - Change
    - Individual
    - Institutional/media
    - Interface
- Technology
- Text organization
- Text type
- Written language

Chapter 3 (Just the Same Old Story? The Linguistics of Text Messaging and Its Cultural Repercussions) by Alexander Bergs

Communication practices
Complaint tradition
Culture
Genre
Grammatical structure
Hybridity
Informality
Literacy
Mobile phone/cell phone
Norms and conventions
Postcard
Print/digital culture
Private communication
Style
- Individual
- Institutional/media
- Interface

Technology
Telegram and telegraph
Telephone
Text messaging (SMS)
Text organization
Text type
Visual communication
Written language

Chapter 4 (E-mail Play and Accelerated Change) by Charley Rowe

Communication practices
E-mail
Face-to-face communication
Hybridity
Informality
Intimate communication
Language play
Norms and conventions
Private communication
Style
- Change
- Individual
- Interface

Spoken language
Technology
Telegram and telegraph
Telephone
Voice-mail and answering machine
Written language

Chapter 5 (From the Bridal Letter to Written Flirting: Love Letter Text Types From the 19th Century to the Era of Internet and Mobile Phone) by Eva L. Wyss

Communication practices
E-mail
Genre
Hybridity
Informality
Intimate communication
Literacy
Norms and conventions
Private communication
Style
- Change
- Individual
- Interface

Technology
Text organization
Text type
Visual communication
Written language

Chapter 6 (From Postcard to E-Card: Tradition and Change in the Smallest Spaces) by Hajo Diekmannshenke

Communication practices
E-card
E-mail
Etiquette
Genre
Grammatical structure
Hybridity
Informality
Literacy
Multimodality
Norms and conventions
Postcard
Private communication
Style
- Change
- Institutional/media
- Interface
Technology
Telegraph and telegram
Text organization
Text type
Visual communication
Written language

Chapter 7 (History Now: Media History and Textual History of Wikis) by Christian Kohl, Thomas Metten, and Wolf-Andreas Liebert

Collaborative technology
Communication practices
Discussion board/forum
Encyclopedia
Genre
Hypertext
Literacy
Norms and conventions
Public media
Style
- Institutional/media
Technology
Text organization
Text theory
Text type
Theory
Web site
Wiki, Wikipedia
Written language

Part II: Changes in Traditional Mass Media Language and Culture

---

Chapter 8 (Radio Broadcasting: From Secret to Public to Private) by Jürg Häusermann

Communication practices
Democratization
Digital audio broadcasting
Informality
Internet radio
Narrative
News
Norms and conventions
Participation
Podcasting
Public media
Radio
Regulation
Rural radio
Style
- Change
- Institutional/media
- Interface
Spoken language
Technology
Telegram and telegraph
Text organization
Text type
Traditional mass media

Chapter 9 (Is Television Language a Spoken Variety? A Corpus Based Study of Italian Television) by Stefania Spina

Genre
Grammatical structure
Hybridity
Multimodality
Narrative
News
News anchor
Public media
Spoken language
Television
Television news
Text organization
Text type
Traditional mass media
Visual communication
Written language

Chapter 10 (Disclosing and Announcing, Interpreting and Entertaining: A Comparative Study of the History of TV News Presentation) by Martin Luginbuehl

Communication practices
Culture
Genre
Informality
Journalism
Multimodality
News
News anchor
Norms and conventions
Public media
Regulation
Style
- Change
- Institutional/media

Spoken language
Technology
Television
Television news
Text organization
Text type
Traditional mass media
Visual communication
Written language

Part III: Evolving New Media Genre Norms

Chapter 11 (Technologies and Techniques: The (n)etiquette of Telephone and E-mail Communication) by Sharon Millar

Business communication
Communication practices
E-mail
Etiquette
Hybridity
Informality
Literacy
Norms and conventions
Postcard
Private communication
Skype
Style
- Change
- Individual
- Institutional/media

Spoken language
Technology
Telephone
Voice-mail
Written language

Chapter 12 (Diary 2.0? A Genre Moves From Page to Screen) by Laurie McNeill

Audience design
Blog/weblog
Communication practices
Diary
Genre
Hybridity
Informality
Multimodality
Norms and conventions
Private communication
Public media
Style
- Change
- Individual
- Interface

Technology
Text organization
Text type
Theory
Visual communication
Web site
Written language

Chapter 13 (Wikis, Wikipedia, and Collaborative Technology: New Challenges and Emergent Language Styles) by Antonella Elia

Collaborative technology
Communication practices
Democratization
Discussion board/forum
Discussion pages
Encyclopedia
Etiquette
Genre
Grammatical structure
Hybridity
Hypertext
Informality
Lexis
Neologisms
Norms and conventions
Online community discourse
Style
- Change
- Individual

Technology
Text organization
Web site
Wiki, Wikipedia
wikispeak
Written language

Chapter 14 (Hypertext and the Evolution of Web Site Genres) by Eva-Maria Jakobs

Communication practices
Culture
Genre
Hybridity
Hypertext
Interactivity
Multimodality
Norms and conventions
Public media
Situational context
Style
- Interface

Technology
Text organization
Text type
Theory
Visual communication
Web site
Wiki, Wikipedia
Written language

I

Tensions Between Newer and Older Media

Chapter 1

"Inter-Activity"

How New Media Can Help Us Understand Old Media

RODNEY H. JONES

City University of Hong Kong

Abstract

One challenge to understanding the potential effects of new media on linguistic interaction lies in the way it facilitates what has been referred to as *multitasking*: the simultaneous management of multiple activities. In computer-mediated communication, just as important as the interaction between or among particular interlocutors is the interaction between or among different activities or interactions—what is called *inter-activity*. This chapter proposes a model for the analysis of new media communication that facilitates the tracing of moves and strategies across multiple simultaneous interactions. The phenomenon of *inter-activity*, however, is not restricted to new media communication but is also an important but neglected issue in "old media" communication. To illustrate this, the same model is also discussed in reference to face-to-face conversation, telephone interaction, and reading in a public place. One of the most fortunate consequences of challenges to the analysis of discourse brought on by new technologies, it is argued, is that they help to reveal gaps in the way we approach more traditional kinds of interaction.

Ming[1] is a 19-year-old student studying in a university in Hong Kong. Every evening after dinner she switches on her IBM laptop, which is connected to the high-speed wireless LAN in her residence hall, and works on her school assignments. While she is working she intermittently answers e-mail and instant messages (IMs) from her classmates as well as her friends and family members.

> Sometimes she chats with three or four people at a time using a chat and IM software popular with her friends. She also has her mobile phone next to her computer just in case somebody wants to call her. Occasionally, it lights up with an SMS (short message service) from a classmate. She is an avid fan of all kinds of music from indie to hip-hop, and she almost always listens to Mp3 files through her earphones when she's working on the computer and swaps songs with other fans using a "peer-to-peer" file-sharing application. Ming's roommate, Mei Lin, sits on the other side of their small room, typing into her own laptop and watching her small portable Sony television. Mei Lin likes to spend time browsing a bulletin board system (BBS) from China where young people express their ideas about culture and politics, and she likes to talk with Ming about some of the news and opinions she finds on it, and sometimes these comments develop into heated exchanges of opinion between the two roommates. These debates, however, rarely significantly interrupt the ongoing tempo of messaging, chatting, e-mailing, surfing the Web, downloading music files, watching TV, and working on homework assignments.
>
> Ah Lam is a 26-year-old homosexual in Hong Kong. Like many single Hong Kong men under the age of 30, he lives with his parents. Three or four evenings a week after his parents have gone to sleep, he logs on to the chat room at gayhk.com and scrolls down the list of potential chat partners looking for user names and profiles that look interesting to him. Ah Lam usually likes to start four or five chats at once to maximize his chances of finding someone who appeals to him. While he juggles multiple chats, he sometimes surfs web pages containing erotic pictures or reads personal messages that people leave on a local gay BBS. If any of his chat partners seems interesting enough, he'll e-mail or IM them his own picture. If their communication is successful, he might have a phone conversation with them or contact them via Web cam. Ah Lam used to go to bars and saunas to meet partners, but now he prefers using the Internet. He has a better chance of meeting and developing a relationship with someone online since he can chat with so many different men at once. Sometimes Ah Lam invites his friend to his parents' flat and they cruise the chat room together, sometimes playing at inventing false identities. They have even engaged online chat partners in "cam sex" together, one of them showing his body, and the other talking to the person on the telephone—and the people never realize that they are having cyber sex with more than one person simultaneously!

For the past 15 years, computer-mediated communication (CMC) has received considerable attention from linguists and communication scholars interested in how this "new way of communicating" differs from the old ways. Among the aspects they have focused on have been lexical and grammatical features of chat and e-mail (see, e.g., Collot & Belmore, 1996; Davis & Brewer, 1997), interactional features of both synchronous and asynchronous CMC such as turn taking and interactional coherence (see, e.g., Garcia & Jacobs, 1999; Herring, 1999; Rintel &

Pittam, 1997), ways participants present themselves and form impressions of others (see, e.g., Douglas & McGarty, 2001), sociolinguistic issues like the formation of online "speech communities" (see, e.g., Paolillo, 1999), and online intercultural communication (see, e.g., Ma, 1996).

Perhaps not surprisingly, most studies of text-based CMC have chiefly concerned themselves with the analysis of texts of conversations downloaded from users' computers and isolated from the situations in which they occurred. The two scenarios cited in the opening of this chapter are meant to illustrate that this practice of isolating interactions from their physical contexts may result in the analyst missing some of the most important aspects of this kind of communication. What is clear from these examples is that the context of CMC can be extremely rich, and extremely physical. It often involves more than just texts sent from one user to another, but rather complex interactions among different media and different modes of communication from text, to graphic, to auditory form. Similarly, CMC often involves a number of people arranged in various physical spaces in complex patterns of participation from interlocutors to onlookers. Finally, and most importantly, CMC often involves multiple, simultaneous interactions interwoven into a dense fabric of overlapping activities. The face-to-face (FTF) conversations Ming has with her roommate Mei Lin arguably have an effect, sometimes subtle and sometimes significant, on the conduct of the computer-based conversations she is having at the same time, and *visa versa*. Similarly, how Ah Lam responds to any one of his chat partners in the gay chat room inevitably depends on how the conversations he is having with other chat partners (or with his best friend who might be sitting next to him) have developed. Both of these examples aptly point to the truth of Hymes' (1974:54) warning that "the common didactic model of speaker and hearer" which is still so central to much linguistic analysis "specifies sometimes too many, sometimes too few, and sometimes the wrong participants." This chapter argues that understanding the interaction that occurs between or among participants using "new media" (and "old media" as well) requires understanding the interaction between or among the different activities that users are involved in, activities that are carried out with multiple interlocutors interacting through a wide variety of media, interlocutors who also include various kinds of onlookers, "lurkers," and future users who, in the case of archived interactions like blog postings and forums) might read and respond to interactions that occurred well into the past.

Inter-activity: Engaging in Multiple Mediated Activities

The practice of engaging in multiple activities using multiple modes has come to be referred to as *multitasking*. This term, however, can be misleading when applied to CMC, creating the impression that the multiple "tasks" in question are clearly "separate" tasks and that they are being engaged in "at the same time," two assumptions that that I call into question later. It also assumes an underlying instrumentality associated with what we do when we use computers—activities are "tasks" rather than processes, pastimes, or diversions. It is this assumption that lies behind recent studies and media stories warning that multitasking is "inefficient" and leads to "wasted time" (see, e.g., "Wasted time . . ." 2001). Rather than multitasking, I use the term *inter-activity*, which I believe better captures the ways multiple activities can flow together and affect one another. The difference between inter-activity and interactivity is that the latter conventionally focuses on the interaction between or among actors, two humans, or humans and things in their environments like computers, whereas the former focuses on the interaction among the various different activities in which these social actors are engaged.

It is its capacity to facilitate not just new patterns of interaction, but also new patterns of *inter-activity*, I argue, that really makes new media communication different from old ways of communicating. At the same time, however, I also argue that the concept of *inter-activity* is equally important to older ways of interacting using face-to-face conversation, print media, and telephones. How communicative behavior (whether it involves sending an e-mail, talking with a friend on a cell phone, reading a newspaper on a train, or attending a university seminar) is organized in terms of continuously interacting multiple channels has thus far not been sufficiently dealt with in linguistic studies of interaction. As Goffman (1981:131) pointed out, "any cross-sectional perspective, any instantaneous slice focusing on talking . . . necessarily misses important features."

The theoretical framework from which I approach this problem brings together a number of traditions in sociology, linguistics, and anthropology, including Goffman's (1959, 1963, 1972, 1981) microsociological approach to behavior in public places, as well as his development of "frame analysis" (Goffman, 1974), early work done on proxemics, gesture, and "context analysis" in interaction (Kendon, 1978, 1990; Scheflen & Ashcraft, 1976), as well as more recent work in what has come to be known as mediated discourse analysis (Norris & Jones, 2005; Jones 2001b; Scollon, 2001a, 2001b).

The data on which I draw for this discussion come from a number of recent studies focusing on the use of computers by young people in Hong Kong[2] (Jones, 2001, 2003, 2005a, 2005b), and on the ways in which CMC is changing the social and sexual behavior of gay men in Hong Kong[3] (Jones, 2005d, 2005e, 2006) They include in-depth interviews, participant journals, and "screen movies" of participants using their computers in real time. The advantage of "screen movies" over chat logs is that they can better capture the complexity of the interaction, in particular how participants manage multiple interactions.

What made collection of the data possible with the informed consent of all parties involved was the participatory model the studies followed: Primary informants were employed on the projects as "participant-researchers," responsible for helping to set the research agenda and for explaining the research and securing consent from others in their communities, particularly their chat partners.

What u do now ar?

Consider one of the conversations recorded on Ming's computer:

1 **A: hihi :)**
2 B: dim a how are u?
3 **A: good and u?**
4 B: ai . . . I have some problem . . . so not good . . . . .
5 **A: what's up?**
6 B: my stomach have some problem . . . . . .
7 **A: Poor girl**
8 B: ai . . . more my uncle near to die . . . . ai . . . . . so unhappy
9 **A: ur uncle?**
10 B: yup . . . . ai . . . .
11 **A: illness???**
12 B: cancer . . . . .
13 **A: *sigh* r u v close to him?**
14 B: He quite take care for me . . . . .
15 **A: ic I'm sorry to hear that**
16 B: i am so sorry for telling u such thing . . . . .
17 **A: not at all, u know life is like that, painful and frustrating**
18 B: Ai . . . forget it la . . . I dunno why I am so unluck . . . . .
19 **A: dun worry, hard time will come**
20 B: ai . . . dim a . . . what u do now ar
21 **A: downloading Mp3**
22 B: do you have any song recommend to me?

A cursory analysis of this small sample of CMC might yield a number of observations that would mark it as an IM session. One might point out, for example, the short turns, the lack of overlap (impossible with this application), the nonstandard spelling, and the use of inventive punctuation and emoticons to substitute for the contextualization cues normally present in FTF conversation (see Quan-Haase, this volume). One might also point out the ways this conversation is characteristic of a particular community of users (i.e., young Chinese in Hong Kong), such as the use of a second language between interlocutors who share a common mother tongue (since English can be typed more rapidly than Chinese) and the adoption of romanized versions of Cantonese expressions (*dim a* = how are you) and particles (*ai, ar*) to lend expressiveness and to modulate the social distancing usually associated with the use of English.

In many ways, however, this conversation does not seem very different from a conversation that might occur between physically co-present participants (or at least participants talking on the telephone) in terms of its opening, topic development, adjacency pairs, politeness strategies, and use of ritual exchanges. Where the conversation starts to diverge from expectations we might have about face-to-face encounters is when, in turn 20, B asks, "what u do now ar?" ("what are you doing now?). This question is perhaps one of the most common in my data of computer chat and IM sessions among Chinese young people. What is so significant about it is B's untroubled assumption that during this conversation about her uncle's fatal illness and imminent death, A has been doing something other than giving her full attention to commiserating with her friend.

In fact, A has been doing a lot more than that. Immediately after the greeting sequence (turns 1–3), A briefly moves her attention to continuing an exchange she is having in another IM window with C, and then takes the opportunity to check out B's "user details" before B replies (turn 4). In between reading B's message that says she has a problem and replying to it, A types another reply to C, selects files to add to her media software play list and begins playing a song. About 40 seconds later she asks her friend what's wrong, but, while waiting for her answer, which just takes a couple of seconds, she sends an Mp3 file to a fourth (more silent) interactant, another user of the "peer-to-peer" file-sharing software she has open. After her friend replies that her stomach is bothering her, A takes another look at her user details, perhaps searching for clues as to how best to respond. After offering an expression of sympathy, her attention moves to a music web page linked with her file-sharing software. When her friend reveals her more serious problem—the illness of her uncle—A's attention apparently momentarily focuses on the import of what has been said, resulting in a more than 30-second period with no screen action (although we do not know what might have been happening off the screen). During the subsequent discussion of the uncle's cancer and his relationship to B (turns 11–19), however, A is also actively involved in searching for songs by Tori Amos and reading message board postings by other fans of this artist. She also receives a message from C reporting on the progress of their joint task. These other activities, however, do not seem to interfere with her ability to comfort B. After turn 20, when B asks A what she is doing, the two heretofore separate activities of chatting with B and downloading music files merge. The topic of the conversation with B shifts to what kinds of music she likes, and the focus of A's file search becomes B's request for A to recommend some songs she might like. Meanwhile, A is able to accept a file transfer from C, thank her for it and resume discussing the details of the homework assignment.

When looked at this way, this interaction seems much less like a FTF conversation. In fact it is almost impossible to imagine a face-to-face encounter in which, while one person is describing the fatal illness of a beloved relative, the other listens to music on her iPod, chats with another friend on the phone, or browses through a CD rack in a music shop. Such behavior would be considered callous or rude. In the context of CMC, however, A's behavior is considered perfectly normal, evidenced by B's question and subsequent reaction (see also Baron, 2008).

An example from Ah Lam's screen recordings is even more dramatic, showing how he interacts with no fewer than five potential sexual partners at one time, skillfully juggling conversations whose topics range from physical appearance to sexual practices to tennis. One of the interactions even involves a Web cam session, but the video presence of this partner does not deter Ah Lam from continuing his text-based chats with other potential partners. What is particularly striking is how adept he is not just at moving easily from one conversation to another without mixing them up, but also at modulating his style and topic to fit the demands of his different audiences, projecting slightly different versions of himself to each interlocutor—professing to one, for example, his desire to find a loyal and sincere mate rather than quick sex, while arranging with another exactly the kind of "quick sex" he professes to deplore.

Taking just one conversational thread from the matrix of communication of either of these examples and attempting to analyze it would be grossly distorting. In order to understand completely what is happening in each separate conversation in these situations, one must attend to all of the different attentional tracks and how what occurs in each affects what occurs in the others. Sometimes these cross-influences are quite dramatic, whereas at other times the user is able to maintain the impression that each track is occurring absolutely independently from any other activity. I stress, however, that this is an *impression* that the media facilitates, not a reality, for it is impossible for the activity in one track not to affect the linguistic and interactional features of others. Some of these effects are more subtle: changes in the timing and conversational mechanics of interactions, the shortening of turns to accommodate the rapid pace of multiple interactions, and the increased use of formulaic exchanges or of hedging or other delaying tactics. Other effects are more obvious, as when topics from one interaction find their way into others, when two communicative activities merge into one, when the attention demanded by one interaction causes others to be somehow compromised or abandoned altogether, or when users accidentally mix tracks by sending to one interlocutor a reply intended for another.

The power of computers to foster interactivity is something that has often been pointed out. Barnes (2003) claimed that electronic communication involves three kinds of interactivity: interpersonal interactivity (the relationship between or among senders and receivers), informational interactivity (the relationship between or among the different bits of information that are available through the screen), and human–computer interaction (the relationship between the user and the machine). *Inter-activity* shows how none of these kinds of interactivity can really be considered separately, that the relationships among information, users, and humans and machines all occur simultaneously and affect one another.

At the heart of the concept of *inter-activity*, then, is an attempt to understand how people in circumstances of multiple interaction balance and prioritize their interactions, and the linguistic and nonlinguistic strategies they use to do this. In examples like those analyzed here, however, it is often difficult to decide which of the various interactional threads are "dominant" and which are "subordinate"; indeed, the status of each thread *vis-á-vis* the others tends to change throughout the occasion. In fact, *inter-activity*, especially in situations where it is facilitated by new media technologies, challenges the whole concept of the "focused encounter" so central to most of our assumptions about language and communication (Jones, 2003).

Participants as Percipients

In order to understand how participants in CMC, or any communicative situation, manage and strategize multiple interactions, it is first helpful to review what constitutes a "communicative situation." In 1951, Jürgen Ruesch and Gregory Bateson attempted to lay down a basic "matrix" through which human communication is possible. In their view, the most basic component of communication is *perception* or, more accurately, "the perception of having been perceived" (15) Communicative situations are created by participants through the "mutual recognition of having entered each other's mutual perception" (23). All signals transmitted from one person to another carry the message of this mutual recognition. In this sense, all communication is about communication—specifically, about maintaining the channel. This view of communication has the dramatic effect of shifting our focus away from the message or the message producer to the *relationship of mutual monitoring* between or among participants. In fact, from this perspective, participants might be more accurately referred to as *percipients* (Kendon, 1978), perception being at the foundation of all participation.

Goffman (1981) provided a similar perspective on the social situation, defining it as "the full physical area in which persons present are in sight and sound of one another" (136). Of course, in CMC, the notion of one person "being in sight or sound" of another needs to be altered to include the perception of electronic signals on a screen. The importance of physical perception, however, remains the same. At the heart of most of Goffman's work on interaction is the examination of how "mutual monitoring" alters behavior. In his classic *The Presentation of Self in Everyday Life*, for example, Goffman speaks of how we divide physical space into regions ("backstage," "frontstage"), characterized by the kinds of information we make available. Physical space allows for what Goffman called "audience segregation," the ability to present different "selves" to different people.

Meyrowitz (1985) built on Goffman's ideas of how social situations are based on the possibilities for directing or obstructing the flow of information between participants. Social situations are best thought of, he said, as "information systems." "The patterns of information flow, whether direct or mediated, help to define the situation and the notions of appropriate style and action . . . when we find ourselves in a given setting we often unconsciously ask, 'Who can see me, who can hear me?' 'Who can I see, who can I hear?' The answers to the questions help us decide how to behave" (39). Meyrowitz claimed that media change the configuration of physical channels and barriers that govern patterns of mutual monitoring, and thereby change what is regarded as socially acceptable behavior (Licoppe & Heurtin, 2001).

Where Meyrowitz, with his grounding in the McLuhanesque tradition, missed the mark in his evaluation of new media effects, however, is in his one-sided emphasis on how electronic media act to *remove* barriers to perception and thereby to "decrease the significance of physical presence" (Meyrowitz, 1985: vii). Meyrowitz claimed that electronic media are causing previously distinct social situations to merge and diminishing our ability to engage in "audience segregation." Taking up this stance, Dresner (2002) argued that Meyrowitz's idea of social situations as information structures helps us to go beyond Goffman's essentially spatial metaphor for communication, allowing us to account for new patterns of information flow that go beyond the laws of physical space. This notion that electronic media render physical space less important can be traced back to McLuhan (1964:3) himself, who claimed that media have been "abolishing both place and time as far as our planet is concerned" causing us to lose our ability to fragment our roles.

Although in some ways this may be true, the two examples provided here suggest that in other ways the opposite is true. What allows these users to so effectively manage *inter-activity* is not that the tools they are using break down geographic barriers, but that they leave them very much intact, giving users greater control over the flow of information and creating *more* opportunities for audience segregation, not less.

Thus, what is often regarded as a *constraint* of text-based CMC—the absence of visual and auditory modes—is, in regard to possibilities for *inter-activity*, actually one of its most important *affordances*, allowing users to distribute their attention over a wider range of encounters and more carefully calibrate the information flow in each one. This is why videoconferencing, no matter how convenient and inexpensive it becomes, will likely not replace text-based CMC, and why more and more people are using their mobile telephones for text-based communication rather than voice-based communication (on the latter, see Bergs, this volume).

Ironically, then, what makes CMC so attractive, especially for activities in which users desire greater control over their information preserve, is not just the sense of "connectedness" it fosters, but also the sense of "disconnectedness." Computer screens have a dual function—they allow us to project information onto them, but they also allow us to hide information behind them. These "involvement screens" (Goffman, 1963) are extremely versatile, allowing us not only to selectively include or exclude participants from situations, but to partially include or exclude participants by selecting which modes of communication we wish to make available. Baron (2008) used the metaphor of "volume control" to express this concept. New communication technologies, she noted, allow users to selectively modulate their availability to other users in the network, altering patterns of interpersonal communication as well as notions of communal propriety.

Thus, what characterizes the mediational means used in newer media in general, and in CMC in particular, is a pattern of affordances and constraints that make people simultaneously more available and less available to one another, and more or less available in different ways. As Cavanaugh (1999) said, "what emerges from the relocation of social interaction in the online environment is a curious tension between intensification of control over the territories of the self and its dissolution" (para. 5).

Paying Attention

One of the trendier theories to be inspired by the new media is the notion of "the attention economy" (Goldhaber, 1997): The idea that, in an age of information overload, what gives value to information is the amount of attention it can attract. The real currency of the information age is not information, but attention. Whether or not this theory holds true on the macro-economic level, it is and has always been an important principle in the micro-economics of situated interaction. All interpersonal communication is fundamentally based on attention: getting attention and "paying attention." In order for successful communication to take place, we must not only pay attention to the content of our interlocutor's words, but we must also monitor the ways in which the interlocutor distributes attention throughout the situation. Thus, not only is attention organized around behavior, but behavior is organized around attention.

According to Kendon (1978:306), in interaction, "participants attend to one another's behavior differentially, and . . . they act on this differential attention in a differentiated way." The way we distribute our attention and interpret how others distribute theirs, he said, is organized through "systems of action and systems of perceptual analysis which are mutually attuned to one another."

Underlying this observation is the notion that when we speak of "attention" we are not really speaking of one thing, but rather of two separate but interrelated systems: one governing actions, and the other governing how actions are interpreted. The first of these is the *cognitive attention system*: the system individuals use to distribute their cognitive attention across the various activities they are involved in. The second is the *social attention system*: the system participants use to *display* attention and to interpret displays of attention by others (Jones, 2005a). What is particularly interesting about the examples given so far in this chapter is that the participants are able to pay attention to so many different interactions at once. What is equally amazing, however, is that they are able to communicate to so many interlocutors at once that they are paying attention. Whereas the cognitive attention system is more of a psychological construct (although it is also affected by training and socialization), the social attention system is based on linguistic patterns and norms of communication that develop within speech communities.

The role of the cognitive attention system in facilitating *inter-activity* lies in its ability to divide up experience into different "tracks"—different paths or routes along which consciousness can travel—and the existence of these multiple tracks is itself is a function of the differentiated character of consciousness. In other words, although it seems in situations of *inter-activity* that people are "paying attention" to more than one thing at a time, this is actually not the case. We are only able to pay conscious attention to one thing at any given moment (Baars, 1988; Raskin, 2000). What is actually taking place, then, is a *choreography* of attention, in which the focus of conscious awareness moves rapidly from one track to another. Mantovani (1996:10) compared this dance of attention to opening and closing windows, a metaphor that is particularly suited to the graphical user interfaces of personal computers. "Actors simplify situations," he wrote, "by opening variously sized windows through which they can focus on what interests them and intervene on the aspects of the situation nearest to their goals. . . . In . . . orienting attention, interests establish a reciprocal hierarchy (of activities) and find an order, precarious though it may be."

One thing that facilitates this process of distributing cognitive attention is the fact that all of the tasks that we perform do not demand the same kind of attention. Tasks that have become familiar to us through repetition do not require focal attention and are regulated by the cognitive unconscious (Baars, 1988; Raskin, 2000). Such tasks may include, for example, operating many aspects of the computer's interface, eating, and, to varying degrees, participating in conversational rituals. Here, of course, is where training and competence play a large role. The more activities can be regulated by the cognitive unconscious through practice, repetition, and training, the more flexibility users have in directing the focus of their cognitive consciousness.

Another important aspect of consciousness is the existence of *peripheral consciousness* (Chafe, 1994): the ability of the mind to hold in awareness that which is in spatial or temporal proximity to its focus of attention. Thus, when we are interacting with multiple interlocutors in CMC, we are aware to varying degrees of being "in the state of talk" (Goffman, 1981) with those with whom we are not immediately engaged in conversation. In many ways, the computer interface augments these mental faculties to facilitate cognitive coherence. The most obvious way is by transforming Mantovani's (1996) metaphorical windows into actual ones, which users can minimize, maximize, and arrange on the screen in ways that help them to "keep track of" the various goings on. Many applications also have methods of alerting users when particular events like an incoming message demand their attention. Finally, because computers can keep a record of the various ongoing actions that accumulate to form activities, users always have the opportunity to review past moves and resume activities where they left off. Herring (1999) pointed out that the persistence of a textual record of computer-mediated conversations, often structured by topics and threads, and now by tags and links, is one of the main reasons users are able to maintain interactional coherence even after long interruptions or in situations where the normal flow of turn-taking is not maintained.

The social attention system, the processes through which interlocutors convey to each other a jointness of attentional focus, is much more culturally and situationally dependent (see Erickson, 1979). Central to these processes is feedback (or "backchannel"), the various signals we send which have as their main purpose providing a display of attention to our interlocutor (although they often also convey important information about our attitude toward what is being said and our relationship to the speaker). Also important is the appropriateness of our responses to what has occurred previously, which is related to conversational goals and plans and the functional organization of interaction (McLaughlin, 1984). What counts, however, as the right kind or right amount of feedback or as the appropriate response to what was last said varies widely among people from different cultures and in different kinds of communicative contexts. Furthermore, the ways joint attentional foci are maintained are not constant throughout interactions, but change as encounters move through various stages (Kendon, 1978).

Feedback in CMC, of course, is very different from that in face-to-face conversation, which depends heavily on visual and auditory cues: gaze, posture, gestures, nonverbal sounds, and minimal response (e.g., "I see . . ."). It is a mistake, however, to think that such "backchannel" is nonexistent or less important in CMC. In fact, despite claims to the contrary (see, e.g., Black, Levin, Mehan, & Quinn, 1983), many utterances in synchronous and asynchronous CMC serve the function of backchannel. In my data, for example, many turns consist of single words or short phrases like "o," "ic," and "cool." Nevertheless, because visual and auditory modes are "muted" in most CMC, the function of other modes to convey listenership become "denser" (Norris, 2004) or more amplified, the most important being the temporal mode. The most common way users convey their attention is by responding within an appropriate time span (Jones, 2005c).

The most important point for the understanding of how *inter-activity* works is the fact that these two systems—the cognitive attention system and the social attention system are reciprocal, that achieving coherence in one system depends on the achievement of coherence in the other. Being able to maintain interactional coherence by, for example, producing an appropriate response to the appropriate interlocutor within the appropriate span of time, is based on being able to either keep various activities running along separate tracks, or to merge them in logical ways. Similarly, being able to make decisions about how to distribute our cognitive attention depends on us being able to make sense of the feedback we receive from others (see Fig. 1.1).

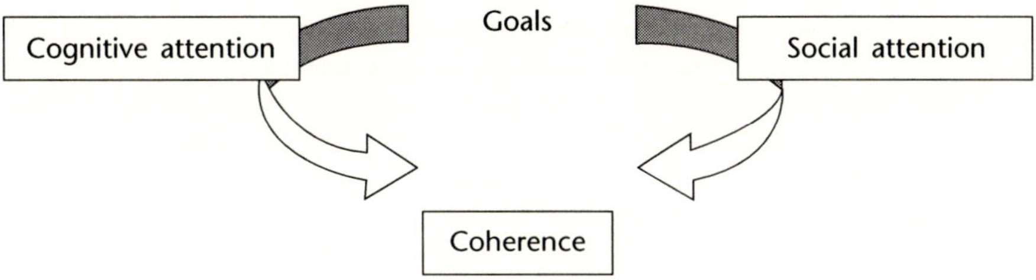

Fig. 1.1. Relationship between cognitive and social attention systems.

What drives this system are the goals of the communication as determined by our own interests and as negotiated with those with whom we are communicating. These negotiated goals are also part of the dynamic system and often change dramatically as interactions progress. As Mantovani (1996:11) wrote, "As our goals supply the criteria to select and interpret situational aspects, so the opportunities or threats identified in situations—together with the responses that the environment makes to our initiatives—supply feedback on our system of goals, encouraging some and discouraging others."

So What's New?

The key point I make here is not so much that computer-mediated interaction is "different" from other types of interaction, but rather the opposite—that the processes I described here are part and parcel of all interaction. Whenever people interact with specific interlocutors, they are *always* doing much more than participating in that single interaction. Social situations commonly involve participants not just shifting from one encounter to another, but also merging different activities in strategic ways. Thus, as Goffman (1981) reminded us, all interaction is marked to some degree by a certain "instability of participation."

Inter-activity is not an idea totally unfamiliar in sociolinguistics and linguistic anthropology. There has been, for example, substantial work done on the ways interlocutors in face-to-face interaction coordinate the different modes (voice, gesture, posture, and proxemics) available to them (see, e.g., Kendon, 1991), as well as on the sequential juxtaposition of differing activities and participant structures within a single interactional occasion (Erickson, 1979; Scheflen, 1974). Perhaps the most famous is Goffman's (1974) description of the way people divide up and compartmentalize their experiences, which he developed in *Frame Analysis*. All encounters, he observed, involve multiple "goings on," and so require participants to manage various attentional "tracks" that define some activities as central or "focal" and others as secondary. Secondary tracks, however, may themselves become focal or conceal covert focal activities or collusion. Thus, every interaction takes place at the nexus of many overlapping, internested, or "laminated" *frames*. Elsewhere, Goffman (1981) pointed out how interactions are often characterized by various "subordinate forms of communication" such as *byplay* (exchanges between subgroups of ratified participants), *cross-play* (exchanges between ratified participants and bystanders), and *side-play* (exchanges between bystanders).

Perhaps the most obvious place to look for examples of *inter-activity* outside of CMC is in telephone communication. Like computers, telephones facilitate *inter-activity* by altering spatial boundaries, muting modes, and providing users with "involvement screens" that can shield certain activities from the perception of interlocutors. Although there has been no lack of linguistic studies of telephone conversations, there have been very few that take this characteristic of the mediational means into account. One reason for this is that most studies of telephone interaction confine themselves to examining the words that are exchanged between those talking on the phone, ignoring the ways in which the introduction of a telephone into a physical place changes the social situation in that place, altering patterns of mutual monitoring. One study that begins to take account of these issues is Rae's (2001) analysis of participation frameworks in counseling sessions in which he looks at how counselors coordinate interacting with clients and making and receiving telephone calls. Central to his analysis is how participants use their voices and bodies to strategically signal attentional orientation. Accounts of such situations, he concluded need to be "rooted in participants' orientation to each other as a matter of their analysis of the attentional structure of the situation at hand and the accountability of their conduct within it" (722).

The written text is another medium which facilitates *inter-activity*. Because writing traditionally involves asynchronous communication, senders and receivers of messages may engage in all manner of activities while composing or decoding messages, some of which affect how messages are transmitted and participants positioned. Considerable attention has been paid to how writing transforms both the language and the person in communication (see, e.g., Ong, 1982), and how practices of reading and writing are embedded in multiple social practices (see, e.g., Heath, 1983). Relatively less attention has been given to how the act of reading alters the communicative architecture of situations. One attempt to do this is Scollon's (1998) study of reading in public in which he conceives of reading as a kind of social "watch" that confers on readers a certain status in the situation in terms of how they are expected to distribute their social attention, and how others present are expected to distribute theirs. A book, magazine, or newspaper used in a public place can act as an involvement screen, protecting the reader from unwanted approaches from strangers or, as in many an old detective film, allowing readers to direct their attention covertly towards others in the situation.

Writing can also serve similar purposes, as for example when students appear to be busily taking notes in a lecture when they are actually drawing cartoon figures or writing messages to their classmates.

Despite the ways technological media like computers and phones, and books and newspapers facilitate *inter-activity*, FTF communication may in the end provide some of the richest and most interesting examples of it. Studies that contribute to how *inter-activity* might be approached in FTF communication are numerous, including Kendon's (1990) work on the coordination of verbal messages with gaze and gesture in conversation, Scheflen and Ashcraft's (1976) exploration of how social actors physically manage multiple simultaneous relations and orientations, and Norris' (2004) examination of how people mix multiple modes in the construction of social identity.

One study that provides an apt contrast to the situation discussed in relation to Ah Lam is my own work on gay men searching for friends and sexual partners partners in parks in China and the strategies AIDS educators use to conduct outreach in such situations (Jones, 2002). The activity of "cruising" (known in Chinese as "fishing") in public places is one that not only involves but requires *inter-activity*. Participants generally engage in one activity (such as taking a stroll in a park) as a cover for searching out and inviting others into relationships. These "laminated" frames of activity result in a situation in which displays of attention are carefully calibrated to both reveal and conceal interest, and thus are able to communicate different messages to the initiated and the uninitiated. Whereas in the example of online "fishing" given previously, the visual mode is muted, in the FTF variety of this activity, the verbal mode is often muted, participants initially communicating through a "vocabulary" of the gaze (Jones, 2001b; Kendon, 1990). In both cases, however, interactions seem to consist chiefly of attempts by participants to negotiate shifts in mode: online participants negotiating how and when to move from a purely textual mode to the exchange of pictures, the use of a Web cam, or a telephone call, and FTF participants negotiating how and when to move from a purely visual mode to verbal communication (Jones, 2005b). The study I refer to, however, involves a third kind of participant, a young man who must to some degree conceal his AIDS education activities within the activity of "fishing," being careful to calibrate his displays of attention in ways which neither alienate nor unduly encourage his potential clients.

Conclusion

I have argued that methods of analysis that extract single streams of talk from the social situations in which they occur can never truly account for all of the variables affecting linguistic production and reception. What is needed are methods that focus on the whole range of behaviors within situations and the continuous dynamic relations these behaviors enter into across multiple modes and multiple frameworks of participation. As I suggested here, I believe the key to this approach lies in understanding how participants organize various levels of joint attentional focus throughout and across situations.

Here I have provided only a brief sketch of the issues that I believe to be important in the study of communicative situations involving multiple interactions. At the heart of these issues, I have suggested, are the ways perception and action function together in a mutually transformative fashion, and how this affects the ways users create what Turkle (1995:13) called *distributed presence*. Much remains to be done in terms of developing tools for the fine-grained analysis of how *inter-activity* affects linguistic and interactional behavior. At the same time, progress in this area will necessitate the use of more ethnographic approaches which gather data from a given situation from multiple perspectives and in multiple modes (Jones, 2001a).

One area that deserves particular attention, especially in analyzing situations like those I presented here, is the *affective* aspect of *inter-activity* (see Beavis, 1998; Danet, Ruedenberg, & Rosenbaum-Tamari, 1997; Herring, 1999), that is, how the "challenge" of managing the cognitive and interactional complexity of multiple interactions and the uncertainty involved can add to and/or detract from users' enjoyment of or engagement in communication, and ultimately how this might subsequently affect their relationships with those with whom they interact.

Another area needing study is the role of rhythmic organization and timing across timescales in *inter-activity*, particular in technologically mediated interaction (Jones, 2005c). Work on these issues in FTF interaction by Lemke (2000) Erickson (1992) and others provides a good theoretical foundation.

Finally, future research should consider how new media affect how old media are used, and how all the mediational means available in a situation work together to constitute an individual's "communicative surround," or to use Goffman's (1972) term, *umwelt* (Jones, 2003). A focus on *inter-activity* may eventually lead us away from studying CMC or FTF communication to studying how, increasingly, all communicative situations involve multiple media (both new and old) and multiple modes that interact with and influence one another.

Notes

1. All names of research participants used in this chapter are pseudonyms.

2. Computers and Youth Literacy in Hong Kong, a grant from the Research Grants Council of the Hong Kong Special Administrative Region, China (Project # CityU 1310/03H) (http://personal.cityu.edu.hk/~en-cyber/Cyberkids/Home.htm)

3. An Ethnographic Study of Computer Mediated Communication among Gay Men in Hong Kong. City University of Hong Kong Small Scale Research Grant #9030988 (http://personal.cityu.edu.hk/~en-cyber/home.htm)

References

Baars, B. J. (1988). *A cognitive theory of consciousness.* Cambridge: Cambridge University Press.

Barnes, S. B. (2003). *Computer-mediated communication.* Boston: Allyn & Bacon.

Baron, N. S. (2008). Adjusting the volume: Technology and multitasking in discourse control. In J. E. Katz (Ed.), *Mobile communication and social change in a global context* (pp. 117–198). Cambridge: MA: MIT Press.

Beavis, C. (1998). *Computer games: Youth culture, resistant readers and consuming passions.* Paper presented at the Australian Association for Research in Education Annual Conference, Adelaide.

Black, S., Levin, J., Mehan, H., & Quinn, C. (1983). Real and on-real time interaction: Unraveling multiple threads of discourse, *Discourse Processes, 6,* 59–75.

Cavanagh, A. (1999). Behaviour in public: Ethics in on-line ethnography. *Cybersociology 6.* Retrieved February 14, 2005, from http://www.cybersociology.com/.

Chafe, W. (1994). *Discourse, consciousness and time.* Chicago: University of Chicago Press.

Collot, M., & Belmore, N. (1996). Electronic language: A new variety of English. In S. Herring (Ed.), *Computer-mediated communication: Linguistic, social and cross-cultural perspectives* (pp.13–28). Amsterdam: John Benjamins.

Danet, B., Ruedenberg, L., & Rosenbaum-Tamari, Y. (1997). "Hmmm . . . Where's that smoke coming from?": Writing, play and performance on Internet relay chat. In S. Rafaeli, F. Sudweeks, & M. McLaughlin (Eds.), *Network and netplay: Virtual groups on the Internet* (pp. 41–75). Cambridge, MA: AAAI/MIT Press.

Davis, B. H., & Brewer, J. P. (1997). *Electronic discourse: Linguistic individuals in virtual space.* Albany: State University of New York Press.

Douglas, K. M., & McGarty, C. (2001). Identifiably and self-presentation: Computer mediated communication and intergroup interaction. *British Journal of Social Psychology, 40,* 399–416.

Dresner, E. (2002, May 10–12). *No sense of global place: Information technologies and political globalization.* A paper presented at the Media in Transition Conference (MiT2). Cambridge, MA.

Erickson, F. (1979). Talking down: Some cultural sources of miscommunication in interracial interviews. In A. Wolfgang (Ed.), *Nonverbal communication* (pp. 99–126). New York: Academic Press.

Erickson, F. (1992). They know all the lines: Rhythmic organization and contextualization in a conversational listing routine. In P. Auer & A. di Luzio (Eds.), *The contextualization of language* (pp. 365–398). Amsterdam: John Benjamins.

Garcia, A. C., & Jacobs, J. B. (1999). The eyes of the beholder: Understanding the turn-taking system in quasi-synchronous computer-mediated-communication. *Research on Language and Social Interaction, 32,* 337–350.

Goffman, E. (1959). *The presentation of self in everyday life.* New York: Doubleday.

Goffman, E. (1963). *Behavior in public places.* New York: The Free Press.

Goffman, E. (1972). *Relations in public: Microstudies of the public order.* New York: Harper.

Goffman, E. (1974). *Frame analysis: An essay on the organization of experience.* Boston: Northeastern University Press.

Goffman, E. (1981). *Forms of talk.* Philadelphia: University of Pennsylvania Press.

Goldhaber, M. (1997). *The attention economy: The natural economy of the net.* Retrieved August 17, 2003, from http://www.well.com/user/mgoldh/natecnet.html.

Heath, S. B. (1983). *Ways with words.* New York: Cambridge University Press.

Herring, S. C. (1999). Interactional coherence in CMC. *Journal of Computer-Mediated Communication 4*(4). Retrieved July 30, 2004, from http://www.ascusc.org/jcmc/vol4/issue4/herring.html.

Hymes, D. H. (1974). *Foundations in sociolinguistics.* Philadelphia: University of Pennsylvania Press.

Jones, R. H. (2001a). *Beyond the screen: A participatory study of computer-mediated communication among Hong Kong youth.* Paper presented at the annual meeting of the American Anthropological Association.

Jones, R. H. (2001b). *Mediated action and sexual risk: Discourses of HIV and sexuality in the People's Republic of China.* Unpublished doctoral dissertation, Macquarie University, Sydney.

Jones, R. H. (2002). A walk in the park: Frames and positions in AIDS prevention outreach among gay men in China. *Journal of Sociolinguistics, 6,* 575–588.

Jones, R. H. (2003). The problem of context in computer mediated communication. In P. LeVine & R. Scollon (Eds.), *Georgetown University Round Table on Languages and Linguistics 2002: Discourse and technology: Multimodal discourse analysis* (pp. 20–33). Washington, DC: Georgetown University Press.

Jones, R. H. (2005a, April). *Attention structures and computer-mediated communication among Hong Kong secondary school students.* Paper presented at the annual meeting of the American Educational Research Association, Montreal, Canada.

Jones, R. H. (2005b, July). *Navigating the emic and etic in participatory research on online literacies.* Paper presented at the 14th World Congress on Applied Linguistics, Madison, WI.

Jones, R. H. (2005c, July) *Rhythm and timing in computer-mediated communication.* Paper presented at the ninth International Pragmatics Conference, Riva del Garda, Italy.

Jones, R. H. (2005d). Sites of engagement as sites of attention: Time, space and culture in electronic discourse. In S. Norris & R. Jones (Eds.), *Discourse in action: Introducing mediated discourse analysis* (pp. 144–154). London: Routledge.

Jones, R. H. (2005e). "You show me yours, I'll show you mine": The negotiation of shifts from textual to visual modes in computer mediated interaction among gay men. *Visual Communication, 4,* 69–92.

Kendon, A. (1978). Differential perception and attentional frame in face to face interaction: Two problems for investigation. *Semiotica, 24,* 305–315.

Kendon, A. (1990). *Conducting interaction: Patterns of behavior in focused encounters.* Cambridge: Cambridge University Press.

Lemke, J. (2000). Across the scales of time: Artifacts, activities, and meanings in ecosocial systems. *Mind, Culture, and Activity, 7,* 273–290.

Licoppe, C., & Heurtin, J. P. (2001). Managing one's availability to telephone communication through mobile phones: A French case study of the development dynamics of mobile phone use, *Personal and Ubiquitous Computing, 5,* 99–108.

Ma, R. (1996). Computer meditated conversation as a new dimension of intercultural communication between Asian and North American college students. In S. Herring

(Ed.), *Computer-mediated communication: Linguistic, social and cross-cultural perspectives* (pp. 173–185). Amsterdam: John Benjamins.

Mantovani, G. (1996). *New communication environments: From everyday to virtual.* London: Taylor & Francis.

McLaughlin, M. L. (1984). *Conversation: How talk is organized.* Beverly Hills, CA: Sage.

McLuhan, M. (1964). *Understanding media: The extensions of man.* New York: McGraw-Hill.

Meyrowitz, J. (1985). *No sense of place: The impact of electronic media on social behavior.* New York: Oxford University Press.

Norris, S. (2004). *Analyzing multimodal interaction: A methodological framework.* London: Routledge.

Norris, S., & Jones, R. H. (Eds.). (2005). *Discourse in action: Introducing mediated discourse analysis.* London: Routledge.

Ong, W. J. (1982). *Orality and literacy.* New York: Methuen.

Paolillo, J. (1999). The virtual speech community: Social network and language variation on IRC. *The Journal of Computer-Mediated Communication 4*(4). Retrieved February 15, 2005, from http://www.ascusc.org/jcmc/vol4/issue4/paolillo.html.

Rae, J. (2001) Organizing participation in interaction: Doing participation framework. *Research on Language and Social Interaction, 34,* 253–278.

Raskin, J. (2000). *The humane interface: New directions for designing interactive systems.* New York: Addison-Wesley.

Rintel, S., & Pittam, J. (1997). Strangers in a strange land: Interaction management on internet relay chat. *Human Communication Research, 23*(4), 507–534.

Ruesch, J., & Bateson, G. (1951). *Communication: The social matrix of psychiatry.* New York: Norton.

Scheflen, A. E. (1974). *How behavior means.* New York: Anchor Books.

Scheflen, A. E., & Ashcraft, N. (1976). *Human territories: How we behave in space-time.* Englewood Cliffs, NJ: Prentice-Hall.

Scollon, R. (1998). Reading as social interaction: The empirical grounding of reading. *Semiotica, 118,* 281–294.

Scollon, R. (2001a). Action and text: Toward an integrated understanding of the place of text in social (inter)action. In R. Wodak & M. Meyer (Eds.), *Methods of critical discourse analysis* (pp. 139–182). London: Sage.

Scollon, R. (2001b). *Mediated discourse: The nexus of practice.* London: Routledge.

Turkle, S. (1995) *Life on the screen: Identity in the age of the Internet.* New York: Simon & Schuster.

Wasted time: Study says "multi-tasking" inefficient. (August 7, 2001). ABC News.com. Retrieved Janurary 29, 2004, from http://abcnews.go.com/sections/living/DailyNews/multitask010807.html.

Chapter 2

Text-Based Conversations Over Instant Messaging

Linguistic Changes and Young People's Sociability

ANABEL QUAN-HAASE

The University of Western Ontario

Abstract

This chapter critically reviews existing literature on the types of linguistic changes instant messaging (IM) has brought about among young people and discusses the impact IM has had on standard writing. The chapter argues that IM is a hybrid form of communication, blending written, spoken, and computer-mediated communication language conventions. Although IM integrates various conventions, it has also developed unique communication practices, forms of expression, and linguistic characteristics. The chapter examines to what extent technological features of IM influence these linguistic practices. As such, IM represents a unique forum of communication for young people, where they can creatively reinvent language use. Implications of linguistic changes in IM are drawn for theories of digital literacies.

Young people have embraced instant messaging (IM) at a rapid rate, making it a part of their everyday lives (Lenhart, Rainie, & Lewis, 2001; Lewis & Fabos, 2005). Of Canadian university students, 69% IM on a daily basis, and 29% IM weekly (Quan-Haase, 2007). A large majority of Canadian students who IM, log into the system every time they are online (Quan-Haase, 2007). The heavy reliance on IM for communication has raised important questions about the types of linguistic changes effected by the medium and its impact on how young people communicate. Because IM has gained its popularity among young people only in the past decade, our understanding of how IM affects young people's literacy is still limited.

This chapter contributes to the body of work emerging around new media and linguistic changes by examining young people's use of IM and the emerging linguistic practices that are developing. The first goal is to provide an overview of existing literature on the types of linguistic practices brought about by IM. The following research question is of central importance: To what extent are IM users employing deviations from written and spoken conventions? The focus is on young people—teenagers and university students—because they are one of the largest user groups, and most importantly, the effect of IM will be most prevalent in young people, who are in the process of acquiring written language skills. The majority of studies on IM are based on U.S. teenagers or university students because they are heavy users of IM (Lenhart et al., 2001; Lenhart, Madden, & Hitlin, 2005; Nielsen/NetRatings, 2002; The Radicati Group, 2005). Studies on how young people use text messaging are more prevalent in Europe and Asia, where a large proportion of young people own their own mobile phones and prefer text messaging to IM (Ling & Pedersen, 2005; Thurlow, 2002). Because the majority of studies focus on linguistic changes in the English language, this is the primary focus of this review. However, a number of studies have examined linguistic changes in IM in other languages, and these studies are also integrated here.

The second goal of the chapter is to examine the extent to which possibilities and constraints in the design of IM have an influence on the nature of linguistic changes. The linguistic characteristics in IM are expected to be distinct from those in other forms of computer-mediated communication (CMC) because of the unique usage conditions. For example, IM supports near-synchronous communication among users who are known to one another. The following research question is addressed: How do features of IM influence the development of language conventions among IM users?

The third goal of this chapter is to examine the current debate about the effects of IM on standard English writing. Much popular media, and many teachers and parents, view IM as a negative influence on young people's ability to write standard English, claiming that IM exchanges are obscured by the use of symbols, "reducing the text to a series of grins and grunts" (Anderiesz, 2003:14). Additionally, they argue that what makes IM messages difficult to decipher is the overuse of abbreviations, acronyms, and contractions (Anderiesz, 2003; Fresco, 2005; Hogan, 2006; J. Lee, 2002). Although the popular press has reported on linguistic changes effected by IM, until recently there has been little systematic analysis (Baron, 2004; Jacobs, 2004; Lewis & Fabos, 2005). To shed light on the debate about the effects of IM on written English, studies analyzing IM conversations between young people are reviewed. The chapter examines to what extent IM messages deviate from standard English and the reasons for these variations.

From the current state of knowledge, some preliminary conclusions are drawn and some implications for theories of digital literacies are provided.

Background Information on Instant Messaging

IM is part of the available array of CMC technologies. One form of CMC is the exchange of text-based messages via the Internet (Baron, 2004). IM differs from e-mail in that its primary focus is the immediate delivery of messages: Users send a message to their interlocutors and receive replies almost instantly, thus engaging in an almost real-time text-based exchange (Faulhaber, 2002). To accomplish this goal, IM applications include, among other features (see Fig. 2.1), (a) a "pop-up" mechanism to display messages the moment they are received; (b) a visible "buddy list" of contacts compiled by the user; and (c) a method for indicating when "buddies" are online and available to receive a message (Alvestrand, 2002). Another unique feature of IM applications is that it makes it possible for multiple

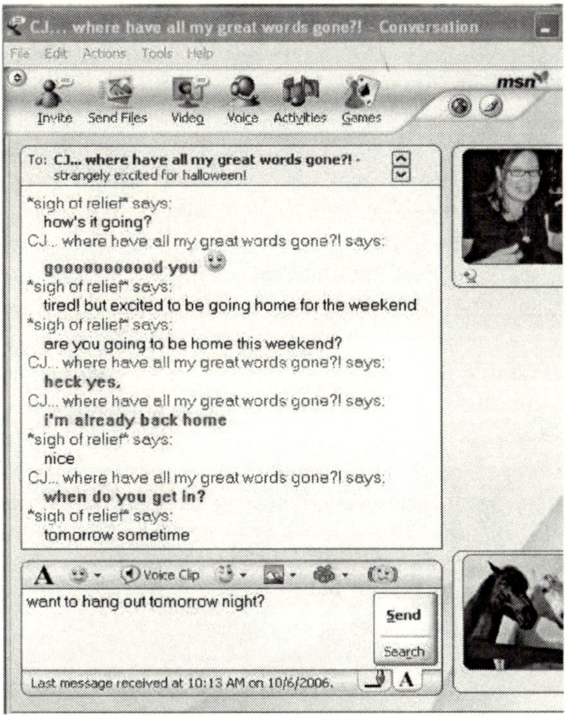

Fig. 2.1. Dyadic IM conversation via MSN Messenger.

conversations to take place simultaneously. IM applications allow not only for dyadic conversations, but also group chats (Grinter & Palen, 2002; Hu, Wood, Smith, & Westbrook, 2004).

IM is enormously popular among young people. In the 18 to 27 age group, 46% reported using IM more often than e-mail, whereas only 18% of 28- to 39-year-olds reported using IM more (Shiu & Lenhart, 2004). As part of the 18 to 27 age group just cited, university students use IM extensively. A study of Canadian university students showed that IM was enormously popular: 97% of respondents were users (Quan-Haase, 2007). When asked how many hours they use IM on a typical day, 28% reported more than 3 hours of use, 41% reported 1 to 2 hours, and 31% reported using IM for less than 1 hour. IM occupies a unique niche in the communication spectrum of young people because of its low cost for long distance exchanges (Hu et al., 2004), its speed, its ease of use, and its ubiquitous availability (Farmer, 2005; Ramirez, Dimmick, Lin, & Feaster, 2004), as well as its functionality, allowing users to detect whether or not their contacts are available (Farmer, 2003).

Linguistic Characteristics of Instant Messaging

IM is a form of CMC that supports interactions over the Internet in almost real time. Although a body of work has emerged documenting the types of linguistic changes effected by other forms of CMC, such as e-mail (Baron, 1984, 2000; Rowe, 2001, 2007, in press; Wyss, 2007), Internet Relay Chat (IRC)[1] (Werry, 1996), and Multi-User Dungeons (MUDs)[2] (Collot & Belmore, 1996; Herring, 1996), little is known about the types of linguistic changes resulting from IM (Baron, 2004; Lewis & Fabos, 2005). Despite the similarities between IM and other forms of CMC, "we need to be aware that each type of CMC has its own usage conditions and therefore, each needs to be analyzed in its own right" (Baron, 2004:398). For example, IM consists of interactions between dyads[3] or a small group, whereas other forms of CMC, such as listservs and MUDs, consist of messages that a larger group will read. E-mail, as with IM, is often used to exchange messages between dyads or a small group, but e-mails are generally longer and the communication is asynchronous. Differences in the linguistic characteristics of various forms of CMC are expected to result from these different usage conditions (Baron, 2004; C. Lee, 2002).

Textual Analysis of Instant Messaging

IM exchanges rely primarily on the transmission of text by two or more interlocutors.[4] Because of the reliance on text for communication, many of the same linguistic variables utilized in textual analysis can be examined in IM. For example, the extent to which users employ acronyms can be studied, as well as the types of acronyms employed. The textual analysis reveals a complex picture of IM linguistic characteristics because it often follows formal writing conventions, such as properly spelling words and writing phrases in full instead of using abbreviations and acronyms. Although these conventions are often known to the producers of IM messages, it does not determine their use of language in IM (Lewis & Fabos, 2005). Instead, producers are creative and often intentionally depart from the conventional use of language (Crystal, 2001; Lewis & Fabos, 2005). Using nonstandard English has meaning for IM users because it signals to others that the producer understands the particulars of the linguistic characteristics of IM, and therefore, understands the IM culture. In this section, a number of key linguistic variables—lexical shortenings—drawn from textual analysis are examined in IM, including acronyms, abbreviations, and contractions.

Lexical shortenings consist of linguistic variations in language use that allow for the speedy transmission of information. Although IM is characterized by the use of lexical shortenings, this phenomenon is not new or exclusive to IM (see, e.g., Elia, this volume). In academic articles, for example, lexical shortenings are commonly used to simplify the writing and communicate more quickly; in particular, this occurs when acronyms replace a string of complex, technical terms. Common in the field of communications is the use of CMC instead of computer-mediated communication and, of course, IM for instant messaging. Although lexical shortenings are infrequently used in formal writing, their use is widespread in IM. In fact, the use of lexical shortenings is characteristic of Internet communication in general (Merchant, 2001; Paolillo, 2001; Raymond, 1996) and is subsumed under the label "Netspeak" (Crystal, 2001), "Internet slang," or "NetLingo" (Jansen, 1994). Netspeak is widespread online, and several dictionaries compile the most commonly used words, lexical shortenings, and writing conventions, making Netspeak more accessible to a larger online audience (Jansen, 1994; Raymond, 1996). Furthermore, the use of Netspeak is not exclusive to a North American or European context: Research conducted in Japan and China shows similar trends in linguistic changes (J. Lee, 2002; Yukiko, 2003). Next, three types of lexical shortenings are examined: abbreviations, acronyms, and contractions (Baron, 2004).

Abbreviations. Short forms for words called abbreviations are commonly used in IM conversations to substitute for words that appear frequently in the conversation. A typical abbreviation in IM is, for example, the use of 'bc' for 'because.' Some abbreviations frequently used in IM have been adapted from informal writing, such as 'nite' for 'night' (see also Bergs, this volume, for short message system). Moreover, the use of abbreviations in IM stems from a long tradition in CMC of relying on lexical shortenings; many of the abbreviations found in IM are typically used in other forms of CMC. However, what is unique to IM is the diversity of abbreviations used and the innovativeness with which new types of abbreviations have been introduced (C. Lee, 2002).

Through the use of a questionnaire, Lotherington and Xu (2004) investigated the language conventions of English and Chinese IM users. They asked respondents to report the types of abbreviations they commonly used in their IM exchanges. The authors identified five different types of abbreviations, including "homophonic spellings" (e.g., 'u' for 'you'), "truncated homophonic spellings" (e.g., 'k' for 'ok'), "borrowed shorthand" (e.g., 'w' for 'with'), "reduced spellings needing a gloss" (e.g., 'ft' for 'faint'), and "simplified but recognizable spellings" (e.g., 'nite' for 'night'). Although both English and Chinese participants reported in the questionnaire the use of a wide range of abbreviations, the study did not examine to what extent users relied on these abbreviations when actually communicating via IM. Carmen Lee (2002) reported similar findings in her examination of IM users in Hong Kong. Participants in a questionnaire reported using similar types of abbreviations as reported in the study by Lotherington and Xu (2004). This reflects similarities in terms of the reliance on abbreviations in IM across languages and cultures. However, IM users in Hong Kong employed a much more diverse set of abbreviations because they relied on English, Chinese, and Cantonese to compose their IM messages (C. Lee, 2002).

In a study of university undergraduate students' linguistic patterns in IM, Baron (2004) investigated, among other things, the types of abbreviations employed and the extent to which abbreviations were used. In the data (12,000 words), she found 31 abbreviations (less than 1% of the corpus). These include: five instances of 'bc' for 'because,' two instances of 'bf' for 'boyfriend,' seven instances of 'cya' for 'see you,' 16 instances of 'k' for 'ok,' and one instance of 'y?' for 'why?'. The types of abbreviations identified in the data set are common in IM exchanges and reflect typical writing conventions in CMC. However, the low number of abbreviations present in the data set is surprising. Similar findings were reported by Tagliamonte and Denis (2006), who examined Canadian teenagers. Based on reports in the media, widespread and extensive use of abbreviations would have been expected (Alcoba & Leong, 2006; J. Lee, 2002). One possible explanation for the low usage of abbreviations is that respondents, aware of their participation in the study, may have altered their writing style. Another explanation could be that university students represent a different user group in comparison to teenagers or young people who do not attend university.

Lewis and Fabos (2005) examined a group of teenagers' use of IM and their linguistic properties. Two younger participants continued to use abbreviations, whereas older participants tended to use abbreviations less. Jake, one of the mature participants, reported on his use of abbreviations: "I used to [use abbreviations]. . . . Now I've started to type out 'you' and 'are' . . . I don't know. It looks better. . . . You just seem smarter" (483). This suggests that the use of abbreviations is more appealing for younger teenagers, whereas older users prefer to adhere to standard forms of writing. This could explain why Baron found few abbreviations in her data set, the participants she examined were college students who may no longer find it as appealing to deviate from written language conventions. A different explanation offered by Baron (2008) is that students spent a lot of time at their computers writing standard English for school and work and hence it is less effortful to write in a formal style in IM than it is to be continuously switching.

Acronyms. In CMC, acronyms are also common, further aiding in speeding up the conversation because they shorten entire phrases. As with the use of abbreviations, Baron (2004) reported that the number of acronyms found in the data was also relatively small. The body contained about 90 tokens, which is less than 1% of the total words in the corpus. The most frequently used acronyms in the data corpus were 76 uses of 'LOL' for 'laughing out loud,' five uses of 'ttyl' for 'talk to you later,' and three uses of 'brb' for 'be right back.' In the IM corpus examined, the most frequently used acronym was by far 'LOL'; which is one of the best-known acronyms on the Internet and therefore, it is not surprising that participants used it the most in their interactions.

In Yukiko's (2003) study, Japanese CMC users did not use exact translations of such English-language abbreviations as 'LOL,' but did employ shortened versions of common Japanese phrases. In her study of IM users in Hong Kong, Carmen Lee (2002) found that participants relied on various forms of code-mixing: (a) standard written English, (b) standard written Chinese, (c) Cantonese represented through characters, (d) Cantonese Romanization, and (e) various forms of translations. Moreover, her participants also combined elements from English, Chinese, and Cantonese to creatively form acronyms.

Contractions. These represent the third method of speeding up communication. Baron (2004) found 498 contractions in 11,718 words, such as, 'I'm' for 'I am.' When comparing the total number of contractions to possible contractions, in 65% of instances, people preferred the contracted form. The study reveals that contractions, similar to abbreviations and acronyms, are an integral part of IM communication. However, it is surprising that participants in Baron's study were "not taking advantage of possible lexical shortcuts" (413) that would speed up writing and reflect a more informal communication style, but instead participants frequently chose to follow the formal writing convention.

The studies reviewed show that lexical shortenings are an integral part of IM communication. IM users employ a great variety of lexical shortenings in their communication, demonstrating creativity and innovation in language use. Few studies have directly compared the use of lexical shortenings in IM and other forms of CMC; a study by Ling and Baron (2008) shows that IM use of abbreviations and acronyms is similar to that of texting. Considering the similarity of IM with MUDs, IRC, and texting, these parallels are not surprising. Additionally, IM users develop new lexical shortenings in their conversations with friends; these are creatively developed in the context of the interaction. The cross-cultural work is especially interesting because it reveals similarities in the changes that IM has effected across languages and cultures, as well as some interesting differences in terms of how languages are creatively combined in IM. What is surprising is the finding that lexical shortenings are not as prevalent as popular writings have suggested (Hogan, 2006; J. Lee, 2002). The low usage of lexical shortenings, on the one hand, could reflect participants' intentional use of standard writing as a response to being part of a study. On the other hand, it could reflect age differences, with younger participants showing greater reliance on nonstandard writing in comparison to older participants. Further research into these discrepancies is needed.

Conversational Analysis of Instant Messaging

IM systems are designed so that interlocutors can send messages back and forth, thereby engaging in an almost real-time conversation. In comparison to other asynchronous forms of CMC, such as e-mail and listservs, the immediacy of IM allows for instant feedback and creates a natural flow in the exchange. Because of its interactivity, IM closely resembles face-to-face (FTF) conversations, despite the lack of nonverbal cues and tone of voice, and the reliance on text. For example, in both IM exchanges and FTF conversations, interlocutors need to negotiate turn-taking so that the exchange remains coherent (Sacks, Schegloff, & Jefferson, 1974). Additionally, in both forms of communication interlocutors engage in conversation openings and closings. Because of these parallels, similar variables can be studied in IM exchanges as when analyzing spoken FTF discourse (Baron, 2004). This section examines turn production, turn-taking, conversation openings and closings, and unique aspects of conversation management in IM.

How turns are crafted and how turn-taking occurs is essential in IM conversations to keep the exchange coherent. However, examining turn-taking in IM, and in CMC in general, is not straightforward because it is unclear when a turn starts and ends (Baron, 2004; Herring, 1999a; Jacobs, 2004). Moreover, in IM, messages are often deliberately broken up into multiple transmissions, with conversation partners inserting comments between single transmissions of a message (Baron, 2004; Lewis & Fabos, 2005). To simplify the analysis of turns, Jacobs (2004) suggested defining a turn as a single transmitted message. In this way, a turn has occurred when the <enter> button is pressed on the keyboard, thereby posting the text, even though following transmissions may be produced by the same interlocutor and directly linked.

Despite the conversational similarity with turn production and turn-taking in spoken FTF discourse, this process occurs differently in IM. In FTF conversations,

> . . . overwhelmingly, one party talks at a time, though speakers change, and though the size of the turns and ordering of turns vary; [that] transitions are finely coordinated; [that] techniques are used for allocating turns; . . . and

> [that] there are techniques for the construction of utterances relevant to their turn status. (Sacks et al., 1974:699)

Because of the computer-mediated nature of IM conversations, new challenges arise in negotiating turn-taking. As Jacobs (2004) noted, turn-taking in IM conversations is not equivalent to turn-taking in verbal communication, because thoughts "in progress" are not visible to the other conversant. Moreover, many of the techniques employed in FTF conversations are no longer applicable because of the lack of physical presence. For example, in FTF conversations partners stop their utterance when they realize someone else is talking simultaneously (Schegloff, 2000). With a lack of voice and nonverbal cues, conversation partners need to use other techniques in IM exchanges to "hold the floor" (Sacks et al., 1974). A second problem in IM results from interlocutors' engagement in several simultaneous conversations or activities. Therefore, conversation partners need to not only negotiate turn-taking, but also need to sustain the attention of their interlocutors during an IM exchange through, for example, backchannel behavior, such as adding short reactions—"yeah" and "right"—to the conversation as it unfolds (McLaughlin, 1984).[5]

In the study reported previously, Baron (2004) also examined turn-taking. Although there was considerable variation in turn length, the average turn was short, comprised of five words. The longest IM turns consisted of 42 words, yet a high proportion of turns were comprised of only one word. Similar results are reported by Palfreyman and Muhamed (2003) in their examination of Arabic IM exchanges, with the majority of turns being short and consisting of about four words. The brevity of turns in IM serves two purposes. First, short messages help to hold the floor because conversation partners are constantly provided with pieces of information that sustain their attention (Werry, 1996). Second, interlocutors use short messages as post-turn minimal responses or backchannels (McLaughlin, 1984). Baron found that 22% of the IM corpus was comprised of one-word turns (e.g., "yep," "okay," and "cool"). Backchanneling is common in text-based CMC environments because they "heighten the interactivity of CMC by signaling listenership and encouraging others to continue" (Herring, 1999b:14). Providing text-based feedback is important because interlocutors lack nonverbal cues as well as spoken cues typical of FTF discourse to show agreement and encouragement, such as, nodding, and saying "aha."

Crafting short messages and engaging in backchanneling is typical for IM users. In the study by Baron, focus group participants reported that parents often did not understand the concept of using IM, and tended to write long messages that resembled letters or e-mails. This reflects normative differences in the two forms of communication. Although parents often are familiar with how e-mails are composed, they do not realize that IM is used in a fashion that resembles FTF discourse, and hence, the norm is to write short messages back and forth, keeping a constant flow of exchanges. Writing short messages is also possible in e-mail, but is not customary. Moreover, e-mail does not provide any status information on communication partners making it difficult to infer if they are online, and whether or not they are currently paying attention to the conversation, which is signaled in IM through presence and typing status information at the bottom of the screen (see Fig. 2.1; Quan-Haase, 2007). Writing longer messages in IM is also possible from a technical standpoint, although until recently, a limit existed in terms of the number of characters one can use in a single transmission. However, long messages do not serve the purpose of keeping an interactive conversation going. Therefore, the differences between IM and e-mail are not only grounded in the technology, but also exist at the normative level.

Not only did communication partners in Baron's (2004) study write short messages to keep IM conversations flowing, they deliberately broke up messages into multiple transmissions. That is, they pressed <enter> (posting the text) even though they had not completed their sentence. In the IM data set, 42% of all exchanges consisted of multiturn transmissions, indicating that this is a common activity. Focus group data collected by Baron suggested that one purpose of breaking conversational turns was to keep communication partners' attention and prevent them from moving on to other activities or conversations. By providing a constant stream of information, the interaction resembles a "delayed" FTF conversation.

What is interesting about IM interactions is that the delay between turns is flexible. Baron found that participants sent an average of 22 words per minute, which is much less than half the typing speed of most students. Baron also reported a gap of 15 minutes between turns in one IM conversation. Hence, although IM supports synchronous communication, it is often used for "delayed synchronicity." The delay can result from users' engagement in several IM exchanges that simultaneously require their attention. It can also result from users' simultaneous engagement in other online activities, such as surfing the Web, listening to music, or engaging in offline activities, such as watching television and talking on the telephone (Quan-Haase, 2007). This is an important distinction between CMC and FTF discourse conventions. In IM, the norms are such that delayed responses are acceptable and part of the discourse, whereas in FTF conversations a prompt response is expected.

One strategy employed to negotiate turn-taking in IM was the use of ellipses. Lewis and Fabos (2005) observed participants using ellipses (. . .) at the end of messages to indicate to conversation partners that they were reflecting on the message and had not yet finished writing. This would help keep the coherence of the IM interaction because conversation partners would not start writing a response, but would wait for the interlocutor to finish typing the message. In Lewis and Fabos' (2005) study of IM, another technique used to negotiate turn-taking was to withhold relevant information from conversation partners that would then be revealed in the next turn.[6] This withholding of information helped to negotiate turn-taking by signaling that the message was not yet completed. Additionally, it added tension to the conversation because partners had to wait for the following turn to receive the withheld information. In this way, their communication partners would not switch attention to another online or offline activity, but would wait for the message to be completed. This technique is similar to what Baron (2004) described in her study, where participants would break up messages into multiple transmissions with the intention of holding the floor.

The feature displaying a user's typing status is also helpful in conversation partners' negotiating of turn-taking. When a conversation partner is typing a message, the system signals this activity. This information can help keep the conversation coherent as the other person will wait to first read the message before starting to type a new one. Despite attempts by conversation partners to signal turn-taking, posts can be out of synch. Herring (1999b) found in her analysis of text-only CMC that the negative effect of incoherence was offset by the availability of a persistent textual record that could help communication partners make sense of messages. This is also true of IM, where conversation partners can scroll up to see text previously typed to help them make sense of the conversation.

Similar to FTF encounters, IM interactions rely on conversation openings and closings. Conversation openings usually consist of short, informal salutations, such as "Hi" or "How is it going?" By contrast, conversation closings in IM tend to be longer and more elaborate than IM conversation openings. Baron (2004) found that closings in IM took on average 7 turns and 32 seconds. She also found a gender difference: Closings in female–female conversations took 41 seconds with 10 turns,[7] whereas male–male closings took 16 seconds with 4 turns. Baron suggested that the gender difference reflects a female writing style identified in previous studies, where females use longer sentences than males in informal essay writing (Mulac & Lundell, 1994, cited in Baron, 2004).

In IM, turn production, turn-taking, and conversation closings and openings are important linguistic dimensions that conversation partners can deliberately manipulate to hold the floor, retain their conversation partners' attention, and keep the conversation flowing. Therefore, deviations from standard sentence construction may not necessarily indicate an inability to communicate, but on the contrary, they may reflect creative ways of keeping conversation partners "in the loop." Although turn production, turn-taking, and opening and closings serve a social function in IM, as they do in FTF exchanges, different techniques are employed. This is in part a result of the limitations imposed by the medium. For example, as noted earlier, interlocutors cannot rely on nonverbal cues to negotiate turn-taking. Technological constraints are, however, not the only factor leading to differences in language conventions—different norms are also at work in IM exchanges. For example, IM partners are willing to wait for a response from their conversation partners because it is expected that people are multitasking. Moreover, Lewis and Fabos (2005) found in their analysis that delaying a message is often intentional because it carries social meaning. The delay reflects engagement in other activities, including other conversations, which reflects a person's popularity in their social circle.

The Nature of Instant Messaging Literacy

A central debate in the linguistic changes of IM literature has been how IM affects young people's ability to communicate (Berg, this volume; Hogan, 2006; J. Lee, 2002; Lewis & Fabos, 2005; Matthewman, Blight, & Davis, 2004; Tagliamonte & Denis, 2006; Tenopir, 2006). The debate is part of a larger concern of parents, educators, and policymakers about the influence of digital literacies on print literacy. Digital literacies (The New London Group, 2000), also referred to as new literacies (Henry, 2006; Matthewman et al., 2004) or information and communication technologies literacies (Alvermann, 2004), consist of reading and writing via Internet-based tools, such as Web sites, e-mail, and IM (Kellner, 2002). The popular press has reported that the English language is deteriorating as a result of IM literacy, also often referred to as "AOL-speak" or "IM lingo" (Hogan, 2006). According to J. Lee (2002), young people will not learn how to write standard English as a result of this "inappropriate" use of language (see also Bergs, this volume).

It is the case that instead of emphasizing poetic writing and proper spelling and grammar, IM stresses speed, efficiency, and brevity (Hogan, 2006; Jacobs, 2004). Additionally, IM simplifies written language, reducing complex ideas to their bare-bones skeletal structure (Hogan, 2006). Thus, it is argued that as a result, language use will be sloppy and the quality of writing will deteriorate, and young people will not learn how to express themselves in a proper way. Finally, it is argued that the heavy use of abbreviations, acronyms, and contractions will create a barrier to communication, making messages indecipherable and obscuring the content (Anderiesz, 2003). This is of particular concern to teachers and educators who are not trained to read and comprehend "AOL-speak" and hence cannot make sense of messages that include CMC-specific lingo (J. Lee, 2002). For a summary and eloquent discussion of these arguments see the content analysis in Thurlow (2006).

Despite the large number of articles published in the popular press, until recently few systematic analyses existed on this topic (Thurlow, 2006). Without systematic analysis of actual IM conversations, it is impossible to understand how IM literacy diverges from standard language use. Moreover, empirical evidence that elucidates the nature of the relation between IM literacy and print literacy is lacking. Does IM literacy have a positive, negative, or indiscernible effect on print literacy? The studies discussed here provide a better understanding of the types of linguistic changes effected by IM. Furthermore, they provide a picture of the complex relationship between IM and print literacy, providing a framework for teachers and policymakers to work from. The studies also point toward future avenues of research.

A recent study on the effect of IM on students' ability to write was conducted by linguists at the University of Toronto and examined IM exchanges of 71 young Torontonians, aged 15 to 20 (Tagliamonte & Denis, 2006). The results showed that many of the fears articulated in the popular press are unfounded, with students showing a strong command of the language. These results corroborate findings from other studies. The studies by Lewis and Fabos (2005), and Baron and Ling (2003) suggest that spelling mistakes are not as prevalent as the popular press had originally suggested. Moreover, participants in Lewis and Fabos' study monitored their spelling and tried to correct errors if they occurred, often due to fast typing. Participants would indicate to conversation partners that they realized they had made a spelling mistake by typing the correct spelling preceded by an asterisk (see also Baron & Ling, 2003). Even though the spelling mistake does not always hinder the comprehension of the message, the effort made to correct the mistake suggests that users engage in reflections on their use of language.

These studies are in part problematic because participants knew they were participating in research, making it difficult to judge the extent to which this influenced their behavior. For example, one of Lewis and Fabos' (2005) participants mentioned that he paid more attention to writing properly when he communicated with teachers. Taking into account the low frequency of abbreviations, acronyms, and spelling mistakes and grammatical errors, this could be a result of "social desirability." One problem is that users usually do not save their interactions, and former IM systems did not automatically save exchanges in the same way that e-mail systems save messages. With more recent versions of IM this may be changing as it provides users with the option of keeping a copy of their IM interactions. By contrast, one can argue that if participants communicated with friends, the fact that they were participating in research may not have influenced their linguistic characteristics because they would have felt comfortable expressing themselves.

Lewis and Fabos (2005) suggested that spelling represents another level of communication, deliberately manipulated by communication partners. Two of their participants made very few spelling mistakes when they were conversing about a serious topic. By contrast, when they engaged in casual exchanges, communicating quickly often was more relevant than employing correct spelling. This discrepancy suggests that IM users adapt their use of language to the specific social context. Participants in Lewis and Fabos' study also indicated that when they wanted to impress a conversation partner, they paid special attention to spelling. This was also the case when communication partners were less well known to one another because then they wanted to make a good impression. Familiarity, thus, led to more colloquial, informal language. In a study of chat conversations, Lotherington and

Xu (2004) found that most conversations were "unparallel," with one conversation partner making use of relatively standard language while the other employed more innovative language.

Although IM users do not always rely on conventions from written discourse, some educators see IM as having a positive effect on students' print literacy, namely, because it gets students to write on a regular basis. J. Lee (2002) reported that students write as much or more using IM than they do for schoolwork. Thus, IM represents an alternative form of expression for young people, in addition to the writing assignments given in school or university (Lewis & Fabos, 2005). Through IM, they practice how to express their thoughts through written symbols. Students also learn to type quickly and to maintain conversations by using creative forms of expression. Communicating well through CMC and IM will increasingly be an important skill in a networked society where many jobs will require employees to be technologically savvy (Lewis & Fabos, 2005; Matthewman et al., 2004) and where communicating via digital networks will be normalized (Quan-Haase, Cothrel, & Wellman, 2005). Indeed, one study found that one-third of faculty surveyed in American universities felt student writing abilities had improved because of the Internet, with only 6% feeling that student writing had worsened (Jones & Johnson-Yale, 2005).

Taken together, the current body of work shows that language is used creatively in IM. Norms, either from written or spoken language conventions, are not always followed. What is interesting is that producers of messages often make use of these deviations purposefully. To what extent they follow certain conventions depends on the social context of the interaction. The writing style, therefore, becomes another linguistic variable with which the message is given meaning. For example, a formal writing style signals the importance and seriousness of the message, whereas an informal writing style signals a casual exchange. Contrary to fears expressed in the popular press, linguistic analyses report fewer deviations from standard language use than expected in IM exchanges. And this is so despite the fact that IM language production occurs at a fast pace with users multitasking. The current evidence suggests that the age of producers of IM messages may be an important factor in the extent to which deviations occur from standard writing, with younger age groups showing more deviations than older age groups.

Conclusions

IM has become an integral part of young people's lives (Boneva, Quinn, Kraut, Kiester, & Shklovski, 2006; Grinter & Palen, 2002; Jones & Madden, 2002; Lenhart et al., 2005; Quan-Haase, 2007). Through the regular use of IM, young people have developed unique communication practices, forms of expression, and linguistic characteristics.

Linguistic Changes in Instant Messaging

What linguistic changes are occurring as a result of young people's use of IM? IM has unique linguistic characteristics because users employ text-based exchanges to emulate FTF conversations. These exchanges usually occur between interlocutors who know one another. Because typing takes longer than speaking, the speed with which a message is transmitted is an important component of the interaction. For interlocutors, conversations that evolve slowly are not as fulfilling as those that unfold quickly (Lewis & Fabos, 2005). Lexical shortenings, such as abbreviations, acronyms, and contractions, are used to speed up the communication (C. Lee, 2002; Lotherington & Xu, 2004). Other linguistic conventions employed in IM to speed up communication include the use of lowercase letters and the omission of proper punctuation (i.e., colons and commas). Instead of using proper punctuation, users press <enter> to substitute for these formal pauses and create a natural flow in the exchange (Lewis & Fabos, 2005).

The motivation behind young people's use of lexical shortenings is as much about speed, convenience, and familiarity as it is about following established norms around Netspeak. By following these norms, young people signal to others that they are a member of the IM culture and understand the unique practices of IM communication. Hence, by employing AOL-speak users feel part of the in-group (see Boneva et al., 2006; Eckert, 2003). For example, IM systems allow users to type longer messages, but the majority of turns in IM are short (Baron, 2004; Palfreyman & Muhamed, 2003). This reflects the conventions of communication established in IM, where exchanges occur quickly and language use promotes speed. Although AOL-speak is fairly standardized, differences in what lexical shortenings are employed, and how often, can be observed between groups and cultures. For example, as discussed earlier, research examining users in Hong Kong showed unique combinations of English, Chinese, and Cantonese to form lexical shortenings (C. Lee, 2002).

Despite the reliance on text for communication and the use of the formal written style, IM is more akin to FTF conversations in terms of the users' norms and practices because it is interactive and casual. To make conversations dynamic, statements are broken down into various turns and each turn is usually short (Jacobs, 2004; Lewis & Fabos, 2005). This practice in IM differentiates it from other forms of writing, such as essays and reports, where the text is usually long and there are no breaks in the statements. It also differentiates IM from other social forms of writing, such as letters and e-mails, because interlocutors communicate synchronously in IM and are expected to consistently switch turns. This frequent turn-taking makes IM interactive, resembling in-person conversations. To end conversations, participants rely on similar conventions as those used in FTF conversations, taking several turns before closing the IM window (Baron, 2004; Boneva et al., 2006).

Turn-taking in IM, like in FTF conversations, is a complex process requiring negotiation between interlocutors (Sacks et al., 1974; Schegloff, 2000). In IM, interlocutors tend to write a series of short messages to "hold the floor," often withholding information to keep their conversation partners' attention. This can be particularly difficult when communication partners juggle various IM conversations and activities simultaneously. This creates delayed synchronicity, where gaps of varying length occur between messages. Moreover, often unrelated topics are inserted in conversations, distorting the coherence of conversations. Similar to findings with other forms of CMC (Herring, 1999b), this incoherence does not affect the positive experience of the interaction because through the persistent log of the messages, interlocutors can make sense of the conversation.

Although thinking about the types of linguistic changes effected by IM is central to our discussion, social and contextual factors also need to be taken into account (Jacobs, 2004; Lewis & Fabos, 2005). The context of IM conversations and different usage conditions will influence IM use (Jacobs, 2004). Moreover, IM is part of a larger digital revolution that has seen the integration of a wide range of digital communication technologies into the lives of young people (Lenhart et al., 2001; Ling & Pedersen, 2005; Rheingold, 2002). Hence, the types of linguistic changes occurring in IM need to be examined in relation to the emergence of digital literacies in general, which are multimodal, interactive, and integral to the culture of the Net generation.

Implications for Theories of Digital Literacies

The most salient linguistic change in IM is the seamless integration of different genres. An analysis of linguistic characteristics reveals conventions that stem from written discourse and others that stem from spoken discourse. These two distinct language conventions are integrated in IM, creating a hybrid form of communication. This mix of genres results from the use of text-based messages to communicate in real time. Additionally, analyses of IM conversations reveal characteristics that are unique to its digital format. IM linguistic characteristics clearly have much in common with conventions prevalent in other forms of synchronous CMC, such as MUDs and IRC.[8] These media share in common interactivity and immediacy in communication, which results in unique language conventions. Despite the numerous similarities with other forms of synchronous CMC, IM linguistic characteristics also show distinct conventions. These result from the nature of the IM interaction, where communication partners, usually with close ties, communicate synchronously over a long period of time.

IM is—unexpectedly, and unlike text messaging via cell phones—very close to the standard literacy. Recent studies question IM's negative reputation in the popular press. No evidence has been found as yet supporting claims that conversations consist only of Netspeak, AOL-speak, or IM lingo: a collection of incomprehensible symbols (Baron, 2004; Lewis & Fabos, 2005; Tagliamonte & Denis, 2006). On the contrary, IM conversations among young people integrate a number of written conventions; for example, employing few acronyms, abbreviations, and contractions (Baron, 2004; Lewis & Fabos, 2005). Spelling mistakes and grammatical errors are rare as well, with users often correcting errors that result from fast typing (Baron & Ling, 2003; Jacobs, 2004; Lewis & Fabos, 2005). Users also deliberately adapt their language use to different social contexts and audiences, showing great flexibility (Lewis & Fabos, 2005). Additionally, IM conversations reflect inventiveness and a good understanding of the various linguistic conventions (Tagliamonte & Denis, 2006). IM provides young people with a positive, unique space of their own separate from the adult world, where they can create and engage in a community of practice around the creative development of conversations through written text (Wenger, 1998; Wenger, McDermott, & Snyder, 2002).

In terms of interactivity, casual use of language, and expressions of playfulness, IM is similar to informal FTF conversations in a number of ways. First, IM includes phatic fillers and back-channeling, such as 'LOL' or 'yes,' that are employed in a similar way in FTF conversations, such as 'OK' and 'uh-huh,' to signal pauses or agreement (Baron, 2004). Second, conversation partners engage in flexible turn-taking (Baron, 2004; Lewis & Fabos, 2005). While the techniques employed in IM are different from those in FTF exchanges, negotiating turn-taking is an important component of IM. Third, similar to disengagements in FTF conversations, users engage in multiple turns to close an IM conversation.

In addition to integrating conventions from both written and spoken discourse, IM shows linguistic characteristics that are unique to its digital format. In IM, conversations are broken down into multiturn sequences because it maintains the attention of communication partners (Baron, 2004; Lewis & Fabos, 2005) and it is visually easier to read on the screen (Baron, 2004). Because the digital, text-based nature of the medium lacks nonverbal cues, such as tone of voice and gestures, it forces users to employ available written resources to express emotions, seriousness of the conversation, and the formality of the message. For example, one way of expressing emotions is through the use of emoticons. Although emoticons are prevalent in all forms of CMC, what is unique to IM is the heavy reliance on emoticons, and the wide range of emoticons available. Finally, IM conversations represent a distinct form of interaction, because IM users seldom engage in a single conversation, usually switching between multiple windows, each representing a different conversation. Moreover, IM users customarily engage in multiple activities, both online and offline, such as assignment activities, telephone conversations, and television watching. Multitasking requires proficient reading and typing skills, as well as the ability to swiftly shift between tasks.

Overall, the evidence has implications for our understanding of digital literacy. First, there is not sufficient evidence available about the nature of the relationship between IM and print literacy because the large majority of studies to date has focused on examining linguistic changes in IM. This would require fine-grained longitudinal analyses of both young people's language use in IM and their print literacy. Second, digital literacy should not be seen as separate from, and necessarily harmful to, print literacy (Kellner, 2002; Manovich, 2001). In fact, digital literacy draws directly from reading and writing skills learned in school. As a consequence, young users can flexibly switch between different types of literacies as they communicate in different contexts with different communication partners. This is not unlike verbal communication, where people use different types of vocabularies and ways of talking in different social settings (Lewis & Fabos, 2005).

On the other hand, digital literacy represents an alternative form of written expression because symbols are employed to express complex ideas. Through IM conversations, users gain familiarity and practice with the composition of digital text. Third, digital literacies are complex in themselves as each type of CMC is different in nature and has unique linguistic characteristics. Hence, studies of digital literacies need to examine similarities and differences of the linguistic characteristics of each form of CMC. Through this comparative analysis, we can gain a better understanding of how various types of digital literacy relate to one another and to print literacy. This type of comparative work is, however, time and resource intensive. Fourth, IM represents a creative blend of written, spoken, and CMC conventions. Users need to be able to integrate these different conventions while engaging in IM conversations. Therefore, conversations in IM and other forms of digital literacy require great flexibility across genres and modes of communication.

Acknowledgments

This chapter was supported by a Social Science and Humanities Research Council of Canada Grant. This chapter benefited from the advice and assistance of Jessica L. Collins and Michael R. Brundin. I also thank Naomi S. Baron, Ron Rice, Charley Rowe, and Eva L. Wyss for their thoughtful comments on an earlier draft.

Notes

1. IRC is a form of real-time chat that primarily supports many-to-many interactions, but also supports dyadic exchanges through private channels. For more information consult http://en.wikipedia.org/wiki/Internet_Relay_Chat.

2. A MUD is a multiplayer computer game that typically runs on an online bulletin board system. For more information consult http://en.wikipedia.org/wiki/MUDs.

3. The term *dyads* in this chapter refers to pairs of people.

4. Although the focus of this chapter is on the exchange of text-based messages, IM also integrates audio, video, and file and application sharing (Shiu & Lenhart, 2004). Although IM in the past decade was primarily used for text-based exchanges, the introduc-

tion of Voice over Internet Protocol has increased the reliance on voice-based interactions. The most frequently used IM application for voice-based exchanges is Skype.

5. Backchannel behavior consists of giving positive feedback in the form of statements, such as 'Uh-huh' or 'Yes,' to encourage interlocutors to continue talking or to signal listenership (Shiffrin & Schneider, 1977).

6. This is a technique commonly used in oral narrative style (Goodwin, 1984; Jefferson, 1978); it is also found as a rhetorical strategy in television; often referred to as *cliffhanger* (Grant & Starks, 2001).

7. Turns were measured in average turn length in words. Other turn measures were also employed, but they are not reported here (see Baron, 2004).

8. Although it is difficult to determine the origins of Netspeak conventions, it is important to note that early precursors to current IM systems existed before many of the current CMC systems. "talk" was developed in the 1970s to support real-time communication between users running the Unix operating system. Probably many of the early online language conventions have their origins in the talk system.

References

Alcoba, N., & Leong, M. (2006, August 2). Instant messaging gets bad rap, linguists say. *National Post.*

Alvermann, D. E. (2004). Media, information communication technologies, and youth literacies: A cultural studies perspective. *American Behavioral Scientist, 48*(1), 78–83.

Alvestrand, H. (2002, November 11). *Instant messaging and presence on the Internet.* Retrieved November 22, 2002, from http://www.isoc.org/briefings/009/briefing09.pdf.

Anderiesz, M. (2003, April 20). Cultural capital: F u can red dis, youre spending too lng n d chatroom. *The Independent on Sunday,* p. 14.

Baron, N. S. (1984). Computer mediated communication as a force in language change. *Visible Language, 18*(2), 118–141.

Baron, N. S. (2000). *Alphabet to email: How written English evolved and where it's heading.* London: Routledge.

Baron, N. S. (2004). See you online: Gender issues in college students' use of instant messaging. *Journal of Language and Social Psychology, 23*(4), 397–423.

Baron, N. S. (2008). *Always on: Language in an online and mobile world.* New York: Oxford University Press.

Baron, N. S., & Ling, R. (2003, October). *IM and SMS: A linguistic comparison.* Paper presented at the fourth International Conference of the Association of Internet Researchers, Toronto, ON, Canada.

Boneva, B. S., Quinn, A., Kraut, R. E., Kiesler, S., & Shklovski, I. (2006). Teenage communication in the instant messaging era. In R. E. Kraut, N. Brynin, & S. Kiesler (Eds.), *Computers, phones, and the Internet: Domesticating information technology* (pp. 201–218). Oxford: Oxford University Press.

Collot, M., & Belmore, N. (1996). Electronic language: A new variety of English. In S. C. Herring (Ed.), *Computer-mediated communication: Linguistic, social and cross-cultural perspectives* (pp. 13–28). Amsterdam: John Benjamins.

Crispin, T. (2006). From statistical panic to moral panic: The metadiscursive construction and popular exaggeration of new media language in the print media. *Journal of Computer-Mediated Communication 11*(3).

Crystal, D. (2001). *Language and the Internet.* Cambridge: Cambridge University Press.

Eckert, P. (2003). Language and adolescent peer groups. *Journal of Language and Social Psychology, 22*(1), 112–118.

Farmer, R. (2003). *Instant Messaging—Collaborative tool or educator's nightmare!* Retrieved July 7, 2006, from http://naweb.unb.ca/proceedings/2003/PaperFarmer.html.

Farmer, R. (2005). Instant messaging: IM online! RU? *EDUCAUSE Review, 40*(6), 48–62.

Faulhaber, G. (2002). Network effects and merger analysis: Instant messaging and the AOL-Time Warner case. *Telecommunications Policy, 26*(5/6), 311–333.

Flanagin, A. J. (2005). IM online: Instant messaging use among college students. *Communication Research, 22*(3), 175–187.

Fresco, A. (2005, November 1). Standard English out, text creeps in. *The Times (Australia),* 3.

Goodwin, C. (1984). Notes on story structure and the organization of participation. In J. M. Atkinson & J. Heritage (Eds.), *Structures of social action: Studies in conversation analysis* (pp. 225–246). Cambridge, UK: Cambridge University Press.

Grant, L., & Starks, D. (2001). Screening appropriate teaching materials: Closings from textbooks and television soap operas. *International Review of Applied Linguistics, 39*(1), 39–50.

Grinter, R. E., & Palen, L. (2002). *Instant messaging in teen life.* Paper presented at the ACM Conference on Computer-Supported Cooperative Work (CSCW '02), New Orleans, LA.

Henry, L. A. (2006). Searching for an answer: The critical role of new literacies while reading on the Internet. *The Reading Teacher, 59*(7), 614–627.

Herring, S. (Ed.). (1996). *Computer-mediated communication: Linguistic, social and cross-cultural perspectives.* Amsterdam: John Benjamins.

Herring, S. C. (1999a). Interactional coherence in CMC. *Journal of Computer-Mediated Communication, 4*(4).

Herring, S. C. (1999b). The rhetorical dynamics of gender harassment online. *The Information Society, 15*(3), 151–167.

Hogan, P. (2006, April 14). AOL-speak is destroying language's beauty. *Chicago Maroon.*

Hu, Y., Wood, J. F., Smith, V., & Westbrook, N. (2004). Friendships through IM: Examining the relationship between instant messaging and intimacy. *Journal of Computer-Mediated Communication, 10*(1).

Jacobs, G. (2004). Complicating contexts: Issues of methodology in researching the language and literacies of instant messaging. *Reading Research Quarterly, 39*(4), 394–406.

Jansen, E. (1994). *Netlingo.* Retrieved October 27, 2006, from http://www.netlingo.com/.

Jefferson, G. (1978). Sequential aspects of storytelling. In J. Schenkein (Ed.), *Studies in the organization of conversational interaction* (pp. 219–248). New York: Academic Press.

Jones, S., & Johnson-Yale, C. (2005). Professors online: The Internet's impact on college faculty. *First Monday, 10*(9).

Jones, S., & Madden, M. (2002). *The Internet goes to college: How students are living in the future with today's technology* (Report). Washingon, D.C.: Pew Internet and American Life.

Kellner, D. (2002). New media and new literacies: Reconstructing education for the new millenium. In L. A. Lievrouw & S. M. Livingstone (Eds.), *Handbook of new media: Social shaping and consequences of ICTs* (pp. 90–104). Thousand Oaks, CA: Sage.

Lee, C. K. M. (2002). Literacy practices in computer-mediated communication in Hong Kong. *The Reading Matrix, 2*(2).

Lee, J. (2002, Sept. 19). I think, therefore IM: Text shortcuts invade schoolwork, and teachers are not amused. *The New York Times,* pp. 1 & 4.

Lenhart, A., Madden, M., & Hitlin, P. (2005, July 27). *Teens and technology: Youth are leading the transition to a fully wired and mobile nation.* Retrieved August 2, 2006, from http://www.pewinternet.org/pdfs/PIP_Teens_Tech_July2005web.pdf.

Lenhart, A., Rainie, L., & Lewis, O. (2001). *Teenage life online: The rise of the instant-message generation and the Internet's impact on friendships and family relationships.* (Report). Washington, D.C.: Pew Internet and American Life.

Lewis, C., & Fabos, B. (2005). Instant messaging, literacies, and social identities. *Reading Research Quarterly, 40*(4), 470–501.

Ling, R., & Baron, N. S. (in press). Text messaging and IM: A linguistic comparison of American college data. *Journal of Language and Social Psychology.*

Ling, R., & Pedersen, P. E. (Eds.). (2005). *Mobile communications: Re-negotiation of the social sphere.* London: Springer-Verlag.

Lotherington, H., & Xu, Y. (2004). How to chat in English and Chinese: Emerging digital language conventions. *ReCALL, 16*(2), 308–329.

Manovich, L. (2001). *The language of new media.* Cambridge: MA: MIT Press.

Matthewman, S., Blight, A., & Davis, C. (2004). What does multimodality mean for English? Creative tensions in teaching new texts and new literacies. *Education, Communication & Information, 4*(1), 153–176.

McLaughlin, M. L. (1984). *Conversation: How talk is organized* (Vol. 3). Beverly Hills: Sage.

Merchant, G. (2001). Teenagers in cyberspace: An investigation of language use and language change in Internet chatrooms. *Journal of Research in Reading, 24*(3), 293–306.

New London Group. (2000). A pedagogy of multiliteracies: Designing social futures. In B. Cope & M. Kalantzis (Eds.), *Multiliteracies: Literacy learning and the design of social futures* (pp. 9–38). London: Routledge.

Nielsen/NetRatings. (2002, June 17). *Instant messaging used by more than 41 million home Internet surfers, according to Nielsen/NetRatings.*

Palfreyman, D., & Muhamed, A. K. (2003). A funky language for teenz to use: Representing Gulf Arabic in instant messaging. *Journal of Computer-Mediated Communication, 9*(1).

Paolillo, J. (2001). Language variation on Internet Relay Chat: A social network approach. *Journal of Sociolinguistics, 5*(2), 180–213.

Quan-Haase, A. (2007). University students' local and distant social ties: Using and integrating modes of communication on campus. *Information, Communication, & Society, 10*(5).

Quan-Haase, A., Cothrel, J., & Wellman, B. (2005). Instant messaging as social mediation: A case study of a high-tech firm. *Journal of Computer-Mediated Communication, 10*(4).

Radicati Group. (2005, July 1). *Instant messaging market, 2005–2009.* http://www.radicati.com.

Ramirez, A. Jr., Dimmick, J., Lin, S. F., & Feaster, J. (2004). *Revisiting interactive media competition: The gratification niches of instant messaging, e-mail, and the telephone.* Paper presented at the annual meeting of the International Communication Association, New Orleans, LA.

Raymond, E. S. (Ed.). (1996). *The new hacker's dictionary* (3rd ed.). Cambridge: MA: MIT Press.

Rheingold, H. (2002). *Smart mobs: The next social revolution.* Cambridge, MA: Perseus.

Rowe, C. (2001). *Genesis and evolution of an email sibling code.* Chapel Hill: University of North Carolina Press.

Rowe, C. (2007). Building code: Development, maintenance, and change in a private language. *American Speech, 82*(3), 235–261.

Rowe, C. (in press). Whatchanade?: Rapid language change in a private email sibling code in computer-mediated conversation. In S. Herring (Ed.), *Computer-mediated conversation.* Cresskill, NJ: Hampton Press.

Sacks, H., Schegloff, E. A., & Jefferson, G. (1974). A simplest systematics for the organization of turn-taking in conversation. *Language, 50*(4), 696–735.

Schegloff, E. A. (2000). Overlapping talk and the organization of turn-taking for conversation. *Language in Society, 29,* 1–63.

Shiffrin, R. M., & Schneider, W. (1977). Controlled and automatic human information processing: II. Perceptual learning, automatic attending and a general theory. *Psychological Review, 84,* 127–190.

Shiu, E., & Lenhart, A. (2004, September 1). *How Americans use instant messaging.* Retrieved April 13, 2006, from http://www.pewinternet.org/pdfs/PIP_Instantmessage_Report.pdf.

Tagliamonte, S., & Denis, D. (2006, July–Aug.). *LOL for real! Instant messaging in Toronto teens.* Paper presented at the Linguistic Association of Canada and the United States annual meeting, Toronto.

Tenopir, C. (2006). What chat transcripts reveal. *Library Journal, 131*(4), 34.

Thurlow, C. (2002). Generation txt? Exposing the sociolinguistics of young people's text-messaging. *Discourse Analysis Online.*

Thurlow, C. (2006). From statistical panic to moral panic: The metadiscursive construction of popular exaggeration of new media language in the print media. *Journal of Computer-Mediated Communication, 11*(3), article 1. http://jcmc.indiana.edu/vol11/issue3/thurlow.html.

Wenger, E. (1998). *Communities of practice: Learning, meaning, and identity.* Cambridge, UK: Cambridge University Press.

Wenger, E., McDermott, R. A., & Snyder, W. (2002). *Cultivating communities of practice: A guide to managing knowledge.* Boston: Harvard Business School Press.

Werry, C. C. (1996). Linguistic and interactional features of internet relay chat. In S. C. Herring (Ed.), *Computer-mediated communication: Linguistic, social and cross-cultural perspectives* (pp. 47–63). Amsterdam: John Benjamins.

Wyss, E. L. (2007). From the bridal letter to online flirting: Changes in text type and writing practice from the 19th century to the Internet era. *Journal of Historical Pragmatics, ##.*

Yukiko, N. (2003). Linguistic innovations and interactional features of casual online communication in Japanese. *Journal of Computer-Mediated Communication, 9*(1).

Chapter 3

Just the Same Old Story?

The Linguistics of Text Messaging and Its Cultural Repercussions[1]

ALEXANDER T. BERGS

University of Osnabrück

Abstract

This chapter investigates the general sociolinguistic effects of short message service text messaging available through mobile phones. The popularity of text messaging has led to some very interesting phenomena, which seem to parallel those in other media to a certain extent: (a) the formation of certain linguistic structures and uses specific for the medium, mostly due to technical constraints and possibilities; (b) the extensive use of the new medium in youth culture, with concomitant youth specific and age-graded language use (e.g., "secret code"); and (c) the (re-)awakening of the self-styled guardians of language, culture, and proper style. This chapter describes and discusses data from text messaging in English and German and evaluates these in the light of comparable earlier and present-day forms of communication (e.g., postcards and e-mails). It also shows that the negative evaluations of text messaging, particularly in teenage users, are mostly misguided and unnecessary comments in the well-established "complaint tradition." The latter aspect underlines the idea that we may find parallel, cyclical developments in the (socio-)linguistic evolution of almost all new media, ranging from the book through the telephone to text messaging and instant messaging/Internet Relay Chat, which will be topic of the final section of this chapter. In particular, it will have to be asked in how far the dichotomy of synchronous and asynchronous communication gradually fades away with crossover phenomena in texting (see Rice, 1999).

One of the latest media innovations in bidirectional communication is so-called text messaging, or short message services (SMS). SMS are sent either from one mobile phone or from a computer to a mobile phone. As one of their major characteristics, one single message may not be longer than 160 characters, including spaces. The first SMS was sent from a personal computer to a mobile phone on the D2-Vodafone Network in 1992—probably as a practical Christmas joke. In 1994, SMS were presented at the CEBIT fair in Hannover, Germany. Despite a few initial problems—mobile phone companies simply could not see why anybody would want to clumsily type up to 160 characters as a means of fast and efficient communication—this new way of communicating has become a great success in most parts of the world. The number of short messages has been growing constantly: Statistics from the Global Systems for Mobile Communications Association show that the number of messages in the Vodafone Network in Great Britain have increased from approximately 600,000 messages per month in January 1998 to about 22 million in May 1999, 257 million in September 2001, and finally 2.8 billion in December 2005. The number of messages in the United States was approximately 150 billion in 2004. By way of comparison, globally, Germany, Italy, and Great Britain are the leading SMS countries, followed by Australia and Singapore (see Fig. 3.1). Last, both in terms of current as well as projected use, is the United States.

Although most countries still show a rapid increase in messaging, the leading countries gradually show signs of market saturation, in terms of a decline or at least stagnation in growth rates. This may be because in these contexts, messages are no longer cheap ways of communicating, but rather expensive; that the mobile communication market as a whole is saturated to some extent; and that traditional short messages are being replaced by newer technologies, such as Multi Media Services (MMS) or Push to Talk (PTT).[1]

With regard to the age of SMS users, A. T. Kearney's (2001) Mobinet studies show that SMS are mostly used by younger (teenage) speaker/writers, but that there also is a general increase of at least about 10% across all age groups. In fact, the strongest increase was found in the 35 to 54 age group, but those 55 years and older showed an astonishing increase of approximately 14% (see Fig. 3.2). These studies also show that SMS are now gradually penetrating into the middle income groups with higher education.

In summary, short messages are not a marginal fashion trend that only affects uneducated teenagers from low-income groups. Rather, they are quantitatively as well as qualitatively well established. Simultaneously, SMS are being used for a growing number of different message types, ranging from short practical questions and answers (e.g., "Wanna tlk?"—"No!"), official reports,[2] shopping lists, poems, proposals for marriage, and even bible quotations and prayers.[3]

The introduction of this new means of communication has had at least two different, although interrelated effects. First, as this new medium of bidirectional

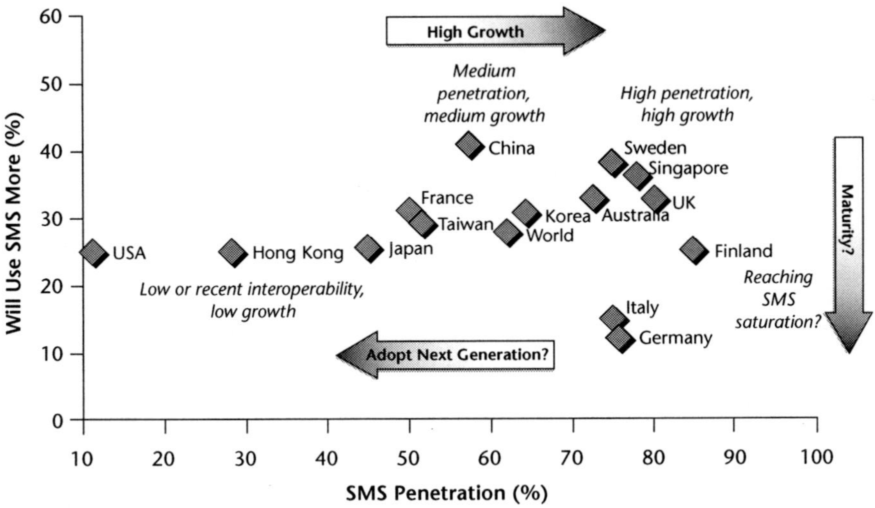

Fig. 3.1. Scatter of SMS penetration and proportion of users saying they will use SMS more (Source: Mobinet 4, http://www.atkearney.com/shared_res/pdf/Mobinet_4_S.pdf, retrieved July 20, 2003)

communication traditionally allows for only 160 characters per message, including spaces, we can expect to find certain linguistic structures, functions, and styles that are at least in part motivated by this restriction.[4] These, of course, warrant some investigation, and serve as the topic of the first part of this chapter. Second, from a sociolinguistic point of view, the introduction and rapid spread of text messaging has led to very interesting reactions from the general public and the self-proclaimed language "experts." Comments similar to those made during the

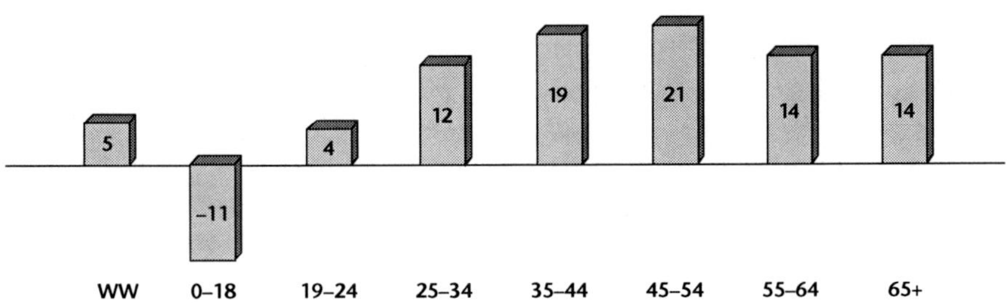

Fig. 3.2. Growth/decline in SMS actual usage by age group from Mobinet 2 to Mobinet 3 (Source: Mobinet 3, http://www.atkearney.com/shared_res/pdf/Mobinet_3_S.pdf, retrieved July 20, 2003)

rise of e-mail communication were made (i.e., many saw good style and general linguistic skills at bay, particularly among teenagers and young adults, who are among the most prolific users of computer-mediated communication [CMC], and SMS in particular).

Thus, the aim of this chapter is twofold. First, it seeks to describe some of the linguistic structures, uses, and styles that can be found in both German and English text messaging. Second, it evaluates the comments made about the sociolinguistic effects of text messaging. To that end, the introduction and actual consequences of text messaging will be compared to those of traditional written media, such as telegrams, postcards, and personal letters. It is argued here that there are parallel developments in the (socio-)linguistic evolution of almost all media, ranging from the book to the telephone to text messaging, which may lead to significant changes in the awareness and evaluation of linguistic norms (see, e.g., Millar, this volume).

The Linguistics of Text Messaging

In the following, some of the linguistic structures typically found in short messages in German and English are described in some detail. Although many phenomena are dealt with, the following discussion cannot claim to be comprehensive, of course. For present purposes it will suffice to discuss some of the likely most salient structures on each level.

Text messaging is, technically speaking, a written/graphic way of communicating, despite its being conceptually rather spoken- and proximity-oriented, or possibly even belonging to "a third kind" (Koch & Österreicher, 1994; Stein, 2003). Some of the most typical and salient morphological and lexicological elements in SMS (and Internet communication in general) are probably so-called emoticons or "smileys"[5] (i.e., ideograms with more or less iconic meaning) and abbreviated orthography, involving the deliberate omission of certain letters or the substitution by homophonous number symbols (e.g., 2, 4, 8), as in *ltl* ('little'), *bk* ('book'), *4* ('for,' 'four'), *2* ('too,' 'two,' 'too'), *b4* ('before'). As can be seen in Table 3.1, the ideograms in SMS communication range from quite transparent and well established (such as the smiley and the winkey) to very opaque (all of the following examples are taken from *wan2tlk? ltl bk of txt msgs,* 2000). It may be interesting to note that a similar diversity is not typical in German (and Swiss) SMS communication (cf. Androutsopoulos & Schmidt, 2002); whether this is a linguistic or cultural phenomenon, or perhaps both, remains to be seen.

Similarly, the extensive use of more or less opaque acronyms and alphabetisms like AFAIK ('As Far As I Know'), AWGTHTGTTA ('Are We Going To Have To Go Through This Again'), ROFL ('Rolling On the Floor Laughing'), ADS ('Alles Deine Schuld'), *g* ('grins'), FUDHUK ('Fall Um Den Hals Und

Table 3.1. Ideograms in SMS

Ideogram/Symbol	Meaning
:-)	Smile (Smiley)
;-)	Wink (Winkey)
:-(	Sad
:-0	Surprized/"oh!"
:-o zz z z Z Z	I'm bored
@:-}	I've just been at the hairdresser's
#(,'%/)	I've slept too long on one side and I didn't have time to wash my hair.
L8er	Later

Knuddel') is often seen as typical of text messaging. Phenomena like these seem to be so salient and popular that little dictionaries and usage guides were published (e.g., *WAN2TLK :-) ltl bk of txt msgs*, 2000). A number of studies, however, have shown that graphemic-orthographic features like these are actually much rarer in actual SMS communication in both English and German than is commonly assumed (Androutsopoulos & Schmidt, 2002; Döring, 2002b; Dürscheid, 2002; Frehner, 2004; Kesseler & Bergs, 2003). On the one hand, we find age-grading in these features (i.e., the quality and quantity of their use depends on the age of speakers—teenagers and young adults show the strongest use). On the other hand, some emoticons, like the smiley, the winky, the frowny, and some acronyms like ROFL or LOL have actually been conventionalized in average language use. (Note, even in older speakers this does not mean that these elements are being used productively.) More baroque sequences of elements, such as #(,'%/ . . . ('I've slept too long . . .') or AWGTH ('Are we going to have . . .') rather belong to realm of language myths than to actual, daily language use, despite the fact that they save precious time, space, and money.

What is more common among many SMS users is the substitution of letters and letter sequences by homophonous number symbols, especially in certain, more frequent, fixed phrases like 4eva ('forever'), str8 ('straight'), and L8R ('later'). This also applies to the use of asterisks as metalinguistic "stage directions" or pragmatic contextualization clues (e.g., *smile* where the asterisks underline the performative, meta-communicative nature of this element). It can be speculated that the rise and demise of these elements is closely connected to issues such as opacity and transparency, economy, and expressivity (i.e., the higher the time and energy that is needed to learn specific forms and their combinations, the more useful they had better be in the long run). This principle can be seen in traditional evolutionary theories and economy: If consumers have two roughly equivalent products (e.g., two mobile phones), they will always buy the cheaper one unless the more expensive one clearly is more useful (i.e., has a better cost–benefit ratio, including immaterial benefits, such as coolness). Even the most expressive new form does not work if the interlocutors cannot decode it, since it might simply be too opaque (see Hard af Segerstad, 2002).

Another issue in the graphemic-orthographic domain relates to sound–spelling correspondences. Users have to type their messages; they have to use letters and symbols, not their voice. While phonological considerations are generally somewhat problematic, certain interesting phenomena can be observed. In informal SMS communication, forms of phonetic spelling, morphosyntactic contractions typical of informal spoken discourse, representations of nonstandard varieties, and the written representation of proximity related forms like interjections can be found.[6]

- Semi-phonetic spellings and eye dialect: *enuf* ('enough'), *nite* ('night'), *u* ('you'), *no* ('know'), *l8r* ('later').
- Morphosyntactic contractions representing spoken informal language: *isnt* 'is not,' *cant* 'can not,' *gonna* 'going to,' *wanna* 'want to').[7]
- Written representations of nonstandard forms (e.g., dialectal or sociolectal variants): English *guid* 'good,' *gerrof* 'get off,' *havin* 'having' (see Schwitalla, 2002).
- Interjections and discourse markers that are usually common in written representations of spoken language: <oh> <ah> <yikes> <hehehe> <yeah> <boo>, <boah>, <yippie>, <ey>, <mmm>.

Another feature that deserves to be mentioned in this context is the use of graphological means to emulate suprasegmental phonological features, such as emphasis and stress. Since **boldface**, underlining, or *italics* are technically unavailable, speakers can employ capitalization on the one hand and special characters on the other. Capitalization (<I SAID NO!>) is often interpreted as shouting and therefore generally considered to be rude (Crystal, 2001). The use of special characters ranges from traditional punctuation marks and the multiplication thereof (<no!>, <what????>) to newly created symbols such as the asterisk representing emphasis and importance (<it's vital that you send it *today*>). Also, traditional spellings may be modified to represent some pronunciation variants or possible pragmatic effect: <soooooo?>, <heeeeloooo!>, <grrreat!>.

Table 3.2. Syntactic Features of SMS Communication

Omitted Element	Example
Subject pronouns	<can't come 2nite> <will kekse> 'want cookies'
Determiners	<seen O kitty?> 'Have you seen the cat?' <O Hund is tot> 'the dog is dead'
Copula/auxiliary	<house O on fire. cant come> 'the house is on fire, I cannot come'
	<rechnung O nicht bezahlt.> 'the bill has not been paid.'
Larger constituents (e.g., subject and verb)	<time off?> 'have you got some time off?' <Geld leihen?> 'can you lend me some money'

Adapted from Döring (2002).

In terms of syntax, short messages are subject to the same limitations as telegraphese (i.e., users often strive for cost-effective communication, leaving out whatever is not crucial for the message itself). This is one of the main features separating SMS communication from IM/Internet Relay Chart (IRC) (for comprehensive surveys, see Androutsopoulus & Schmidt, 2002; Döring, 2002b; Schwitalla, 2002; Hard af Segerstad, 2002).[8] This results in the frequent omission of function words (see Table 3.2).

Gerhadinger (2001, cited in Schwitalla, 2002) demonstrated that in (German) "dialogic" SMS communication (e.g., in adjacency pairs such as Speaker 1: <wanna come 2nite?> Speaker 2: <Sure. When?> Speaker 1: <the cave at 8 CU>) the average message length is no more than 11.06 words; Schwitalla added his own statistics which show an average length of 18.31 words per message (in a corpus of 100 messages) with significant peaks at 8–12, 15–17, and 20–27 words (Schwitalla, 2002). Naturally, extremes may also occur. Döring (2002a) pointed out that the shortest message in her corpus contained only one "word" consisting of two letters: <ok>. Messages of up to 160 characters are also not unusual. Döring's (2002a) empirical analysis generally confirms the first impression: 38% of the messages in her corpus of 1,000 texts contained at least one omission; multiple omissions occurred in 6% of the texts. Subject pronouns (particularly first-person singular) were the most frequently omitted elements, especially when followed by auxiliaries. Nevertheless, personal pronouns were still among the most frequent words in the corpus (<Ich> 'I' occurred 372 times, <Du> 'you,' second-person singular T-pronoun, occurred 329 times). This again correlates with the claim that SMS communication is generally an informal, spoken and proximity-oriented variety of language, as first- and second-person pronouns also figure, for example, in Biber's INVOLVED dimension of register variation (Biber, 1988, 1995).

The Evaluation of Text Messaging in the News

It did not take long before the general public reacted regarding this media innovation and its alleged consequences. Some examples include the following:

> I think it's a weird way to communicate, inevitably it'll affect the way people talk to each other. There are already problems with university students with their ability to write English. The more people use it, the less they'll be aware of different styles of communicating. (Ken Lodge, *The Guardian*, January 22, 2001)

> As text messengers discover and share new abbreviations and acronyms, the language becomes familiar to a growing population of cell phone users. And as more people use the lingo for text messaging, Sheidlower said, it is more likely to spill into speech or writing. That worries American University linguistics professor Naomi Baron, who said text messaging is another example of a trend in written communication. "So much of American society has become sloppy or laissez faire about the mechanics of writing," Baron said. Problems arise when people use the casual language in other forms of written communication, such as e-mail, in which the sender may not receive the message for some time, or writings in which the reader may not even know the author, she said. (Naomi Baron, *CBS News*, Feb. 13, 2003, http://www.cbsnews.com/stories/2003/02/13/tech/main540529.shtml)

> **The Sydney Morning Herald, April 23, 2004**
> **Technology marches ahead, grammar gets worse**
> Text messages, e-mail and the push for faster and more efficient communications are taking their toll on grammar, language experts believe. . . . Macquarie Dictionary editorial committee member Sue Butler said grammar and punctuation have been increasingly underused in Australia and around the world. "We now do get a failing in standards of punctuation that can be disconcerting," she said. http://www.smh.com.au/articles/2004/04/23/1082616314126.html

> **Girl writes English essay in phone text shorthand**
> By Auslan Cramb, Scottish Correspondent
> Education experts warned yesterday of the potentially damaging effect on literacy of mobile phone text messaging after a pupil handed in an essay writ-

ten in text shorthand. The 13-year-old girl submitted the essay to a teacher in a state secondary school in the west of Scotland and explained that she found it "easier than standard English." Her teacher, who asked not to be named, said: "I could not believe what I was seeing. The page was riddled with hieroglyphics, many of which I simply could not translate." (Telegraph.co.uk, Filed: 03/03/2003)

Text messaging is spoiling teenagers' sleep
Roger Dobson, *British Medical Journal* 13 September 2003
These preliminary findings suggest that mobile telephones may be having a major impact on the quality of sleep of a growing number of adolescents. . . . The threat to healthy sleep patterns is potentially more important than the threat posed by entertainment media. The latter mainly appear to influence time to bed, while mobile phones actually seem to lead to interrupted sleep. http://bmj.bmjjournals.com/cgi/content/full/327/7415/582-c

Other headlines include *Wirtschaftswoche*, "LiDuMiNo—Geheimcode der Jugend," June 8, 2001, *BBC News,* "Attacked by text message," April 15, 2002, *Daily Mail,* "Texting and emailing 'fog your brain like cannabis,' " April 22, 2005.

Some voices point out the advantages of text messaging: "The fact that e-mail and text messaging are both short and quick is a big part of the attraction. Many people find them more informal than making a phone call or writing a letter, and so simpler to use" (S. Gaines, "Now we're getting the message," *The Guardian,* February 15, 2001). Most commentators concentrate on salient aspects like opaque acronyms and alphabetisms. In November 2000, the very popular German tabloid *BILD* started a campaign with the heading "Wer etwas wichtiges zu sagen hat, macht keine langen Worte: Die große SMS Aktion gegen lange Sätze" ("Those who have to say something important don't use many words: The great SMS campaign against long sentences") and rewarded the most innovative SMS acronym send-ins, for example LIDUMINO "Liebst Du mich noch?" ("Do you still love me"). This, naturally, does not lead to positive associations related to SMS communication. Other worries and warnings could be heard regarding the loss of literacy skills, or even general communicative competence, the secret language of teenagers, and the sociopsychological dangers of bullying at school and the workplace via text messages. Text messaging was even seen as cause of sleep deprivation and dwindling IQ points (*Daily Mail* April 22, 2005; British Medical Journal, September 13, 2003).

But are these scenarios actually true and real? Do text messaging and similar media innovations impose some kind of threat? Can these developments trigger "negative language change" (i.e., language decline or language impoverishment)? The following section uses a historical-comparative perspective in order to evaluate these claims.

Text Messaging in a Historical Context: A Comparative Perspective

In their study *Authority in Language*, Lesley and James Milroy (1999) identified a so-called "complaint tradition" regarding language (i.e., some kind of tradition to complain publicly about the state of the language and the misuse of linguistic elements). Part of the tradition, in many cases, is to denounce and ostracize those people whose language use does not correspond to the ideas commonly held about certain "standard language norms" (see Jensen, 1990). L. Milroy and Milroy distinguished between two types of complaints:

> Type 1: This type is implicitly legalistic and essentially deals with correctness. These complaints usually deal with the "wrong" use of certain phonological, orthographic, grammatical, or lexical elements.
>
> Type 2: These complaints are implicitly moralistic. They usually deal with clarity of expression and ambiguous or misleading use of language.

Complaints of both types can be found in virtually all epochs and cultures, albeit to different degrees and concerning different linguistic elements. Moreover, many actually vacillate between Type 1 and Type 2. Some of the most striking examples include the fear in the later Middle Ages that the newly introduced "silent reading allowed for day-dreaming, for the danger of *accidie*—the sin of idleness" (Manguel, 1996: 51) or the 19th-century death sentence for the personal letter: "Seine eigentliche Geschichte liegt hinter uns, und es scheint, als ob es mit seiner weiteren Entwicklung überhaupt völlig vorbei sei" (Steinhausen, 1889, "Its true epoch lies behind us, and it seems as if its further development has come to an end altogether.") Jonathan Swift, the author of *Gulliver's Travels* complained in his famous proposal: "My Lord, I do her, in the name of all the learned and Polite Persons of the Nation, complain to Your Lordship as *First Minister,* that our Language is extremely imperfect; that its daily Improvements are by no means in proportion to its daily Corruptions; that the Pretenders to polish and refine it, have chiefly multiplied Abuses and Absurdities; and, that many Instances, it offends against every Part of Grammar" (Jonathan Swift, 1712, *Proposal for Correcting, Improving and Ascertaining the English Tongue*; cited in Bolton, 1966: 108). As far as we can tell, neither has silent reading led to a significant increase in idleness nor has the personal letter died out. Often, declining language standards are associated with declining standards in culture and ethics: The headmaster of Westminster School, John Rae was worried about the introduction of creative writing "The overthrow of grammar coincided with the acceptance of the equivalent of creative writing in social behaviour. As nice points of grammar were mockingly dismissed as pedantic and irrelevant, so was punctiliousness in such matters as honesty, responsibility, gratitude, apology and so on" (Rae, 1982, cited in L. Milroy & Milroy, 1999:41). And again, sociolinguistically speaking, this link is anything but obvious and warrants some discussion.

In the following, the linguistic features of text messaging discussed in the previous section are compared with those of other means of communication especially in a historical perspective. In particular, telegrams and postcards are discussed, but also the language of real personal letters and informal personal notes as they can be found in the 19th century. The hypothesis to be checked is that most of the features that give rise to critical comments today can also be found in other previous media and in earlier informal discourse. If this is the case, then we currently will not witness a decline in language and culture, but rather a cyclical development that probably takes place every time a new means of communication is introduced (see also Quan-Haase, this volume).

Let us begin with the use of complex and opaque acronyms and alphabetisms. These are by no means new and innovative as such. In fact, they have always been available to language users, even in formal writing: <etc.>, <NATO>, <RSVP>, <UAWG>, <mfG>, <uvm.>, <z.B.>, <e.g.>, <snafu>[9] are just some examples. The same applies to some extent to the use of ideograms and other non-orthographic symbols. The "x" representing a kiss or hug can be traced back at least to the 17th century (see Bergs, in preparation-b; Kesseler & Bergs, 2003), similarly, some people, especially female teenagers and young adults also use little hearts as i-dots, for example. In a word, similar phenomena can also be found in traditional written language and are not specific for text messaging or Internet communication (see Hard af Segerstad, 2002). Also, it needs to be pointed out again that both the use of acronyms/alphabetisms and ideograms is generally very restricted, and that the more complex forms are mostly found as an in-group code in close-knit networks.

Opaque linguistic forms of all kinds seem to fulfill one important function in these mostly teenage groups (just like in military groups, one might add): They support a linguistic group identity and allow for private, "secret" messages.[10] And again, this is nothing new. Historically, we find "Pig Latin" and "Löffelsprache"[11] (Applegate, 1961; Halle, 1990) for the same function. However, that Döring (2002a) did not recognize any specific identity function of short form usage per se in SMS communication. She pointed out that SMS communication generally only happens between members of (externally) well-established groups, which do not require any additional means of identification, and that the opaque and special short forms are generally much too rare to fulfil such a function (in her corpus there was only one opaque short form in 1,000 messages). Although this appears to be generally true, it seems also plausible to assume that teenagers in particular seek to establish special, secret in-group codes (Stenstrom, Andersen,

& Hasund, 2002). These group codes, however, appear to be extremely volatile and are certainly age-graded so that they have hardly (or rarely) any lasting impact on the linguistic structure of the language community as a whole.

The comparison of text messaging with telegram language is also informative, both in terms of external, sociocultural features and internal, linguistic phenomena—not in the least because both types seem to share certain normative expectations (in other words, they are probably not subject to the same normative pressures of standard language ideologies as other forms of communication). Telegrams are surely one very special and marked means of communication that is hardly ever used today. Nevertheless, before the rapid spread of the telephone in the 1950s, telegrams were among the fastest means of long distance communication and enjoyed quite some popularity; in 1960 there were more than 21.6 million telegrams in Germany alone, in 1980 the number was down to approximately 6.6 million (Schwitalla, 2002). Telegrams, in contrast to short messages, are principally not limited in the number of characters, but impose constraints on their use as their price depends on the number of words used. Therefore, we do not necessarily find misspellings and short forms, but rather modified syntactic structures, also known as "telegraphese." Tesak and Dittmann (1991) showed that a prototypical experimental German telegram on average contains five words and two utterances, Brandstetter (1968) established an average of 8.62 words in 93 private telegrams, Schwitalla (2002) counted 1.46 "sentences" on average in 60 German private telegrams until 1910. The sentences contained 5.35 words on average; telegrams on the whole contained 7.8 words, although extremes ranged from two (*mutter tot* "mother dead") to 88 words. Schwitalla (2002) also showed that the omission of constituents and single words mostly affects articles, personal and possessive pronouns (mostly "I"), auxiliaries and copula verbs, and prepositions.

As pointed out earlier, these are the same elements omitted in short message communication. The two media, although technically unrelated and used for different purposes (brief, expensive, mostly monologic messages containing vital information in telegrams versus brief, rather inexpensive messages sent for informative, phatic, or playful purposes, sometimes even in a truly dialogic situation in SMS), exhibit very similar linguistic structures and uses. And neither of them has or will seriously endanger existing linguistic structures or uses. In other words, people will continue to use subordinate clauses in the editorial of *The New York Times*.

The second type of texts that warrants comparison with short messages are postcards. Again, these impose severe restrictions on their users, simply because of limitations of space. The second similarity is that they are also rather inexpensive and useful for brief, informal messages, such as birthday cards or holiday greetings (see Diekmannshenke, 2002, and this volume). They do not, of course, lend themselves to use in dialogic communication (as, e.g., notes would), as sending and receiving postcards is a rather time-consuming process. Postcards, at least in informal style, exhibit also those features that are associated with proximity-related language use, such as contractions, short forms, nonstandard spelling, nonstandard linguistic structures and vocabulary (dialectal, sociolectal). Interestingly, they often also contain small hand-drawn pictures or emoticons, such as hearts, smileys, and so on (Diekmannshenke, this volume; Elspaß, 2002). It might also be speculated in how far postcards also foster the use of "secret code" as their level of privacy is also fairly limited (see Diekmannshenke, this volume; Rowe, this volume; 2007). But this question needs further investigation. And, once more, nobody today would suspect that postcards seriously threaten the literacy of their users. People seem to be able to differentiate between different media and text types, and their associated usages.

Finally, informal personal letters and notes might also be used for comparison at this point. As pointed out earlier, it is important not to compare apples and oranges. Comparing letters of highly literate and educated authors with everyday short messages will inevitably lead to the impression that the general educated written style is going down the drain. What should be compared are everyday notes and letters by the general public of all ages and backgrounds vis-à-vis short messages. The problem with such an approach is that contemporary everyday letters are rarely if at all available for linguistic investigation (see Elspaß, 2002). However, three studies can be mentioned in this context. Kesseler and Bergs (2003), Bergs (in preparation-a), and Wyss (2000, 2002, this volume) analyzed and compared "love letters" in the 19th-century corpus of London Foundling Hospital Letters (see Bergs, in preparation-b) to present-day amorous writings and showed that basically the same kind of imagery and stylistic devices can be found now and then. For example, in the 19th century letters, people often represented kisses by (surreptitiously) using *X*.

1. I wish that you were living there with me . . . you affectionate lover, John J.

   XXXXXXXXX

The same can be found in present-day e-mails and SMS:

2. Som1 tell sugarlips I LOVE HER!!! Babychops ;-) xxxxxxx

Hearts are also frequently used in letters (as i-dots, for example) but also in SMS (note the interesting mixture of verbal and iconic signs):

3. Jon—u are the * in my life! I'll always be yours. <3 <3 <3 love Cath

Similarly, the early letters analyzed also exhibit frequent orthographic "errors," jumbled syntax, and nonstandard morphology:

4. I accept the kisses you sent in your note with pleasure and will return with interests on Friday night althou I would rather had them from your lips than your hands . . .

Elspaß (2002) compared the language of early 20th-century personal letters in German with that in the new media (chat, e-mail, etc.). He concluded that there are no fundamentally new forms of language use, but only different attitudes toward the norms of language use. Just as with the London Foundling Hospital Letters,[12] the letters in Elspaß's study abound in nonstandard forms, structures, and vocabulary, ranging from simple "errors" through the use of interjections (Oh! Ha!) and (consistent) dialectal and sociolectal variants to various ideograms and icons such as crosses (+) and hearts (♥). In summary, Elspaß argued that if there are differences between traditional everyday personal letters and informal modern media, these are often differences in degree rather than in kind, such as in the use of interjections (Elspaß, 2002). To what extent, however, this is due to competence and performance, then and now, remains to be seen.

A final comment can and should be made concerning the status and development of literacy and communicative competence. To begin with the latter: Some commentators have suggested that text messaging (and other forms of electronic communication) endanger the communicative competence of speakers in general. In other words people don't talk to each other anymore (or they don't know how to do so). This is clearly not the case. As was shown in the first sections of this chapter, the use of text messaging has actually led to an increase in communicative acts. In a similar vein, Wellman and Haythornthwaite; Katz and Rice; and others (e.g., Quan-Haase, this volume) have shown generally positive correlations with use of new media and use of traditional media, as well as meeting with friends, and so forth (i.e., those who communicate more use more media in general). Text messages are now often used for purposes which did not require or allow for communicative acts beforehand. One of the main purposes today is brief phatic or ludic communication (the "how are you" or "just to let you know" type of message) and short information bites (the "bring x" or "dog dead" type of message). Döring also reports that in text messaging speakers are more willing to talk about more intimate and face-threatening issues (see Döring, 2002a, 2002b). But text messaging actually leads to more daily communication and interaction, not less. As a matter of fact, it may even be suggested that text messages do not even oust other forms of communication such as phone calls, e-mails, and letters, but that they actually supplement them. They fill a gap in communicative possibilities.[13]

Second, it needs to be pointed out that many of the communicative doomsday scenarios as they are reported in the media cannot be confirmed in any way. Nobody has actually seen the mysterious essay in text message form and the teacher "asked not to be named." That said, exact empirical studies that measure *declining* literacy are still lacking. The fact is, however, that never before in English and

German history have so many children attended schools and acquired basic literacy skills. The Programme for International Student Assessment studies on the quality of schooling certainly show significant accumulated needs and gaps, and they should be taken seriously. But we should not forget that in England and Wales, for example, an astonishing 31% of all bridegrooms and 46% of all brides in 1850 could not sign their own name in the marriage register. This rate decreased to 3% in 1900. By way of comparison: Studies of the U.S. National Center for Education Statistics show that 85% of all subjects had at least basic literacy skills. This means that what we compare are the relatively high rate of secondary or functional literacy today with a high rate of complete illiteracy in (relatively recent) historical periods. The group of people whose lack of primary literacy we complain about today probably could write and read even less 100 or 150 years ago. And this means that we should not create some "Golden Age of Literacy" with which we compare our state of affairs today (Pattison, 1982).

Summary and Conclusion

This chapter described some of the linguistic structures and uses to be found in text messaging on mobile phones in both German and English. It argued that few, if any, of the features in SMS communication are genuinely new and that most of them can also be found in other, earlier and well-established means of communication. In particular, it was pointed out that the media and certain language critics base much of their criticism and apocalyptic views on drastically exaggerated information and nonempirical speculation. It appears as if an evolutionary cycle in the introduction of new media can be seen. New media are often introduced as special means of communication for experts and business purposes (even the first letters in antiquity were business letters!). Shortly afterward, the general public gains access to the new means of communication and popularization sets in. Teenagers and younger adults seem particularly susceptible to media innovations of various kinds so that using new technologies becomes not only a matter of prestige but also of developing group identity among younger members of society.

Concomitantly, we generally witness a growing concern in the media about the long-term effects of media innovation on society, language, and culture in general. Rarely ever are the effects of such innovations praised; rather, the complaint tradition with its jeremiads seems deeply rooted in societies' subconscious so that news about declining standards simply sell better than good news about possible advantages (which is well in line, of course, with the general rule in journalism that "good news is bad news," in other words bad news sells better than good news). After a while, then, the saliency and newsworthiness of these items ebb off, and actual facts gain the upper hand. People become accustomed to the new means of communication and start employing them in their everyday life. Once a critical mass has been reached (i.e., the more common these new media become), the fewer worries there are to be found, until communication in these new media finally becomes established practice in everyday life. Simultaneously, the most marked linguistic structures and uses that are associated with the medium are leveled out to some extent and the most scandalous phenomena are often ironed out. And then a new form of communication is introduced and everything starts all over again. This kind of development can be seen in most if not all media innovations, from the book, through silent reading, to the telegram, the fax, telephone, e-mail, chat, SMS, and instant messaging. It appears to be a universal evolutionary cycle of innovations.

Nevertheless, it still seems important and worth the time from a linguistic point of view to look at each of these innovations carefully and individually and to describe and evaluate their structures from a neutral, descriptive point of view, just as established linguistic practice would have it. After all, it's still fun to watch new things grow and develop.

Acknowledgments

I would like to thank Charley Rowe, Eva L. Wyss, and one anonymous reviewer for the many instructive comments and questions. My gratitude also extends to the participants of the workshop on Language and Media at the 16th International Conference on Historical Linguistics, Copenhagen 2003, where a first version of this chapter was presented. Needless to say all errors are mine.

Notes

1. MMS basically work like short messages, but are not limited to 160 characters. Also, they allow for added pictures, audio files, and even videos. Push to Talk (PPT) is a recent innovation for mobile phones that allows users within a certain (limited) geographical range to use their mobile phones like walkie-talkies.

2. Rumor has it that the German chancellor Merkel first heard of her victory in the 2005 elections via a short message on her mobile phone (see http://focus.msn.de/politik/deutschland/amtlich_nid_21749.html).

3. During the World Youth Day 2005 in Cologne, pilgrims could subscribe to a short messages service from and about Pope Benedict (see http://www.katholische-kirche.de/8497_12057.htm).

4. In fact, the latest technology (in early 2003) allows for more than 160 characters. Enhanced-Message-Service allows for 2750 characters, MMS can also include pictures, sounds, animation etc. For an informative overview of the technological background and development, see Le Bodic (2005).

5. The ASCII smiley symbol was probably created by Scott Fahlman in an e-mail thread, where he pointed out that people in the thread may want to use :-) as a metalin-

guistic symbol to mark jokes (or irony), and :-(to mark something that is not a joke (Hard af Segerstad, 2002; see http://research.microsoft.com/~mbj/Smiley/Smiley.html).

6. The fact that these structures occur in text messages does not always imply that they were specifically invented for the medium. English apostrophe omissions, for example, can be found in telegrams in the 1930s (see Rowe, this volume).

7. Apparently, English messengers tend to leave out apostrophes in possessives or contractions, as has also been diagnosed for informal e-mail and chat communication (see Crystal, 2001).

8. It needs to be pointed out that this is only true for certain styles. Formal styles as well as leisurely, playful communication certainly tend to more complete or even redundant information, while informal styles and information-oriented communication (the "who, what, when, where?" type) tend toward greater brevity. Also, as Tesak and Libben (1994) pointed out, economy is not the only factor in the production of telegraphese.

9. <SNAFU> stands for "Situation Normal All Fouled/Fucked Up." This was probably first used in World War II as a joke in five-letter code developed in radio communication (http://www.snafu.com/Snafu/SnafuStory.html).

10. See Rowe (this volume) on one-on-one private "codes."

11. Pig Latin puts the first letter of the word last and ads "ay": "Pig Latin" > "Igpay Atinlay." Löffelsprache adds—lef after every vowel, which is then reduplicated: "Löffelsprache" > "Löleföffelefelspralefachelefe."

12. The "London Foundling Hospital Letters" are documents (mostly letters, notes, postcards and essays) that were archived during the late 18th and throughout the 19th century at the London Foundling Hospital. Women could leave their illegitimate children at the hospital, but had to prove their (moral) innocence. They often did so by presenting letters from their lovers in which these promised to marry them, for example. These documents are a unique source for informal written English in the 19th century from almost all social groups (for a detailed description of the corpus, see Bergs, in preparation-b) and are particularly well suited for comparative studies of 19th and 20th century informal (written) English.

13. Rowe (2007) came to similar conclusions regarding e-mail communication.

References

Androutsopoulos, J., & Schmidt, G. (2002). SMS-Kommunikation: Ethnografische Gattungsanalyse am Beispiel einer Kleingruppe [SMS-Communication: Ethnographic genre analysis of a small group]. *Zeitschrift für Angewandte Linguistik, 36,* 49–80.

Applegate, J. R. (1961). Phonological rules of a subdialect of English. *Word, 17,* 188–193.

Bergs, A. (in preparation-a). Investigating love letters across time: Some thoughts on metaphors, stylistic inertia, and meme theory. In E. L. Wyss (Ed.), *Transformations of passion. The mediatisation of intimacy in the 20th century.* Amsterdam: John Benjamins.

Bergs, A. (in preparation-b). *The London Foundling Hospital Letters: A new corpus of authentic, informal 19th century English.*

Biber, D. (1988). *Variation across speech and writing.* Cambridge: Cambridge University Press.

Biber, D. (1995). *Dimensions of register variation: A cross-linguistic perspective.* Cambridge: Cambridge University Press.

Bolton, W. F. (1966). *Essays by English and American Men of Letters, 1490–1830.* Vol. 1. Cambridge: Cambridge University Press.

Brandstetter, A. (1968). Zur syntax des wetterberichtes und des telegramms [On the syntax of weather reports and telegrams]. In Zur Syntax des wetterberichtes und des telegramms [On the syntax of weather reports and telegrams]. *Duden-Beiträge, 33,* Mannheim, 23–43.

Crystal, D. (2001). *Language and the internet.* Cambridge: Cambridge University Press.

Diekmannshenke, H. (2002). "und meld" dich mal wieder' Kommunizieren mittels Postkarte ["call again soon!" Communication via postcards]. In E. L. Wyss & U. Schmitz (Eds.), *Briefkommunikation im 20. Jahrhundert. OBST (Osnabrücker Beiträge zur Sprachtheorie)* (pp. 93–124). Duisburg: Gilles & Francke.

Döring, N. (2002a). 1xBrot, Wurst, 5Sack Äpfel I.L.D.—Kommunikative Funktionen von Kurzmitteilungen (SMS) [1x bread, spread, 5bags of apples I.L.U.—Communicative functions of short messages (SMS)]. *Zeitschrift für Medienpsychologie, 14*(3), 118–128.

Döring, N. (2002b). *'Kurzm. wird gesendet' Abkürzungen und Akronyme in der SMS Kommunikation.* ['Message sent' abbreviations and acronyms in SMS communication]. *Muttersprache. Vierteljahresschrift für deutsche Sprache.* 2.2002. [Electronic version]. Retrieved July 20, 2003, from http://www.nicola-doering.de/publications/sms-kurzformen-doering-2002.pdf.

Dürscheid, C. (2002). E-Mail und SMS—ein Vergleich [E-mail and SMS—A comparison]. In A. Ziegler & C. Dürscheid (Eds.), *Kommunikationsform E-Mail* [Communication from e-mail] (pp. 93–114). Tübingen: Stauffenburg.

Elspaß, S. (2002). Alter Wein und neue Schläuche? Briefe der Wende zum 20. Jahrhundert und Texte der neuen Medien—ein Vergleich [Old wine in new skins? Letters from the turn of the 20th century and texts from the new media—a comparison]. In E. L. Wyss & U. Schmitz (Eds.), *Briefkommunikation im 20. Jahrhundert. OBST (Osnabrücker Beiträge zur Sprachtheorie)* (pp. 7–32). Duisburg: Gilles & Francke.

Frehner, C. (2008). *Email—SMS—MMS: The linguistic creativity of asynchronous discourse in the new media age.* Frankfurt/M. Peter Lang.

Halle, M. (1990). Phonology. In D. Osherson & H. Lasnik (Eds.), *An invitation to cognitive science. Vol. 1: Language.* (pp. 43–68). Cambridge, MA: MIT Press.

Hard af Segerstad, Y. (2002). *Use and adaptation of written language to the conditions of computer-mediated communication.* Unpublished doctoral thesis, University of Göteborg, Göteborg, Sweden.

Jensen, J. (1990). *Redeeming modernity: Contradictions in media criticism.* London: Sage.

Kesseler, A., & Bergs, A. (2003). Literacy and the new media: Vita Brevis, lingua brevis. In J. Aitchison & D. Lewis (Eds.), *New media language* (pp. 75–85). London: Routledge.

Koch, P., & Österreicher, W. (1994). Schriftlichkeit und Sprache [Literacy and language]. In H. Günther & O. Ludwig (Eds.), *Handbuch Schrift und Schriftlichkeit. Ein interdisziplinäres Handbuch internationaler Forschung* [Handbook of writing and literacy. An interdisciplinary handbook of international research] (Vol. 1, pp. 587–604). Berlin: de Gruyter.

Le Bodic, G. (2005). *Mobile messaging technologies and services: SMS, EMS and MMS* (2nd ed.). Chichester: Wiley.

Manguel, A. (1996). *A history of reading.* London: Penguin.

Milroy, L., & Milroy, J. (1999). *Authority in language. Investigating standard English.* London: Routledge.

Pattison, R. (1982). *On literacy.* New York: Oxford University Press.

Rice, R. E. (1999). What's new about new media? Artifacts and paradoxes. *New Media and Society, 1*(1), 24–32.

Rowe, C. (2007). Building 'code': Development, maintenance, and change in private language. *American Speech, 82*(3), 235–261.

Schwitalla, J. (2002). Kleine Botschaften—Telegramm—und SMS-Texte [Short messages—telegram—and SMS-texts]. In E. L. Wyss & U. Schmitz (Eds.), *Briefkommunikation im 20. Jahrhundert. OBST (Osnabrücker Beiträge zur Sprachtheorie)* (pp. 33–56). Duisburg: Gilles & Francke.

Stein, D. (2003). *Computer-mediated communication: Written, spoken, or a third kind?* Discussion panel held at the 8th IPrA conference, Toronto.

Steinhausen, G. (1889). *Geschichte des deutschen Briefes: zur Kulturgeschichte des deutschen Volkes* [History of the German letter: On the cultural history of the German people]. Vol. 1. Berlin: Gärtner.

Stenstrom, A-B, Andersen, G., & Hasund, I. K. (2002). *Trends in teenage talk: Corpus compilation, analysis and findings.* Amsterdam: John Benjamins.

Tesak, J., & Dittmann, J. (1991). Syntaktische Strukturen und Ellipsen in deutschen Telegrammen [Syntactic structures and ellipsis in German telegrams]. In E. Feldbusch et al. (Eds.) *Neue Fragen der Linguistik* [New issues in linguistics] (pp. 245–251). Tübingen: Niemeyer.

Tesak, J., & Libben, G. (1994). On economy in written language production: Telegrams, short letters, and notes. *Grazer Linguistische Studien, 42,* 115–120.

Wyss, E. L. (2000). Intimität und Geschlecht. Zur Syntax und Pragmatik der Anrede im Liebesbrief des 20. Jahrhunderts [Intimacy and gender: The syntax and pragmatics of address forms in love letters in the 20th century]. In D. Elmiger & E. L. Wyss (Eds.), *Sprachliche Gleichstellung von Frau und Mann in der Schweiz. La féminisation de la langue en Suisse. La femminilizzazione della lingua in Svizzera. L'egualitad linguistica da dunna ed um en Svizra* [Linguistic equality of women and men in Switzerland] (pp. 187–210). Bulletin VALS/ASLA 72.

Wyss, E. L. (2002). Fragmente einer Sprachgeschichte des Liebesbriefs. Texte im Spannungsfeld von Sprachgeschichte, Geschichte der Kommunikation und Mediengeschichte [Fragments of a linguistic history of the love letter: Texts marking the polarity between linguistic history, history of communication and history of the media]. In E. L. Wyss & U. Schmitz (Eds.), *Briefkommunikation im 20. Jahrhundert. OBST (Osnabrücker Beiträge zur Sprachtheorie)* (pp. 57–92). Duisburg: Gilles & Francke.

Wyss, E. L. (2003, August). Love *Correspondences from the 19th century to the Internet era.* Paper presented at the 16th International Conference on Historical Linguistics (ICHL 2003), Copenhagen.

Chapter 4

E-mail Play and Accelerated Change

CHARLEY ROWE

University of Hong Kong

Abstract

Within e-mail communication, modifications in social interaction and language use are observable, particularly between intimates.[1] This chapter examines the influence of the physical and sociopsychological features of e-mail on language play and change between two sisters.

I begin with a conceptual history of two modern media that provided precedents for computer-mediated communication—the telegraph and the telephone. I then discuss the nature of e-mail, and its role in language play between pairs of users. I note that the tensions between delay and immediacy, asynchronicity and (near-)synchronicity, and well-composed and fragmented, are all possible in e-mail, with the result that people and contexts can create very different kinds of e-mail communication. In particular, I examine the play and exceptionally rapid change in the private code of two adult sisters. This sibling code, having originated in spoken interaction as a family code when the sisters were children, evolved rapidly over 4 months' time when used over e-mail. I suggest that the e-mail medium's inherent physical and socio-psychological features provide a ripe environment for play, and that these features provided by the e-mail context furnished the necessary conditions for language play to effect rapid and dramatic change.

With the invention of telegraph (1838) and telephone soon thereafter (1876), interpersonal communication changed radically. Many have argued that the telegraph was the first truly revolutionary modern medium, as it increased the speed and distance many more times over prior media than the telephone did over telegraph (Ron Rice, 11/3/08 pc). Telegraphs were used when near-immediate contact was desired with a recipient who would not be easily reachable, either by phone or by other means. Telegrams were expensive—charged by the word or character—and so messages were kept to maximum brevity. Function words like articles, pronouns, copula, particles, and punctuation unnecessary for basic comprehension were generally omitted for the sake of cost-efficiency:[2]

```
Western Union
Baltimore MD 3 817P 1934 Oct 3 PM 8 29
    THOUSAND CONGRATULATIONS SINCERELY HOPE YOU AND BABY
OK DELIGHTED ITS A GIRL BOTH DOCTORS PITTMAN AND LANEE
SAID OVER PHONE YOU AND BABY OKAY CANT WAIT TO SEE HER
LOVE BILL
```

Functionally, the telegraph is the precursor to the answering machine, and later, e-mail, e-card[3] (see Diekmannshenke, this volume), short message service (SMS),[4] and chat.[5]

Telephone, for its part, allowed for oral communication without time-consuming travel in cases where a written letter would not do. Telephone is unique in that it is real-time like face-to-face (FTF) communication but, like the written medium, it cannot draw on visual paralinguistic cues like gestures and facial expressions. However, unlike the written medium, it offers aural paralinguistic cues like intonation, pausing (hesitation), and the like, alongside other oral–aural effects like misspeeches and backchanneling, all of which can help interlocutors feel that the absence of visual cues does not leave them at a significant psychological and communicative disadvantage. People speak of the advantage of having a "good telephone voice." This refers not only to a mellifluous timber, but also to the ability to use intonation and other nuances to compensate for the loss of visual cues over phone (see Millar, this volume).[6]

Answering machines and voice-mail, companions to the telephone, eliminate the host interlocutor. The absence of the addressee introduces additional responsibility for the caller. In the early days of the popularization of answering machines (~1985), it was common for the caller to become dumbstruck upon hearing that he or she would be recorded instead of being directly connected to the other party. This led to several responses on the part of the caller: hesitate then hang up (or with practice, hang up without hesitation); dive into a long, verbose monologue; litter the message with verbal hesitations (*uh, um*); formulate a cumbersome closing (either a drawn-out series of "wells" and "OKs," as if the caller were awaiting a response from the apparatus, or a resigned and hurried *OK-talk-to-you-later-bye*) (Rowe, 2001).[7]

These responses reflected discomfort with one-on-one conversation that is effectively monologic and asynchronous when the expectation is for dyadic and interactive speech. That is, society had become comfortable with the format that speeches and lectures take (essentially the oral pronunciation of written language style), whereby speakers tailor the speech to the audience, regarding the audience as silent dialogue partners in this situation. Even speakers recording for a later presentation (e.g., in pre-recorded news shows) could always project the interest of the listeners and address their common interest even while presenting a monologue to them. But society was suddenly not comfortable delivering a quasi-monologue for a solitary listener who was not present to impart any cues of understanding, interest, or encouragement. Indeed, callers were acutely aware that the proxy for the listener was an impersonal, unforgiving machine. As callers spoke to the answering machine, they could only imagine the perceptions and thoughts of the listener who would play back the message, and attempted to adjust their message accordingly. Callers' nervousness and discomfort resulted in part from the lack of opportunity to edit smoothly (unlike with written notes), or naturally (as with canonical dialogic speech). Thus, the recording device became, in terms of Bell (1984), both audience (proxy) and referee. Accordingly, it can be argued that in every such call, the caller's speech was transformed, modified to a style that would not have been used with the answering machine's owner.

After a decade or so, callers became quite comfortable talking "to" the answering machine and later, voice-mail, on the other end of the line. Some callers speak, in total comfort, entire paragraphs, happy to be able to avoid potential interruptions from the interlocutor. Some callers even seek out this type of communication, calling precisely when they can be sure that there will be no one available on the other end to answer. Persons have at times answered the phone and heard the caller say "What are you doing there? I wanted to talk to your voice-mail!" This is facilitated even further with internal voice-mail systems, where the caller can choose to leave a message into the voice-mail account, bypassing the option to ring the owner (Rice & Danowski, 1993; Rice & Shook, 1990). Ultimately, caller-to-voice-mail communication is efficient for both parties: small talk and greetings are not expected, and long introductory exchanges and full conversations can be avoided. For anyone craving privacy, voice-mail becomes "e-mail that talks" (Duff, 1998).

In recent years, just before cellular (wireless mobile) phones became affordable and thus ubiquitous, pagers were common. A caller who calls someone on a numeric pager is prompted by an automated message to leave the number to be called.[8] Then, when the pager beeps, the person holding the pager reads the number off the pager and returns the call. Interestingly, the culture of pager-users developed systems of numeric codes for the purpose of efficiency or fun typing in a previously agreed-upon code (e.g., 911 means "urgent," 143 means "I love you," etc., Hafner, 2000).[9] Pagers soon gave way to increasingly affordable mobile phones, all but replaced by them now for the benefit of efficiency and quantity of conversation.

On the written front, new developments arose in the form of faxes. Often, faxes are, quite simply, standard office memos transmitted (essentially "mailed") over telephone line, emitted from a fax machine in the familiar form of a photocopy. In this use, faxes are electronic messages. But even if faxes are transmitted in seconds, the fax mode is a transmission mode only, and not a composition medium; thus the linguistic effects of faxed messages are dependent on the original medium: handwriting, typewriter, or computer-generated text.[10]

E-Mail[11]

Compared with memos that are printed and hand-distributed (such that the computer serves only as a printer-networked word processor), the language of computer-mediated communication (CMC) may appear hastily composed. The speed and immediacy of Internet communication plus the rapid composition speed provided by the computer—much faster than the typewriter because of the low-resistance keys—can result in several dramatic linguistic effects, for example, spelling errors, truncated syntax (omitted words, short or incomplete sentences), and certain lexical effects (neologisms, special lexis, and casual language). Some of these informality effects are undoubtedly largely due to those technical features[12] that apparently neutralize the capacity of the computer for easy error correction (facilitated by easy text manipulability, like cut-and-paste mode, drag-and-drop editing, and spellcheck functions). We surmise this because CMC errors seem more frequent than errors in handwritten text, typewriter-composed text, or computer-generated hard-copy (printed-and-distributed) memos.[13]

Other deviations, however, such as lexical effects, are clearly deliberate. After a brief excursus on the functions and nature of e-mail, this chapter will focus more narrowly on these effects in the e-mail environment.

E-mail (electronic mail), the primary medium of interest in this study, holds unique status in CMC. It was first used internally in U.S. government offices in 1971; by the early 1990s it had extended past the government network and was implemented at universities (Baron, 1998). Now e-mail is pervasive in industrialized societies.[14]

E-mail's advantages over traditional telephone communication include the feasibility for parties to contemplate before composing and responding, and the capacity to leave a paper trail (messages can be easily stored on a server or in designated mail folders on the user's computer hard drive).

As a written device, e-mail covers a multitude of functional domains: formal documents, official memos, informal notes, quick greetings, and so on (Lan, 2000)—although usually not for condolence letters/obituary announcements (see Günther & Wyss, 1996). This may have to do with users' discomfort with newer media, alongside courtesy and formality conventions regarded as appropriate for serious events. This can be seen currently in Skype (and similar) users' avoidance of uncomfortable topics, relying instead on the old comfort of telephone for such functions (see Millar, this volume).

Like faxes, e-mail messages are written; but because they are usually composed in the e-mail editor, they generally do not have the overall appearance of a memo. In fact, most earlier e-mail programs lacked text formatting like tabs, underline, *italic*, **bold**, and font modifications, and it is still the case in some e-mail applications that "rich" formatting is not used due to the demands on server space. Because standard word-processing options were absent until recently (and are still not easily engaged in all e-mail platforms), e-mail messages substituted other features; for example, with expressions that users would normally italicize or bold, users instead flanked words to carry emphasis with double asterisk, *like this*; in partial broken underlines, _like this_; or in single or double quotation marks (avoiding, of course, uppercase writing because of its apparently CMC-specific pragmatics of "shouting"). Without these markings, the writer surrendered a great deal of paralinguistic expression entirely. This is interesting because formal conventional written language often dispenses with special formatting, in favor of emphatic words (e.g., instead of writing "*This* is what the researcher intended," one writes "This is precisely what the researcher intended."). Perhaps now one could say that this style is particularly characteristic of the e-mail-writing culture that has been established in the last 30 years (Eva L. Wyss, 5/7/07 pc), and it is not certain which direction(s) it will take, toward formality, informality, or toward a hybrid register.

Like faxes, formal e-mail—even despite the lack of special formatting—can closely reflect its hard copy counterparts (Gains, 1998). In other words, e-mailed memos may not differ stylistically from hard copy memos, e-mailed contracts do not differ in wording from hard copy versions, and so on. But unlike most written media, e-mail can also serve effectively as a dialogue device (Lan, 2000) despite the technical asynchroneity, and as such, is often a mirror of the writer's spoken style (Davis & Brewer, 1997). As a result, it reflects much of the personality and mental state of the writer: Creativity, carefulness/caution, impatience, perfunctory demeanors, and the like (Rowe, 2001). In e-mail, special uses of punctuation (such as copious use of exclamation marks, question marks, emoticons, etc.) can convey a sense of intimacy (Lan, 2000), as can novel abbreviations (e.g., *BTW* 'by the way') and stylistic hybridizations (e.g., *What the **** does my esteemed colleague mean by that?*).

Particularly interesting for e-mail are the combined tensions between delay and immediacy, asynchroneity and (near-)synchroneity, and well-composed and fragmented. When individual users' preferences are factored in, it becomes clear that people and contexts can create very different and complex types of e-mail communication. E-mail, as a result, then, more versatile and pliable than traditional forms of exchange (both oral and written), frequently shows hybridizations and stylistic innovations—even within sub-genres ("note," conversation, memo, business letter, etc.). Because it is asynchronous, it is not always (or not necessarily) composed hastily, and thus it affords greater control by the user than, say, chat; and e-mail texts are generally not as cryptic or fragmented as chat texts often are. But when interaction and transmission *are* almost instantaneous (messages arriving within seconds or even milliseconds after the sender clicks the "send" button), the speed of e-mail interaction approaches real-time communication; furthermore, e-mail can stylistically appear as handwritten notes might, with abbreviations, truncated syntax, and hurried signatures.[15] Moreover, many e-mail conversants expect an immediate reply—a demand that may also result in less formal "telegraph" style (see Wyss, 2003). In its most casual realizations, e-mail communications are largely the electronic equivalent of the handwritten note, with abbreviations, truncated syntax, and hurried signatures. Interactionally, they are much more.

The Sociopsychology of Computer-Mediated Communication

E-mail, even some 35 or so years on, is in some sectors (e.g. industry, among others) not regarded as an indispensable means of communication, but rather as a compromise between telephone and FTF communication on the one hand, and paper-written communication on the other. Far from supplanting these primary modes of communication, it has become well established as an additional way to communicate (Rowe, 1996, 2001). The stylistics and efficacy of e-mail are still compared—unfavorably—to oral communication and "paper" (canonical written) communication.

Although e-mail use continues to increase, many users still complain of some sociological discomfort with this relatively new medium. Without the pragmatics of vocal intonation, some e-mail users feel at a loss to express themselves naturally. And the absence of standard memo signals (letterhead, embossing, etc.) leaves some users uncomfortable with the guesswork of what degree of formality to invoke (see Sherblom, 1988). Additionally, the time and effort it takes to compose a message, relative to speaking, sometimes results in e-mailers not taking the time to check their messages for politeness and other pragmatic effects; as a result, messages may seem more abrupt than what the e-mailer intended. This (coupled with the "anonymity effect," which affects even familiars who cannot see or hear each other) can result in a backlash by the addressee, who assumes the worst and fires back a furious response that would not have been dared in FTF interaction.[16]

On the positive end, e-mail has become a link for those who would normally not interact—or at least, who would not interact as frequently. Part of the reason is fiscal: while private users in many cases must pay (usually a monthly fee) for Internet connection, e-mail itself is free and unlimited. Many users pay nothing, availing themselves of an internet connection at their institution or place of employment. This makes e-mail in many cases preferable to telephone particularly where discreet interaction is preferred. It may also be more practical than FTF communication—particularly in long-distance relationships or relationships with highly mobile or very busy persons. Seniors who are e-mail users happily report increased contact with their children and grandchildren; college and high school classmates regain contact by locating one another's e-mail addresses over the Web; and many users engage in new e-mail-only interactions that are typically short-lived but nonetheless in many cases socially rewarding. Moreover, with e-mail, the typical psychological constraints that usually bind writing are largely absent. And so for every e-mail user who is uncomfortable with the ambiguous rules of formality and informality connected with e-mail, there is another who finds the ambiguity of this interface liberating (see, e.g., Rice & Love, 1987).[17]

But what specifically causes users to feel so open to this medium? Mihalo (1985), who studied e-mail-based bulletin boards, suggested that when one must interact solely with what is written, some social barriers may be diminished (see also Hiltz & Turoff, 1978). Thus, posits Mihalo, greater intimacy can be achieved, and a new dimension is added to human relationships (see also Walther, 1996). An early study by Kiesler, Siegel, and McGuire (1984) also found that CMC changes interaction patterns. Their study suggests that CMC promotes a less-inhibited style of interaction in users of various ages than does FTF interaction.[18]

Like oral language, language use over e-mail can be quite unrestrained. This psychological "loosening" can also feed creativity: Once the bonds of grammatical rules and format conventions are broken, e-mail users are free to "play" with language and make it their own (Rowe, 1996, 2001).[19] For these reasons, CMC allows individuals to create and assert new social identities to play with, and fictional worlds to play in (Poster, 1996; Yates, 1996b), where "language use is separated from biographical identity" and stable identity is disrupted (Poster, 1990).

Esther Guinell (*Online Seduction*), in an NPR interview on Talk of the Nation (Feb. 11, 1999) expounded on these observations. Guinell, a clinical psychologist, theorized that the lack of facial expression in e-mail results in fewer fears of rejection; thus conversants are more likely to "pour their hearts out" when they can't see each other.[20] Conversing over e-mail, then, according to Guinell, is more akin to journal writing; it is "inner speech," in the sense of Vygotsky (1934, 1986). Yates (1996a), on the other hand, suggested that CMC does not actually "promote" emotional language; rather, users who anticipate seeing conventional written texts are "challenged by the dynamic, interactive and interpersonal forms that CMC takes,"[21] and the style of their messages reflect this psychological experience. Most likely, both observations are correct: Either way, the effect is generally a more intimate interaction over e-mail than what may occur in FTF interactions.

Another important sociopsychological contribution of e-mail to personal interaction is that for the frequent or "heavy" user, CMC has entertainment value. Some describe it as "just plain fun" (Marks, 2000). The reason for this may be the emergence of "short attention spans" that some (e.g., Marks, 2000) have noted;

e-mail messages have to be, according to Al Gore spokesman Chris Lehane, "so over the top that people pay attention" (Marks, 2000; see also Lee, 1999).[22]

How does the context of e-mail explain the "play" aspect of e-mail interaction? When speakers are free to construct their own experience, they often engage in linguistic play: They speak more, expand their linguistic repertoire, and invent new elements of linguistic ritual (Danielewicz, Rogers, & Noblit, 1996). In particular, ritual language and humor can arise quickly in new linguistic situations and become dominant features of speakers' conversations (Corsaro, 1986; Danielewicz et al., 1996). CMC is a new domain; following Danielewicz et al.'s observation, it is not surprising that humorous and ritualized language would emerge here.

Inherent Features of E-mail

Separation of Conversants in Time and Space

In e-mail, conversants technically are spatially separated, but time is compressed as a function of speed of exchange. With e-mail, responses can come within a few minutes or sometimes within a few seconds.[23] Response time may be low because transmission is rapid and utterances may be formulated and received almost instantaneously, reflecting an inherent tension between asynchronicity and simultaneity. As a result, conversational "turns" and the length of the exchanges (in ongoing e-mail conversations) are commonly short (Baron, 1998) and frequent.

Vygotsky explained the importance that speed of exchange has to "oral speech," which he understood as a "dialogue" (with the implication that "monologue" maps onto "written language"):

> The speed of oral speech is unfavorable to a complicated process of formulation—it does not leave time for deliberation and choice. Dialogue implies immediate unpremeditated utterance. It consists of replies, repartee; it is a chain of reactions. Monologue, by comparison, is a complex formation; the linguistic elaboration can be attended to leisurely and consciously. (Vygotsky, 1934/1986:242)

The speed with which many users compose and send e-mail back and forth maps onto Vygotsky's formulation of oral speech.

Other types of users (and the same users in some instances) treat e-mail as a monologue device, slowly, thoughtfully, and deliberately creating missives that have a precise formulation and less directly interpersonal expression. In other words, although e-mail conversants *can* respond almost instantaneously, many e-mailers prefer instead to word their e-mails carefully, structure them logically and tightly, spellcheck them, and check them for etiquette before sending them off. Many users quickly become quite adept at combining the relative separation in time and space from their interlocutors to form a hybrid interaction.

Context Independence

Reliance on extralinguistic context is associated with spoken language (Baron, 1998). Similarly, Vygotsky noted that the psychological context of an (oral) utterance is revealed in gestures, facial expressions, and intonation. Oral speech requires contextual references in part because it is spontaneous (Baron, 1988); speech created "on the fly" lacks efficient resources for self-containment (i.e. "integration"; Biber, 1988). That is, we speak without the luxury of footnotes, appendices, glossaries, and so on. Certainly we may use parenthetical expressions when we speak, but these are usually perceived as digressions by the addressee, which tax the addressee's memory and processing skills. Thus, a person's speech is generally restricted to text whose background is at least in part understood or shared by the addressee. Written language, by contrast, cannot draw on paralinguistic means to reveal context. However, it allows for much speech planning, so that an author of written text has the advantage and, indeed, the duty, of carefully building context into written language, by various means. Without a carefully integrated context, a reader will not understand the text.

The extent to which e-mail is context-dependent (the extent to which it relies on information that must be included for clarity) can depend on the amount of lag time the conversants allow between responses. If response turn-around is rapid, speakers can engage in a conversation flow effortlessly because their memory of the last response is still fresh. In this case, as with oral language, the context is redundant in the conversational situation. Even for speakers who allow more lag time (several hours or days), e-mail can still be context independent, if they consistently use the "include previous message" feature. Clearly, if an e-mailer takes the liberty of a several-hour or several-day lag time and does not invoke the "include previous message" feature, the context is absent and so confusion runs rampant:

```
Date: Sun, 16 Nov 1997 13:13:05 -0500
Subject: Re: addresses
put for what? (if you don't include what you're responding
to, I have no idea what you're talking about!!!!!)
t.
—Begin original message—
> Date: Fri, 14 Nov 1997 16:47:34 -0500 (EST)
> Subject: Re: addresses
>
> Betcha Brother didn't have to put for it!!!!
```

The author of the original message is attempting the most convenient of both worlds: delayed reply and a default setting to "do not include previous message."[24] This combination of conveniences does not promote efficient and effective conversational exchange. Conversation cannot flow time-efficiently if some features of oral exchange are not present. Conversants must either have a speedy exchange (such that memory is not taxed), or else they must have an exchange that includes portions of the conversation partner's previous message, as a reminder of the content and context of the dialogue. Otherwise, the context must be built in deliberately (e.g. "As you wrote in your last letter, . . .").

Conscious or Nonconscious?

In its technical expression, e-mail is written language. Canonical written language is generally accepted to be produced with full consciousness. However, unlike canonical written language, e-mail messages are often composed in a "lower consciousness" style: Some e-mailers fire off messages rapidly, with the primary effect of expressing emotion. With its stream-of-consciousness effects, e-mail can be regarded as a direct outlet for inner speech (Rowe, 2001, 2007). Like inner speech, some e-mail style is truncated, representing Vygotsky's (1934/1986) "thought connected with words." In the following e-mail, the pragmatics are most clear, through use of all capitalization (known as "shouting" in e-mail circles), multiple punctuation marks, and misspellings in the heat of the composition.

```
Date: Wed, 24 May 2000 07:42:53 -0700 (PDT)
Subject: Re: r.
OH GOD!!!!!!
of course p. is going to want to see
that!!!!!!!!!!!!!!!!!!!!!!
well, i guess my friday night is palnned isn't it? do
you 2 want to meet us there??? the usual plan is drop
r. off at 7 and then go to the rodeo near
pleasantvalley and then go to the 8 or 9 show at the
grand. last time the car got broken into: that was NOT
in the plans!
we'llknow better about parking from now on!
kc
```

Other e-mailers are quite conscious of their adherence to and deviations from standard nonelectronic writing patterns. Where the language of e-mail is consciously attended to, it represents strongly self-conscious and self-reflective metalingual activity. This, too, implies a creative process at work when users carefully craft their e-mail messages to humorous or other effects. In terms of consciousness, e-mail combines the best of both worlds: Its nonconscious side reflects inner speech, which connotes intimacy; its conscious side reflects creative linguistic "art."

Creativity

Several scholars have emphasized the creative aspects of CMC—e.g., Rice & Love, 1987, on innovations and Collot & Belmore, 1996, on the frequency of nonstandard spellings, such as *yer*, *fer*, and nonwords, such as *offensivity*. One reason for this stylistic freedom may be that e-mail is driven more by the exigencies of communication and usage trends (Lan, 2000).[25] This assertion is supported by *Wired Style*, the style manual for online communication; it prescribes:

> We encourage you to . . . welcome inconsistency, . . . treat the institutions and players in your world with a dose of irreverence. Play with grammar and syntax. Appreciate unruliness. (Hale, 1996:96)

Creativity is consciously attended to. Conscious creativity within the CMC context is noted by Fry (1995):

> Put the average geek on the telephone and he or she will not be up to much. Put them behind a keyboard, however, and the act of literary composition forces a wit, an integrity, an insight, an emotional and moral honesty that would amaze even an optimist . . .

Creativity in the e-mail context is quite often more pointed or extensive than what would be expected in an oral context, perhaps in part because it is a written medium, which allows more time for creative formulations. For example, e-mailers may invent separate identities for themselves, even if their actual identities are known:

```
Date: Thu, 25 Jan 2001 16:20:01 -0500 (EST)
Subject: No Lunch Today
Today is fend for yourself day.
I have a dual with the Evil Ford Emperor today, and I will
be unavailable for lunch. Two of the Jedi LunchMasters,
myself and Master O. will be in Charlotte today.
May the Lunch be with you,
LunchMaster
```

Among its features, making it in many respects a written–spoken hybrid medium, e-mail provides a very convenient conversational tool not available in any other medium, written or spoken—namely, the "include previous message" feature available in all e-mail software and interfaces. Given the greater lagtime in e-mail responses, many if not most e-mail users engage the optional feature, which allows the respondent the option of including all or part of the previous message in the response, to keep up with the flow of the conversation. If both participants use the feature without exception, the entire message history is included in each mailing. The result is that the e-mail exchange is both conversational and allows for elaborated creativity. In such instances, e-mail becomes the electronic version of clever "note-passing" so common in those FTF situations where oral communication is impractical.

Creativity and Play in E-mailer Pairs

I have discussed one example of "anonymous" character play in e-mail in a one-to-many message (the "Lunchmaster" message). I now show creativity effects between two separate pairs of one-to-one e-mailers. The first set is between long-time friends, who, prior to e-mail, established a "middle name" guessing convention along the lines of "Hey, 'Efficiency' (etc.) is my middle name!," which has now been elaborated in their e-mail signatures:

```
Hey there,
Thanks for the editing. Your changes aren't bad, I have to
admit, so I guess you've still got it!
Thanks again and talk to you soon,
BQPR
```

```
So, just got back from a meeting in Raleigh at the City
Attorney's Office. I've finished setting up a database for
them and—you guessed it—they're tickled pink at all the
new possibilities and time-saving features. It interfaces
with another database I setup for a different city
department and they are starting to realize that I'm the
man. B"P"E.
```

```
Howdy,
[. . .] I'd also like to go back to Spoleto in June this
year, but that still needs to be decided. After the Vien-
nese cultural experience, we may not do that one.
[. . .] OK, we're off to a Mozart concert at Duke Chapel.
Talk to you soon, BCE (too easy)
```

```
Do you know if you can create a script [. . .] "F" in
Word? I need to create the character for some relational
algebra formulations in the database theory class.
B"F"E :)
```

```
Got a quick question: I was trying to reach L. and emailed
him at his hotmail account, but it came back undeliver-
able. Do you know if 1) he's still around and 2) what
another email address for him might be?
Thanks
BPPE (plumbing problems) :(
```

```
>I can't decide who should be decapitated first: Bush or
Rumnsfeld or
>Chaney. Those stupid shits piss me off to no end.
% add ari Fleischer to that list. What a scumbag!
>I think M. mentioned something about your summer deal
[. . .] You might be able to stay with her sister if you
want. [. . .]
% you know the dates—let me know. d. "i.t.b.i.a.s.t" g.
>Ciao for now,
>B STBIAS E
("still thinks Bush is a schmuck")
```

Following are examples of sibling e-mail between two brothers, one of whom[26] creates a pattern into which he weaves "cat commentary" (identifying the pet cat as *Calico, Mini-T, Lil'T, Sweets McT,* etc.), usually relating it to movies or rap music:

```
You rummy lorry! It's the chicken men of Krankor. And as
for the brits love of bad films, didn't they give us that
gem—"Attack of the Bees" [. . .]
P.S. Lil'T licks her fur in waves over here, boy.
RMT
```

```
I just finished "A bridge too Far." I may rent the movie
again. The Nazis loved to use fierce names for their weap-
ons, eg, the Panther and Tiger tanks. Can you imagine a
"Calico" tank. I envision a small tank that consume huge
amounts of petrol and emits a fowl odor.
RMT
```

```
An observation: After Lil't is fed, she reminds me of the
jedi-master Yoda
Both Ice Tea and Lil'T want me to drop some knowledge on
you, so: shotgun shells be bad fo' yo' health.
RMT
```

Mini-T and I saw a TV show were a cat "accidently" killed its owner by knocking a plugged-in hair curler into the tub with her. I hope Lil'T didn't get any ideas!
RMT

Remember the movie 2001? Today, Sweets McT was siting in-front of and staring up at the refrigerator, almost in a trance. It reminded me of that scene where the apes are fascinated by the Monolith. I could almost here the theme song. Daa Daa Dada!
RMT

I got some calls from BB&T this morning concerning a job.
[. . .]
P.S. Here's what Lil't will say to me:
Big T . . . I'm in one of these boxes . . . Find me!
RMT

I just saw the film "Legally Calico." It wasn't that good.
I received your diploma in the mail. Sweets wants to sleep on top of it, but I don't let her. Trust me, when she's yackin' on a bone, you don't want that.
RMT

Lil't says:
"Pack it up pack it in let me begin(to eat)."
and
"Big T you better back up."
or
"You pet me you get burned."
finally
"I came to drop bombs(turds)."
Have a good weekend.
RMT

Here's a recording of Lit' T's frantic 911 tape:
Help! Help! Oh, oh, I shit on myself! I can't stop crying.
I've been lisening to dave chapelle. He's funny.
RMT

```
By the way—Sweets doesn't like it when you yell, "flak gun,
twenty mill. Take it out!," into her face. Quote from
"Saving Private Ryan." They messed up and calld a Marauder
tank killer a Panther tank. Like they're going to slip
that one by me. I think not!
RMT
```

Clearly, rapid turnaround, high text visibility, and the socio-psychological environment, promote a proliferation of this type of creativity, and the brothers' style—if not the friends' style—is almost unimaginable except in the e-mail environment.

An E-mail Sister Code

The next creative e-mail code I discuss is a private code shared between two sisters. The private sibling code is restricted in its purest form to e-mail correspondence. The nature of the code's development is crucial: The first stage of this code emerged when the younger sister (S2) imitated their father's idiolect and dialect in parodied form, for ironic effect. The rest of the family (the sisters' mother and the older sister [S1]) adopted the code in similar fashion, using the code primarily to "quote" the father. The father later began to "quote" himself, using the parodied code. Other features of mixed origin also presented themselves as separate family codes (FCs)—a set of language parodies that included parodies of southern White and southern Black English, "Archie Bunker" variants (from the popular television show about a family in the Queens borough of New York City), and other linguistic renditions that reflected family "bonding." However, it was not until the sisters used the code in frequent e-mail correspondence that the linguistic features of the codes were combined, negotiated, and modified; some features even stabilized to some extent graphemically, phonetically, lexically, and syntactically, much as language forms undergoing conventionalization do.

When the sisters used the codes over e-mail, some striking changes occur. First, the rather chaotic and nonunified features of the family code are merged into one code (the "sister code"); that is, the different family variants now occur within the same messages, and even within the same sentences. Second, the variants are actively negotiated by the two sisters, and some stabilize to an extent. Third, the individual elements of the other family codes evolve to have specialized meanings that they did not bear before their use over the e-mail medium.

Many sociological and linguistic effects of the code's use result from the technological and psychological features inherent in e-mail communication. Technological features that promote the use and development of the code are lagtime, the quoting feature, speed and immediacy, and convenience. In tandem, these technological features provide the optimal environment for high use. One important psychological feature of e-mail that promotes the sisters' use of their code is the "distancing" effect of e-mail, which introduces a certain degree of objectivity with regard to addressee and text.

Moreover, much of electronic communication is game-oriented (e.g., Multi-User Dungeons [MUDs], Mud Object Oriented) or entertainment-oriented (e.g., some listservs); likewise, e-mail, perhaps in part because of the "novelty" effect it seems to have retained, also appears to invite light-hearted and casual communication. These observations are borne out in the sisters' use and development of their private code as a play language over e-mail. Although the sisters often used the code to discuss both professional and informal matters, they reverted to standard English or "standard colloquial" English when resolving disputes or discussing serious issues with each other.

Data following the sisters' changeover to e-mail show both a steady increase in the quantity of communication (Fig. 4.1), as well as notable rapid increase in code use already by the end of the second month of e-mail communication (Fig. 4.2).

By the end of the Month 4 there was also less negotiation of variants, and more stability in common linguistic forms. By late March, as Fig. 4.2 shows, code variants are more frequent (approximately 130 per 100 lines of text):

(1)
3/29/96, S2: Hey **FIBE**!!!!!
In uniquely **Fibe** fashion, I notice you didn't answer my question re: the nail parlor search!!! Well??? I've got more on my plate than I know what to do with: [. . .] support **wib yo daddy** [. . .] **yo daddy** is one **ub dem**! [. . .] support for my buddy and colleague R., the smoking menopausal study?! [. . .] aerobics. AAAAAAH!
bye, my **fabe fibe!**
LUB!
Ya bizzy sista!
p.s. no **tehhhhhm fa min**!

(2)
3/29/96, S1: **Girl, dare's alwa tehhhm fa min . . . ya** make **dat tehhm** if **ya hab** to!
Whibe I wuz ansring ya 2nd maeh, I had an application failure, so it was wiped out; didn't **hab** time to repeat. Still don't **hab tehhm** to talk. . . . finally got some excitement!!
Beh
lub ya!
E.

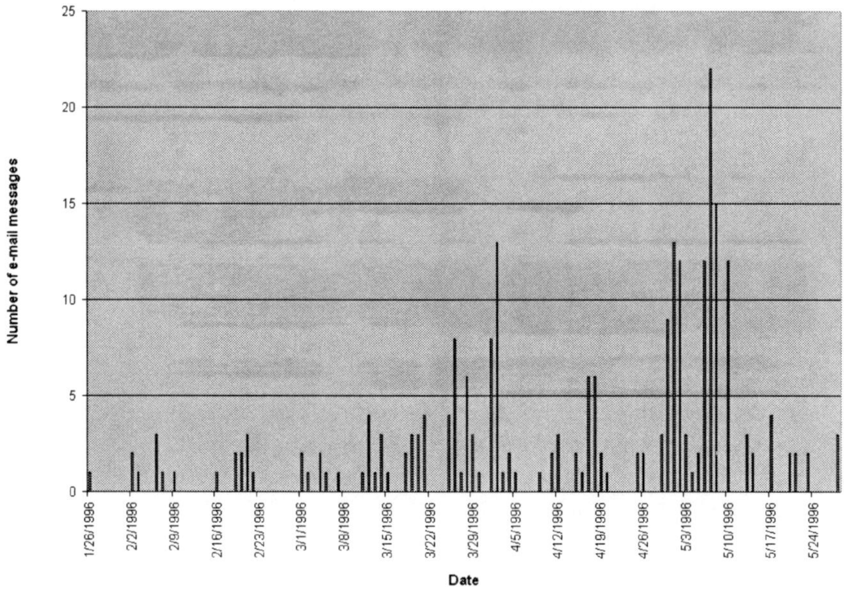

Fig. 4.1. Relative frequency of the sisters' e-mail correspondence over time.

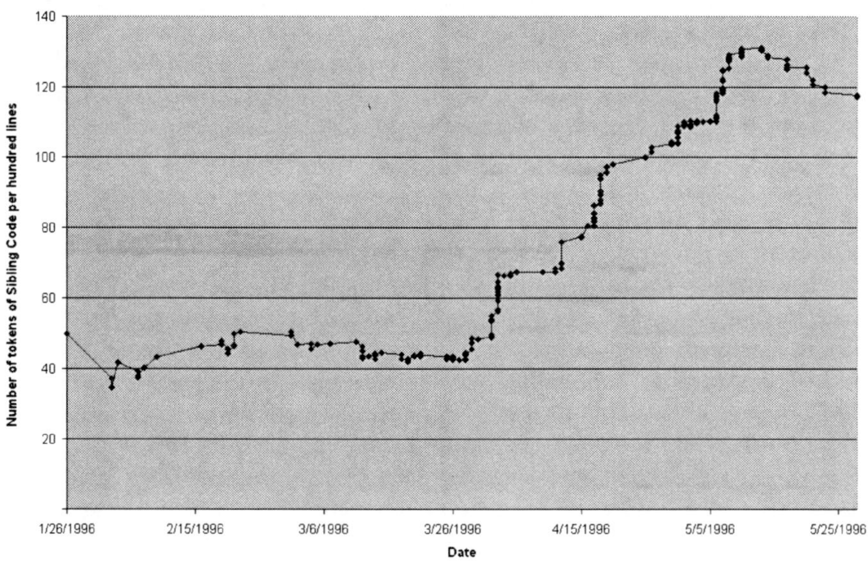

Fig. 4.2. Change in density of the e-mail sibling code over time.

(3)
4/2/96, S2: Hey! **Dat** was a **shote may**! **Yo daddy didn't raise no quiet youngns, naeho**! Have you communicated **wib de** parental unit yet???? [. . .] **Hab day** written you?
S1: **Ya no, dare ah** times when **shote maze is betta dan no maze! Ya lucky ya got dat**!
S2: **Ya na, dat sound jes like yo DADDEH**!

Apparently the electronic medium provided the unique conditions that promoted the development of the sibling code.

The most striking characteristics of the e-mail sibling code are the following: (a) its hybrid nature with regard to dialect (interweaving the dialects of different speakers of White southern English and Black southern English and New York-Queens, along with various individual speech parodies); (b) its hybrid nature with regard to the written/spoken polarity; (c) not only the speed but also the acceleration of its development; and (d) the changes effected in the structure of the code. All of these phenomena in the sibling code take advantage of—and rely on—the special features and conditions of the e-mail environment.

Interestingly, the code's form and development appear to have been calculated at every occurrence. And e-mail, by virtue of certain inherent features, such as the lag-time factor, and the viewability of every utterance, is the optimal medium for planned, self-conscious speech to occur in its most deliberated form. On the other hand, the open, liberating nature of e-mail allows the intrusion of unfettered and unchecked creativity, which is often linked with nonconscious activity. It is this tension between these opposing features that makes e-mail a unique genre, and thus it seems quite logical that new conventions would be constantly negotiated and set in this medium.

It is precisely the nature of the electronic medium that allowed in large part for the accelerated development of the family code to form the sibling code. Unlike traditional written correspondence, the electronic medium offered a speed of interaction approaching real time, which provided for more frequent communication. The sisters report that when they earlier had relied on telephone to communicate with each other, they spoke approximately once per week briefly. When the sisters switched over to fax, they corresponded approximately once daily with brief messages, largely due to issues of privacy and convenience.[27] But when they began to communicate by e-mail, the frequency of their interaction immediately increased to two or more times daily. This is likely attributable to several factors: the intimacy of the sibling relationship coupled with (a) the tendency of e-mail to "draw out" persons, (b) the convenience of e-mail, (c) the near-instantaneous transfer of messages available in e-mail mode, and (d) the "self-reflective" and creative nature of CMC expression already pervasive in e-mail culture. But other, more complex factors are also in play, such as (1) the juxtaposed cooperative-competitive nature of the sisters' relationship; (2) technological features such as the tension between the polarities of lagtime and speed, and the availability of the message history. The latter not only provides the necessary textual memory, but also foregrounds metalingual awareness, which provides the basis for some types of creative linguistic activity (Rowe, 2001, 2007).

Conclusion

There are various features of e-mail that contributed to the creative expression seen here, particularly in the accelerated change of the sisters' code over time. Although the speed of interchange is much greater in spoken communication than over e-mail, the written component of the e-mail mode affords a suspension of time constraints that appears to be indispensible in the development of certain codes: As written communication, e-mail leaves the "flow control" up to the respondent, allowing for careful, thoughtful, and well-formulated responses; e-mail in particular, in the "reply + include previous message" mode, allows the sender and receiver in each instance to view the communication history on responding. These effects are magnified when the respondent inserts her replies at the relevant points in the other participant's text. It was to a large extent the unique combination of high-speed interchange (relative to traditional mail) and suspension of time constraints (relative to spoken communication) that provided the optimal environment for the formation of the sibling code from the family code. On the other hand, the high-speed capability of e-mail encouraged a frequent flurry of communication; as a result one goal of the e-mailers in this chapter was apparently to entertain and "perform" for each other. Moreover, the combination of informality, the rapidity of exchange, and its inherent spontaneity—all mapped onto deliberate production in the written mode—allowed the nonconscious to emerge alongside the conscious creative process, in ways different from what is commonly encountered in prototypical written and oral exchange. As a result, the sisters' e-mail in particular sometimes assumed a chaotic form somewhat reminiscent of inner speech, but with a strongly creative "performance" component.[28]

Like slang, argots, and other special linguistic varieties, e-mail constitutes its own genre. Distinct from prototypical written language and oral language, it has characteristics of both genres. As a written medium, e-mail grants its participants a visual connection to the exchange; it also has a strongly conscious component that can produce deliberate creative activity. Like oral conversation, e-mail grants immediacy of context of the conversation and its text. That is, it is relatively context independent. Crucially, it is strongly dialogic: Like oral conversation, e-mail allows for a rapid exchange and lends itself to informal expression.

Distinct from most forms of both genres, e-mail provides time for message formulation, in tandem with a persistent memory and a vertical dialogue structure. In this sense, the nearest equivalent is classroom "note-passing"; it differs from this subgenre, too, however, because unlike note-passing, e-mail is not used as a last resort (when spoken communication is not possible), but often as the first choice of communication medium, even for times when telephone or FTF communication would be more efficient. Moreover, despite some users' guardedness about the possibility of unauthorized users intercepting their messages, personal e-mail is, unlike "note-passing," felt overall to be more reliably private.[29] As a result of this sense of security, unrestrained creativity is more likely to emerge. Furthermore, e-mail has quickly emerged as an "entertainment" mode, which means that sometimes users—for example, in messages from most of the participant pairs (in addition to the Lunchmaster mailings) shown here—use e-mail largely for the purpose of performing humor (see e.g. Schnurr & Rowe 2008).

E-mail, it can be argued, then, is electronic communication *par excellence*. The cumulative effects of its technical and sociopsychological features operating in tandem create a unique environment for rapid and accelerated change—a development almost unimaginable outside of the e-mail environment.

Notes

1. See Wyss (this volume) on intimate written communication both inside and outside of the e-mail context.

2. See Häusermann (this volume) on the journalistic use of telegraph. (See also Davis, 2006).

3. There is even a nostalgia Web site that provides an interface that mimics the classic telegram: http://www.retro-gram.com/.

4. See Bergs (this volume) for a lengthier discussion of the conceptual connection between telegram and SMS.

5. Of course, the effect of cost on the number of characters used more closely relates to SMS.

6. For some illuminating and often amusing historical accounts of these conflicts, see Fischer (1992). See Rutter (1987) on the cuelessness model; see also Ochsman and Chapanis (1974), and Pool (1977).

7. See Millar (this volume) who analyses instructions for recording messages, as promoted in prescriptive manuals on the topic—the very existence of which reveals a reaction to such cumbersome (and often irreversible) recordings.

8. On a voice pager, callers may leave an ordinary voice message, just as they would with an ordinary answering machine or voice-mail. Some pagers allow both means of "messaging."

9. 911 relates to the code dialed on a landline to contact public emergency services; 143 is "pre-teen note-passing" code (on hard-copy handwritten form, and predating newer

technologies) resolved as follows: 'I' is one character; 'love' consists of four characters; 'you' consists of three characters. Hence, '143'='I love you.'

10. On the functionality of modern faxes, see Rice and Schneider (2006).

11. For treatment of e-mail as a genre, see Yates and Orlikowski (1992), Orlikowski and Yates (1994), and Orlikowski, Yates, Okamura, and Fujimoto (1996).

12. See Millar (this volume) on e-mail style, particularly as a function of technology.

13. For a wide-ranging review of media characteristics, see Rice (1987).

14. For a review of its diffusion, see Rice and Webster (2002).

15. This may also be due the fact that people receive much more daily e-mail than conventional mail (Lan, 2000); this higher volume of mail may contribute to the response urgency, which results in some characteristic e-mail effects like emotionality, terseness, misspellings, and so on.

16. Some e-mailers are characteristically across-the-board blunter and ruder than they would be in any other context. This is perhaps nowhere more evident than in professional bulletin board lists, regardless of field or discipline, or in topic-oriented Web forums.

17. Walther (1996) also discussed how users can be more than personal through CMC by the ability to shape their messages.

18. See Walther (1996) on relationship development over time via CMC, and hyperpersonal communication through CMC.

19. Concerning e-mail and play, see also Trevino and Webster (1992) and Danet, Ruedenberg, and Rosenbaum-Tamari (1998).

20. Weizenbaum's Eliza-Computer showed a similar effect (http://i5.nyu.edu/~mm64/x52.9265/january1966.html) (Eva L. Wyss, 5/7/07 personal communication).

21. Arthur M. Schlesinger Jr., writing for *Brill's Content*, theorized that the challenges of the Computer Age are the result of the traumatic shift from a factory-based economy to a computer-based economy, which happened so rapidly as to deny adequate time to people and institutions to adjust. The impact of the Computer Revolution, Schlesinger wrote, is "dynamic, compressed, and drastic" (Schlesinger, 2000).

22. Lehane is referring to an "e-mail war" between year 2000 presidential candidates. 2000 has been identified as the first election in which the Internet played a significant part.

23. Of course, responses may also take days to arrive, or indeed never arrive at all, due to personal issues (the person chooses not to respond), mechanical issues (e.g., server issues, spam filters, etc.; Cramton, 1997). However, all interaction is hostage to external features, in this sense, e-mail is no different.

24. There are various reasons that some e-mail users do not use the "include previous message" feature when replying. For some, the default reply function in their e-mail application is set to "do not include previous message," and users either do not care to go into the preferences file to reset the preference, do not know how to reset it, or are not aware that it is possible to reset. Some other users describe the "include previous message" feature as annoying and avoid it for that reason.

25. See Millar (this volume) on the development of e-mail conventions.

26. Messages from the other brother in the pair were not available; but the informants recall that these messages were of a similar type.

27. Both sisters described their office fax machines as centrally located, with multiple users.

28. Baym (1995), in fact, asserted that humor in CMC can "define individuality," particularly because of humor's performative nature.

29. Of course, there are some individuals whose workplace practices involve surreptitious forwarding. Ultimately, it is mostly personal e-mails that are reliably private, due to the higher level of trust between participants which disfavors surreptitious forwarding.

References

Baron, N. S. (1998). Letters by phone or speech by other means: The linguistics of email. *Language & Communication, 18*, 133–170.

Baym, N. (1995). The performance of humor in computer-mediated communication. *Journal of Computer-Mediated Communication, 1*(2), 1–24.

Bell, A. (1984). Language style as audience design. *Language in Society, 13*, 145–204.

Biber, D. (1988). *Variation across speech and writing*. Cambridge: Cambridge University Press.

Collot, M., & Belmore, N. (1996). Electronic language: A new variety of English. In S. Herring (Ed.), *Computer-mediated communication: Linguistic, social, and cross-cultural perspectives* (pp. 13–28). Amsterdam: John Benjamins.

Corsaro, W. (1986). Routines in peer culture. In J. Cook-Gumperz, W. Corsaro, & J. Streek (Eds.), *Children's worlds and children's language* (pp. 231–252). Berlin: Mouton.

Cramton, C. (1997). Information problems in dispersed teams. In L. Dosier & J. Keys (Eds.), *Academy of Management best paper proceedings* (pp. 298–302). Cramton, GA: Southern University Press.

Danet, B., Ruedenberg, L., & Rosenbaum-Tamari, Y. (1998). Hmmm . . . Where's that smoke coming from? Writing, play, and performance on Internet Relay Chat. In F. Sudweeks, M. McLaughlin, & S. Rafaeli (Eds.), *Network and netplay: Virtual groups on the Internet* (pp. 41–76). *Journal of Computer-Mediated Communication, 2*(4). Available: http://jcmc.indiana.edu/vol2/issue4. Retrieved February 20, 2009.

Danielewicz, J. M., Rogers, D. L., & Noblit, G. (1996). Children's discourse patterns and power relations in teacher-led and child-led sharing time. *Qualitative Studies in Education, 9*(3), 311–331.

Davis, B. H., & Brewer, J. P. (1997). *Electronic discourse: Linguistic individuals in virtual space*. Albany: State University of New York Press.

Davis, M. (2006, February 3). Last post for the telegram? BBC News. Available: http://news.bbc.co.uk/2/hi/americas/4674782.stm. Retrieved March 5, 2009.

Duff, C. (1998, January 14). Pick up on this: Just don't answer, let freedom ring. *The Wall Street Journal*, Section A, p. 1, column 4.

Fischer, C. S. (1992). *America calling: A social history of the telephone to 1940*. Berkeley & Los Angeles: University of California Press.

Fry, S. (1995, November 13). I fell into the safety net. *Daily Telegraph* (page nr. not available).

Gains, J. (1998). Electronic mail—a new style of communication or just a new medium: An investigation into the text features of email. *English for Specific Purposes, 18*(1), 81–101.

Guinell, E. (1999). Interview [Talk of the Nation]. *National Public Radio*, Feb. 11, 1999.

Günther, U., & Wyss, E. L. (1996). E-Mail-Briefe—eine Textsorte zwischen Mündlichkeit und Schriftlichkeit [Email letters—a text type between spoken and written]. In E. W. B. Hess-Lüttich, W. Holly, & U. Püschel (Eds.), *Textstrukturen im Medienwandel* [Text structures in media change]. *Forum Angewandte Linguistik* (pp. 61–86).

Hafner, K. (2000, March 15). Hi, Mom. Hi, Dad. At the beep, leave a message. *New York Times*, Section G, p. 1, circuits.

Hale, C. (1996). *Wired style: Principles of English usage in the digital age*. San Francisco: Hardwired.

Herring, S. (1996). Introduction. In S. Herring (Ed.), *Computer-mediated communication* (pp. 1–10). Amsterdam: John Benjamins.

Hiltz, S. R., & Turoff, M. (1978). *The network nation*. Menlo Park, CA: Addison-Wesley.

Kiesler S., Siegel, J., & McGuire, T. W. (1984). Social psychological aspects of computer communication. *American Psychologist, 39*(10), 1123–1134.

Lan, L. (2000). E-mail: A challenge to standard English? *English Today, 64*(16), 4.

Lee, M. (1999). Out of the hood and into the news: Borrowed black verbal expressions in a mainstream newspaper. *American Speech, 74*(4), 369–388.

Lin, C., & Atkin, D. (Eds.), (2002). *Communication technology and society*. Cresskill, NJ: Hampton Press.

Marks, P. (2000). E-mail messages to the press have made the Gore-Bush race a cyberwar. *New York Times*. June 1, 2000. http://www.nytimes.com/library/tech/00/06/biztech/articles/01email.html.

Mihalo, W. (1985). The microcomputer and social relationships. *Computers and the Social Sciences, 1*, 199–205.

Ochsman, R. B., & Chapanis, A. (1974) The effects of 10 communication modes on the behavior of teams during co-operative problem-solving. *International Journal of Man-Machine Studies, 6*, 579–619.

Orlikowski W. J., & Yates, J. (1994). Genre repertoire: The structuring of communicative practices in organizations. *Administrative Science Quarterly, 39*, 541–574.

Orlikowski, W., Yates, J., Okamura, K., & Fujimoto, M. (1996). Shaping electronic communiation: The metastructuring of technology in the context of use. *Organization Science, 6*(4), 423–443.

Pool, I. (Ed.) (1977): *The social impact of the telephone*. Cambridge, MA: MIT Press.

Poster, M. (1990). *The mode of information: Poststructuralism and social context*. Cambridge: Polity Press.

Rice, R. E. (1987). Computer-mediated communication and organizational innovation. *Journal of Communication, 37*(4), 65–94.

Rice, R. E., & Danowski, J. (1993). Is it really just like a fancy answering machine? Comparing semantic networks of different types of voice mail users. *Journal of Business Communication, 30*(4), 369–397.

Rice, R. E., & Love, G. (1987). Electronic love: Socioemotional content in a computer-mediated communication network. *Communication Research, 14*, 85–108.

Rice, R. E., & Schneider, S. (2006). Information technology: Analyzing paper and electronic desktop artifacts. In C. Lin & D. Atkin (Eds.), *Communication technology and social change: Theory, effects, and applications* (pp. 101–121). Mahwah, NJ: Erlbaum.

Rice, R. E., & Shook, D. (1990). Voice messaging, coordination and communication. In J. Galegher, R. Kraut, & C. Egido (Eds.), *Intellectual teamwork: Social and technological foundations of cooperative work* (pp. 327–350). Hillsdale, NJ: Erlbaum.

Rice, R. E., & Webster, J. (2002). Adoption, diffusion and use of new media in organizational settings. In C. Lin & D. Atkin (Eds.), *Communication technology and society* (pp. 191–227). Cresskill, NJ: Hampton Press.

Rowe, C. (1996). Genesis and evolution of an email "creole." Unpublished manuscript.

Rowe, C. (2001). *Genesis and evolution of an email sibling code.* Unpublished doctoral dissertation, University of North Carolina, Chapel Hill.

Rowe, C. (2007). Building "code": Development, maintenance, and change in a private language. *American Speech, 82*(3), 235–261.

Rutter, D. R. (1987): *Communicating by telephone* (International Series in Experimental Social Psychology). Oxford: Pergamon.

Schlesinger, A. M., Jr. (2000). The vital center. *Brill's Content*, November, 50–54.

Schnurr, S., & Rowe, C. (2008). The 'dark side' of humor: Subversive humor in Hong Kong workplace email. *Lodz Papers in Pragmatics* 4.1. [Special issue on humour. 109–130].

Sherblom, J. (1988). Direction, function, and signature in electronic mail. *The Journal of Business Communication, 25,* 39–54.

Sudweeks, F., McLaughlin, M., & Rafaeli, S. (1998). *Network and netplay: Virtual groups on the Internet.* Cambridge, MA: MIT Press.

Trevino, L., & Webster, J. (1992). Flow in computer-mediated communication: Electronic mail and voice mail evaluation and impacts. *Communication Research, 19*, 539–573.

Vygotsky, L. (1986). *Thought and language.* Cambridge, MA: MIT Press. (Original work published 1934)

Walther, J. B. (1996). Computer-mediated communication: Impersonal, interpersonal, and hyperpersonal interaction. *Communication Research, 23,* 3–43.

Wyss, E. L. (2003): Metamorphosen des Liebesbriefs im Internet: Eine korpusgestützte, textlinguistische und kommunikationswissenschaftliche Bestimmung des Liebesbriefs und seiner Pendants im Internet [Metamorphosis of the love letter on the internet: A corpus-based text linguistic and communication-scientific designation of the love letter and its internet counterpart]. In J. Höflich & J. Gebhardt (Eds.), *Vermittlungskulturen im Wandel: Brief—E-Mail—SMS* [Changing communication cultures: Letter—e-mail—SMS] (pp. 199–231). Lang: Frankfurt a.M.

Yates, J., & Orlikowski, W. (1992). Genres of organizational communication: A structurational approach to studying communication and media. *Academy of Management Review, 17*(2), 299–326.

Yates, S. (1996a). Computer-mediated English: Sociolinguistic aspects of computer-mediated communication. In J. Maybin & N. Mercer (Eds.). *Using English: From conversation to canon* (pp. 76–83). London: Routledge.

Yates, S. (1996b). English in cyberspace. In S. Goodman & D. Graddol (Eds.), *Redesigning English: New texts, new identities* (pp. 106–133). London: Routledge.

Chapter 5

From the Bridal Letter to Online Flirting

Changes in Text Type from the 19th Century to the Internet Era

EVA L. WYSS

University of Koblenz-Landau

Abstract

Over the past two centuries concepts of love, as well as the nature of intimate relationships, have undergone modifications. Along with these modifications, the language of desire, the text type of the love letter and love letter-writing practice have changed as well. It is therefore surprising that certain elements of the correspondence between prospective brides and grooms of the 19th-century bourgeoisie seem today to be enjoying a comeback on the Internet. Some of the parameters, however, have changed. This chapter explores the similarities and differences of intimate literacy from a historical and pragmatic point of view. On one hand, it discusses the text type in its pragmatic, textual, stylistic, and medial aspects, and on the other, it focuses on a number of distinct writing practices.

Historical Perspective

A love letter is a "metasign" of love, as Roland Barthes (1977) noted in his work *Fragments d'un discours amoureux*. This semiotic term reveals the love letter's function: As a whole it signifies—or, more broadly, expresses[1]—the writer's love. This illocutive force evolves in interaction with the sociocultural knowledge of the text type "love letter" and love discourse in general.[2]

In this chapter, I focus not only on contemporary letters and especially Internet love messages and flirting, but also on letters written more than a century ago. I thus analyze correspondence from a historical perspective focusing on letters, text type, and writing practices as an integral part of evolving textual meaning and social practices. In a contemporary context, readers decode the meaning and the function of an individual text by referring to their more or less tacit knowledge of the "gestalt" (phenomenological form), specific cues indexing the text type, letter content, discursive norms, and performative rules that govern their milieu. A historical analysis, on the other hand, requires the reconstruction of the norms and rules of a sociohistorical context that is different from one's own contemporary milieu. Accordingly, one of the tasks of this chapter is to render comparable not only the text type of the love letter, but also a number of writing practices within their particular historical and media-specific writing contexts. In this sense, the chapter is to be understood as an interdisciplinary part of the more general project of the social history of private life and intimacy (see Aries & Duby, 1994), with a specific focus on the change of text type, language use, and linguistic practice in the tradition of Barton and Hall (2000), Mattheier (1987) and Schikorsky (1990), and and in relation to the change of writing culture in the context of the "evolution" of the culture of literacy following Goody (1977) and Ong (1982).

I begin by exploring letter-writing as a social and historical writing practice of everyday literacy. This means that I describe and analyze historical and contemporary language use following Wittgenstein (1953/2001) and as a transformation[3] of mediatized linguistic material that becomes meaningful in the moment of aesthetical experience, for instance by reading it and in relation to the context and interest of the reader. The gestalt, form, and function of the text as a linguistic transformation are then connected to and discursively prescribed from the author's environment. This discursive prescription is made explicit, for instance, by the models of letter-writing found in the numerous advice books of a certain period (cf. Ettl, 1984; Nickisch, 1991).

Thanks to the widespread alphabetization and the improved postal system around the turn of the 19th century, the letter—and especially the love letter—became a medium of communication for everyone, even for those who had little letter-writing know-how (see Elspass, 2005; Grosse, 1989; Jakob, 2000; Klenk, 1997; Mattheier, 1987; Messerli, 2002; Schikorsky, 1990). I then consider the writing of love letters as a means of indexing and performing social and personal identities (Butler, 1990; Eckert, 2000) within the realm of private, intimate correspondence and, in the case of e-mails, of semi-public[4] correspondence. The last section of this chapter concerns the consequences the changes in performance have for the interlocutors and their relationships.

Methodology

The letters I use for this study come from the Zurich Love Letter Archive (ZLA), a large collection of authentic love letters, cards, telegrams, e-mails, and text messages[5] written in German by women and men from all social classes, and from the predominant religious groups, who lived in Switzerland and Germany from 1880 to 2006.[6]

The fact that there is a preference for writing love letters makes this text type not only widespread, but also especially convenient for research on the interrelations of social, technological, and linguistic changes. Moreover, this text type has emerged in every new medium of communication used for individual communication in the 20th century. The aim of this chapter is to describe the historicolinguistic transformation of the text type "love letter" and to explore whether the changes we can see relate strictly to the text type, or whether something occurs to the performative function of the traditional lover letter when it is transposed into the emerging media of the 20th and 21st centuries. In order to answer these questions, I have sorted the empirical material found in the "uncensored" archive of letters according to the sociohistorical communicative context and three overlapping principal parameters: medium, linguistic categories (style, text type, codes), and writing practice. The data in this study are taken from the three last decades of the 19th century to 2007.

For this reason, the methodology employed here does not take the form of a straightforward comparison. Rather, the analysis is conducted in a dialogical manner, in which one medium is compared to another, and one set of similar letter types to another. This method results in a focus on similarities and differences of text types both between the media and between the text types within the same medium.[7] This generates a crossing of perspectives and an increased insight into the text types for a more profound understanding of private everyday literacy. It does so by focusing on the ongoing and changing uses of various media, and the practices that are "invented," "tested," and "introduced" into certain sociocultural milieus (Durkheim 1894/1967; see also Schulze, 1995).[8]

The Love Letter in the 19th Century

The 19th century was known as the century of correspondence (Chartier, 1997). The love letter must be understood as a part of this social institution. But what does the term "19th century" mean? The mathematical distinction from 0 to 99 is obviously not of historical relevance, as Schikorsky (1996) explained in relation to her discussion of history, which defines the 19th century as lasting from 1815 to 1918.

From a linguistic point of view, as von Polenz (1999) noted, the crucial question is how to define historical linguistic periods and their connection to historical ones. The study of linguistic change shows that in language (and language use) there is no abrupt, clear-cut change of linguistic features. Instead, there are often overlapping strands of different formal and functional linguistic features that evolve in parallel. In addition to formal and functional changes, linguistic pratices also emerge and end.

The Bourgeois Bridal Letter

The exchange of letters between bride and groom during their period of engagement became a "must" within the bourgeois context of the 19th century. This form of letter-writing is referred to as *Verlobungskorrespondenz* or engagement correspondence, which names the ritual it pertains to, or as *Brautbriefe*[9] or bridal letters, which refers to the authors of the letters. This form of letter-writing was so relevant socially that it was integrated into common language use. The fact that the practice was named, of course, also promoted speaking about it. Other forms of correspondence or letter-writing are not named.

In Germany, bourgeois couples who corresponded during their engagement usually did so with the permission of their parents. In fact, because the bride and groom were often chosen by their parents, the exchange letters often began without prior personal contact (Joris & Witzig, 1992). The practice of parents choosing their children's mates began to change in the middle of the 19th century, thus those corresponding would then typically know one another. By the end of the 19th century, the volume of correspondence between partners who were not designated by their parents increased. In other words, the emotional connection had more weight for the choice of partner. In contrast to U.S. 19th-century practice (see Lystra, 1989), German and Swiss bridal correspondence usually ended at the time of marriage. (See Fig. 5.1.)

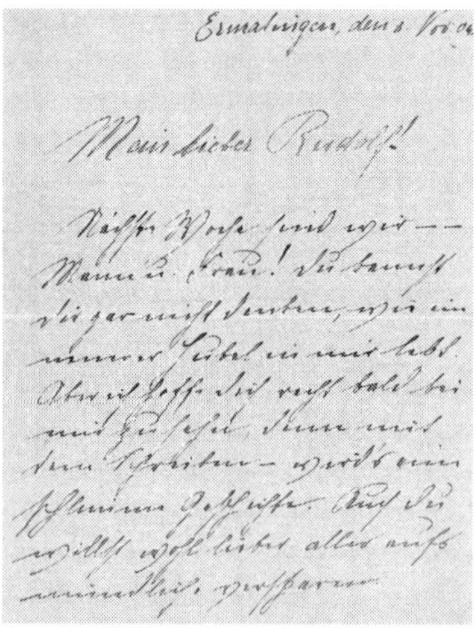

4296 Ermatingen, den 4. Nov 04

Mein lieber Rudolf!

Nächste Woche sind wir—
Mann und Frau! Du kannst
dir gar nicht denken, wie ein
innerer Jubel in mir lebt.
Aber ich hoffe Dich recht bald bei
mir zu sehen denn mit
dem Schrieben – wird's eine
schlimme Geschichte. Auch du
willst wohl lieber alles aufs
müncliche versparen

Fig. 5.1. Letter from Emmy to Rudolf 11-4-1904 (ZLA 4296); the last letter before they were married. (English translation: Next week we are husband and wife. You cannot imagine how jubilant I am within. But I hope to see you soon at my place, because it will be difficult to write. I think that you would prefer to talk when we meet.)

This tradition also is reflected in the semantic description (lexical definition) of the love letter in a well-known German thesaurus at the end of the 19th century, where Heyne (1892) wrote in his language thesaurus that a love letter is "written by engaged people" (650).

Reading the letter in Fig. 5.1, written after a 2-year engagement period, one can see that the bride had already taken on the role of the wife. Her enthusiasm was controlled. The passion discernible in the letters at the beginning of the engagement was lively. Her training as a housewife, as well as the long negotiation process that was undertaken between the parents and the preparations for the actual marriage, had changed the young woman (she was 17 years old). It is also evident that the young woman was writing in a polished style combined with a nice handwriting and good ink on beautiful, expensive brown paper. She not only uses an elaborated vocabulary, but wrote with a certain rhythm, line after line in a very narrow spacing that is—as Ehlers (2004) pointed out—related to informality. She had the competence to put a rhythmic pause (a caesura) in her first phrase to heighten its pathos.

Highly sophisticated norms seem to have governed the writing of this middle-class woman. Such usage can be called a bourgeois language culture, which can be expressed as a highly esteemed language practice that upper-class children learned at home in a set of strict and disciplined, and at times frightening, writing rituals (i.e., the New Year's greeting letter or the Christmas letter for the parents) performed under severe maternal pressure at their homes (Schultz, 2000). In the context in which letter-writing became ritualized, amatory correspondence was a rite of passage from the status of son and daughter to that of betrothed (bride and groom) and finally to husband and wife. In this context of ritualization it was the form of the letter and not its content that had to be emphasized. By writing aesthetically elaborate letters, people demonstrate their belonging to a distinct (Bourdieu, 1979) social and cultural milieu. This practice stands in sharp contrast to the contemporary notion of fixing thoughts by writing.

As Ettl (1984) noted in his work on the German epistolary advice book, the sex and age of the author were linked to a certain letter-writing style. There was a style in which the young bourgeois man showed his passion. The highest textual style, the literary style, showed his highest passion (see Table 5.1). The genre that best befitted this style was the poem.

The old man, on the other hand, was not supposed to feel fire but instead warmth, not sensitivity but boldness; he therefore did not write love poems for his bride. Love letters by the working class and by soldiers were supposed to be written with restraint.

Women—there is only one category of woman mentioned in the literature—were not supposed to show passionate feelings in their letters. They were meant merely to reply to a letter, and to do so with decency; and of course they were never to initiate amatory correspondence (Ettl, 1984). As Lystra (1989) showed, women did write passionately as did men. The ZLA also provides evidence of feminine passionate writing, when the young Lizbeth (1896) writes to her secret fiancé about her lust and desire, which expresses her frustration that they cannot live as she would like to:

> O, wie gerne will ich mich all dem fügen, was vorläufig noch nicht anders sein kann, u. dann später wird's schöner! Wir müssen uns dann nicht mehr auf so manchmal gar schmerzende Art bezwingen, u. möge Gott uns zusammen gefügt haben, dass Du in mir gefunden, was Du mein Teurer begehrtest!—Mein innigst Geliebter es ist wonniglich sich so vereint zu wissen, u. dürfen wir es dann aller Welt mitteilen, dann: "O Liebesglück!"—Mein lieber Rudeli, da ich Dich ja wahrscheinlich bis Sonntags nicht mehr sehen darf, so will ich doch in Gedanken glücklich bei Dir sein u. sage auf freudiges Wiedersehen! Sende Dir indessen mein Liebster die herzlichsten Grüsse u. die innigsten Küsse, Dein treues Liseli
>
> Oh how i would love to have everything that at present cannot yet be otherwise, and then later on, things will be nicer! Then we won't have to force ourselves so painfully, and may God put us toegher, that you found in me, what you my dearest desired! My true beloved, I am delighted to know that we will be unified that much and then we can tell the whole world: "We're in love!"——My dear little Rudy, since I probably can't see you again until Sunday, I will be happy with you in your thoughts, and say a joyful "Good bye," until we meet again! I send you hereby my dearest the warmest greetings and the deepest kisses, Your faithful little Lizbeth. (ZLA 7026 [1896])

So, we can say that there is a gap between the societal norms and factual normality: They do not mirror the writing norms, but rather are deontical valorizations of letter style.[10] If the highest writing style prescribed in letter-writing advisory books is called poetic, that does not mean that every line in a text was actually written in a poetic or literary style. But the named style preference is an anthropological formula that can be sketched as follows: Aside from the standardized salutation and close of a letter, high valuation of the addressee is indicated by means of a socially elevated writing style and writing competence.

An important function of amatory correspondence is the self-representation and self-assurance of the writer as a member of the bourgeoisie. Letter-writing in this sense could thus be understood as a performance of the bourgeois self. This is not merely a narcissistic maneuver because it relies on a certain bourgeois code of virtue and sociability. Very detailed emotional sequences are sometimes entered symbolically into the discourse, or staged as allegorical scenes, so that the phatic function and the construction of relationship fades.

Letter-Writing Without Much Linguistic Competence

Letter-writing in the 19th century was not only practiced by the bourgeoisie, but became more widely practiced among all social classes. This was possible because at the end of the 19th century it was not only alphabetization that promoted letter-writing among different social classes: People who were not used to writing did so nonetheless because they found themselves in uncertain circumstances, such as war and emigration (Elspass, 2005), and wanted to send a sign of life and to stay in contact with their relatives (Wyss, 2003a).

The linguist Isa Schikorsky (2001) analyzed the love correspondence of a lower middle-class (or petty-bourgeois) couple over a period of 11 months in summer 1870, during which time the man was fighting in the Franco-Prussian war. The couple corresponded every 4 to 5 days and this correspondence can be situated in a communicative "state of emergency" (70) in which their distance to literacy in general is expressed. The couple is used to neither reading nor writing in their everyday life. Their writing style, therefore, does not follow any epistolary standard. Indeed, no one style is discernible in this correspondence and the writing oscillates between informality and conformity, the most general model followed being the "pragmatic" business letter. Orality can be discovered in the letters as well: The authors imitate taking turns by organizing the sequential structure with formulas of address (see Assmann, 1995; Koch & Oesterreicher, 1985, 1997). Furthermore, the usual text-structuring devices, such as punctuation marks, paragraphs, and marks of direct discourse, are absent. The letters are characterized by a certain liberty of "orthographic rules, colloquial abbreviations and contractions, deviations of grammatical congruence, syntactical breaks and ellipses and a certain lack of a stringent argumentation" (Schikorsky, 2001: 73f.; cf. Häcki Buhofer, 1985).

At the same time, they show a certain overadaptation in hypercorrect forms. And, like bourgeois couples, they use stylistically highly marked language in sequences that express love and care. The topic of the letters is the everyday life they lead, so the functional focus is on the construction of relationship; the fact that they are writing could be seen as the need to overcome their separation.

Two Poles of Stylistic Variation

Focusing on the textual style, ranging from the letter paper to the written text, one can observe a diversity of stylistic qualities. The most elaborate letters are written in a "literary" style using such things as an elevated vocabulary ("ein Jubel lebt in mir" ['I am all jubilant within']), a rather complex or unusual syntactic structure, employing symbolic expressions of nature ("Heute Mittag stahl ich mich ganz leise ins Dörfchen. Von Konstanz konnte ich gleich mit dem Dampfboot heimfahren. Ein nasser Schirm wirbelte einem um den Kopf und die frische Seeluft erfrischte die Sinne, wie gern liess ich diesen Elementen ihr Spiel" ['At noon I absconded for the small village. From Constance I could take the steamboat home. A wet umbrella whirled around my head and the fresh air of the lakeside refreshed my senses, how much I liked the play of the elements']), an original address ("Meine liebe böse Braut" ['My beloved bad bride']), and an interesting or entertaining occurrence narrated to serve as a plausible reason to end the letter ("Heute habe ich vom Militär ein unangenehmes Schreiben bekommen, ich habe einen Meldungstermin verlümmelt. Nun muss ich mich daher gleich hinsetzen und den Leuten etwas vorlügen, denn Entschuldigungen gibt es bei Militär nicht" ['Today I received an unpleasant missive from the army, obviously a reporting date had slipped my mind. Now I have to sit down and write down a pack of lies, because there are no excuses in the army']).

And of course one also finds the literary topoi of passionate, romantic emotional expressions ("meine Liebe, mein Heimweh ist so gross, ach . . ." ['my love, I am so homesick, alas . . .']); occasionally even the closing was carefully chosen. Female and male members of the bourgeoisie wrote often such eloquent, correct or playful letters, employing an old-fashioned stereotypical literacy usually observed only in very formal literacy—letters with a sense of esprit modeled by the great German writers, whose letters were already being published in edited collections in the second third of the 19th century.[11] The textual aestheticization is not the only stylistic peak. Outstanding writing—the result of high competence—is combined with advanced letter design: "stylish" colored ink, nice handwriting and expensive personal letter paper with initials and decorations.[12]

On the other hand, there were less ritualized or even spontaneous exchanges of love letters. This form of letter-writing is uncommon and therefore is not a ritual, but is rather due to the Franco-Prussian war, or to migration—in these cases, the separation was the true[13] reason for writing. These two poles of aestheticization are not designing homogenous types of two letter styles but they are referring to a variety of stylistic features that are used also in combination with elements positioned on the other end and in between the poles.

To optimize the scholarly distinctions, I will avoid the high-low culture bias that imports unwelcome prejudices, as Elspass (2005) noted.[14] This is important for the analysis of letter practices because the so-called German bourgeois[15] culture is highly connected to Protestant (pietistic) practices, and not to the likewise bourgeois Jewish or Catholic practices. There is a risk of anticipating an implicit high-low assumption because of its connection to literacy in general. In fact, the aestheticization and ritualization can have the effect of a linguistic leveling of writing culture, which establishes a good basis for standardization processes.[16]

Table 5.1. Levels of Aestheticization[a]

High Level of Aestheticization	Low Level of Aestheticization
1. Poetic, literary, or elaborate letter style: eloquent, correct, or playful 2. Letters with esprit expressing the author's passion and emotional experiences within symbolistic impressions and stereotypes including moral reflections on society (letters with esprit); aside from the salutation and the greeting formula at the end, the structure is open/free 3. Literary topoi of passionate, romantic, emotional expression 4. Complex or unusual syntactic structure 5. Coherent and logical argumentation structure 6. In keeping with orthographical and grammatical norms	*1.* (Private) and business letters are the letter models that are orally perceived, learned in school, or copied from advisory books *2.* Stereotypical letter structure (address, thanks for the previous letter, inquiry about health, statement about one's own health, declaration of love, request for a date/present/reply, indication of the reason for not writing, valedictory formula) *3.* Standardized love-declaration formulas *4.* Stereotypical syntactical features *5.* Argumentative discontinuities and breaks *6.* Difficulties in keeping orthographical and grammatical norms: non written (dialectical) lexic, oral-syntactical structures

[a]Elements of styles therefore can be organized along a certain hierarchical scale, but the hierarchies do not mirror social stratification, because styles—as Eckert (2000) shows—are mixed or clustered selections of language elements people use to follow peer-group norms.

Amatory Correspondence in the 20th Century

In the 20th century the forms of correspondence known in the 19th century continued; however, with a number of changes. I distinguish, generally speaking, among three different strands of written amatory communication. They cannot be classified only according to a chronological progression because their application (as a sort of communication habitus) does not occur exactly at the same time in each social milieu, or even for each person with his or her preference for communication. Because of the simultaneity of historically different social practices—for instance, in the 1960s some couples are still writing love letters according to the traditional models, whereas others might be living together without being married—it is difficult to position these strands in a clearly defined period of historical progression.

However, the first strand of letter-writing continues with similar differences between bourgeois and petty-bourgeois couples, as in the 19th century. The second strand is the type of letter-writing that is mixed up with other written and oral communications: the telegram for long-distance and rapid communication, and the telephone for closer distances, are used more and more often in (upper) middle-class society. The postcard is important as an inexpensive communication for lower middle-class couples (see Fig. 5.2 and Diekmannshenke, chap. 6, this volume). Conversation takes up more space and time within cultures relative to writing: correspondence (letters, faxes, e-mails) becomes a supplement to conversation, both face-to-face and by phone calls. The latter became the usual form of communication for couples (see the next section).

The third strand evolves in the Internet that creates new space for, and entails new forms of, amatory correspondence: an adaptation of handwritten practice and very special form of written flirt communication (see section on correspondence on the Internet).

Mixing Media

The phenomenon of mixing media—the second strand—is related to the emergence and easier access to different types of media in the 20th century. Although the telegram was used by members of the upper middle class to announce an arrival, the postcard became a very important medium for love discourse around the beginning of the 20th century. The phone, by contrast, did not become important until the 1960s, when there was a boom of subscriptions that reached the 50% level in Switzerland in the 1970s (see Fig. 5.2, p. 110).

The 20th century also stood for big social changes in Western cultures: for instance the up-and-coming leisure culture, the importance of youth movements, the sexual revolution and women's liberation. These changes affected pick-up communication, dating rituals, communication patterns between dating couples, and perceptions of partnership and intimacy.

A psychosociological study from Döring (2003) on the use of media for amatory communication showed that the media choice for private amatory communication depends on many social and psychological factors: People use the mobile phone because both partners possess one and they travel a lot; they use chat rooms because they both have access to computers with a standing online access and one of them does not like to talk on the phone; others report that they write e-mails because they like to write, and they have access to the Internet at home.

As the ZLA shows, letter-writing was in fact not the sole form of communication between lovers in the 20th century. Correspondence was mixed with other written and oral forms of communication. Nor is written communication limited to love letters. In the ZLA, men often used a telegram to announce to their lovers—who was not in every case a fiancée—when they planned to arrive at the central station. Telegrams also had a status similar to love letters as they too were often considered collectables. The postcard is an inexpensive—and only partly private—means of communication for a short message with the advantage of a tidy design and the ability to speedily celebrate an event of relevance to the couple, such as an anniversary or a holiday (Diekmannshenke, this volume).

Around the 1920s, "other" means of communication were mentioned and entered the love-letter discourse. People would, for instance, refer to the telephone: "—Warst Du heute schlecht gelaunt oder meint' ich's nur am Telephon? Du warst so kurz angebunden" [ZLA:550] ('Were you in a bad mood today, or did I take it the wrong way on the telephone? You were so curt').

In the following note from the early 1930s the whole text focuses on a call. First the problem is introduced, namely that George could not reach Elisabeth by phone, then the anxiety that the call was not welcome or might even have disturbed the other person is expressed, and finally the "solution" is offered, when the author asks his lover to call him at his number.

> Liebstes Lisel!
> Bin nun ganz traurig dass ich Dich
> heute nicht telephonisch erreichen konnte. Aber weist
> Du, ich habe den Eindruck, es wird eben nicht sehr
> gefällig aufgenommen, wenn ich Dir anläute.
> Darf ich Dich deshalb herzlich bitten, mir morgen
> (Freitag) um 11h oder 18h No 34645 anzurufen,
> das heisst natürlich . . . . . wenn Du mich treffen
> magst! Bin ich ein sehr grosser Gauner deswegen?
> Ich möchte Dir sehr sehr viel sagen, aber ich
> kann jetzt nicht, obwohl ich ja so reichlich Zeit hätte.
> Tausend herzliche Grüsse und Küsse
> Georges. [ZLA 3393]

> ('Dearest Lisel! I am very sad that I was not able to reach you by telephone today. But I have the impression my calls are not entirely welcome. May I ask you, therefore, to call me tomorrow (Friday) at 11 a.m. or at 6 p.m. at no. 34645, that is of course . . . if you would like to meet me! Do you think this makes me a very big scallywag? I would like to tell you a lot of things, but now I am not able to, although I have lots of time. I send you my love and a thousand kisses, Georges.')

This example documents not only the problems people had with telephone calls, but also that they dealt with them by means of written communication.

The letter was also used to announce future telephone calls. For instance, on October 23, 1924, Robert wrote a letter to Elisabeth (from Brienz in the Bernese alps to Zurich): "Wenn nicht Du telephonierst so telephoniere ich / Sonntag morgen zwischen 1/2 11 u 11 Uhr und dann / besprechen wir mündlich wegen dem andern Sonntag." [ZLA 3314]. ('If you do not telephone I will/on Sunday morning between 10:30 and 11:00 and then/ we will talk in person about the following Sunday.')

The war situation augmented letter production (Schwender, in press), although in the 1940s it moved a lot of Swiss to purchase a phone subscription, while the boom in the 1960s and 1970s meant that the phone call became a normal thing for most people.[17]

This kind of correspondence is not found in the 1960s, 1970s, 1980s, and early 1990s in the ZLA letters. In the 1960s, the appearance of the phone could be considered an important reason for the abundance of letters. However, because at that time couples could meet freely, the personal exchange of intimacies was more common (Vincent, 1995).

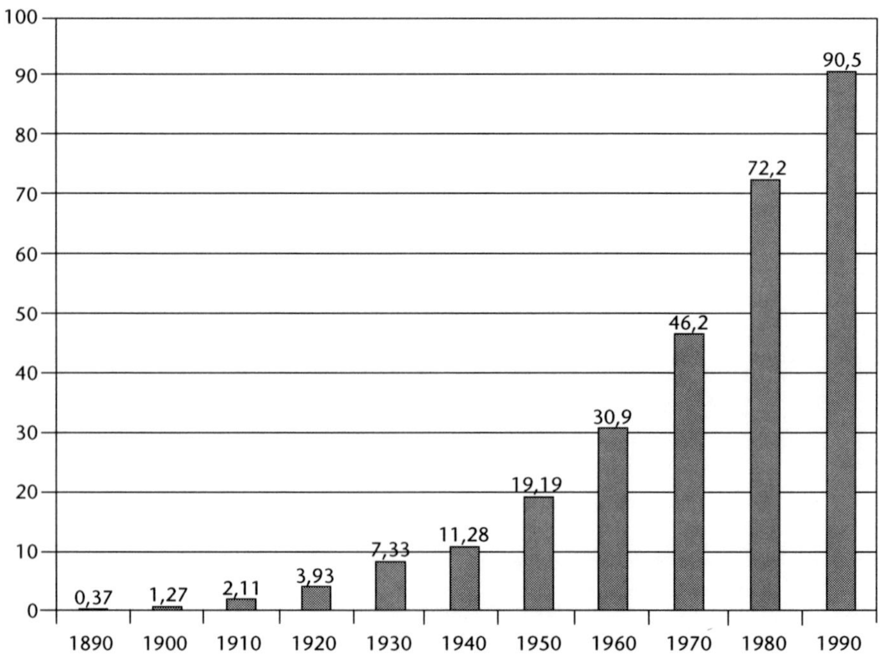

Figure 5.2. Phone stations in Switzerland (per 100 inhabitants since 1890).

Correspondence as a Supplement

When correspondence becomes a supplement, letter-writing is no longer a higly valued (bourgeois) writing practice. Private face-to-face interaction becomes ever more possible with the change in leisure and youth culture, an interaction that will play a more important role in the phases of beginning and establishing a love relationship. (Symptomatic for this change is the myth of the disappearance of the love letter: I am often asked whether people today still write love letters.)[18] Thus, the act of writing letters complements conversations and becomes a supplementary practice. In combination with the growing importance of the telephone and in relation to the content of the letters, there is a tendency to give correspondence a new function: Correspondence becomes a supplementary communication.[19]

In the following example, the writer refers four times to meetings that have either taken place the same week, or that she is expecting by the weekend.

> St. Gallen, 19 Dez. 47.
> Mein lieber Uli! <ein Herzbild>
> Schnell möchte ich Dir mein "lieber" ein/paar Worte schreiben, Wie geht es Dir?/Hoffe gut, was auch bei mir der Fall/ist. Bist Du noch gut heimgekommen, am/Mittwoch? Ich war grad etwa 5 Minuten/in meinem Zimmer, als es Licht gab im/Treppenhaus und die beiden Herren fort/gingen, also hatten wir wieder einmal/Glück. /Nun möchte ich Dir nur noch mitteilen/wegen morgen Abend. Weiss noch nicht ob/Müllers ausgehen. Aber nun habe ich//noch allerlei Weihnachtsabeiten zum fertig/machen, und Päckli machen. Nun muss/ich das einfach <Streichung> Samstag abend in/Angriff nehmen, den nächste Woche ist/nicht mehr viel los, und auf den/letzten Tag kann ichs auch nicht sparen. /Hoffe Du werdest die Zeit schon/verbringen, und bald ist es ja/wieder Sonntag. Vielleicht muss ich/morgen Nachmittag in die Stadt. /Werden wir uns dann wohl sehen? /Nun aber muss ich schnell unter/die Decke, es ist schon wieder ziemlich/spät. /<Ein Herz>/Nun empfange meine liebsten Grüsse/und viele Küsse von Deinem Klärli. [ZLA 1289]

> ('St. Gallen, Dec. 19, 1947.
> 'My dear Uli [short form of Ulrich]! <image of a heart>
> 'My "dear" I want to dash off a quick note to you. How are you ?/ Hope you are well, as am I. Did you get home all right on/Wednesday? I was in my room for only 5 minutes when the light came on/ in the stairwell /and the two gentlemen left/ so we were lucky/again. /Now I just wanted to let you know /about tomorrow evening. Don't know yet whether/the Muellers will go out. But I still /have a number of Christmas preparations to take care of/ and parcels. But now I simply have to/tackle this <cancelled> on Saturday evening, for next week not much will be happening, and I cannot save /it to the last day. Hope you will have a nice time, after all / it will be Sunday soon. Perhaps /tomorrow afternoon I will have to go to town. Do you think

> we might meet? But now I have to get under the/ covers quickly It's quite late already/. <a heart>/Now I send you my most heartfelt greetings /and many kisses from your Klaerli [diminutive of Klara].')

Letters like this no longer serve as a kind of standardized self-presentation of middle-class people: This is more of a caring gesture that performs closeness without much content beyond reflexive remembering and the managing of dates as a way of constructing intimacy.

The following written text—as mentioned earlier—is an example for a face-saving form of communication for "difficult" and "forbidden" topics: In this case the ambivalent feelings about first sexual experiences. A young woman writes the following lines (excerpt) to her fiancé (ZLA 1253: 10–19 [1947]):

> Schau, manchmal finde ich es gar nicht/recht von mir, wenn ich immer nein sage,/wenn Du recht lieb mit mir sein willst./ Du tust mir dann sehr leid, auch ich/meine manchmal es nicht mehr aushalten/zu können, und doch, gell, wir wollen/uns beherrschen. Wir halten ja/noch eine sehr schöne Zeit vor. Wie muss/das schön sein, einmal von Dir Kinder/zu haben, und sie zusammen wieder/zu guten und lieben Menschen zu erziehen. /Ja, auf das freue ich mich sehr.
>
> ('See, sometimes I don't think it is /right of me to say no all the time / when you try to be really loving with me./ I always feel very sorry for you then, I too/ sometimes think I can barely contain myself/, and yet, right, we want to/ control ourselves. We are saving up a beautiful time. How lovely it will be to have children with you and to raise them together to be good and lovely people. Yes, I look forward to that very much.')

Bound up in a complicated ideological and moral argumentation, the author discussed the man's sexual desire from his perspective and called it euphemistically "loving." His desire was described (by her) as a sensual bodily force that makes the man suffer, and makes the woman sad in turn. She did not consider herself fair when she said no. The letter is taken as a means for discreet communication designed to avoid conflict, which was a must for bourgeois interaction.

When they were corresponding (and living in the same village) the correspondence took on a new status as a special practice. It could have been a practice that mimicked bourgeois correspondence, but it lost its very crucial function of marriage introduction. If the people knew each other, fell in love, and met, the virtuality of the written letters in which written fantasies could be experienced together no longer held, because it was reduced by face-to-face communication. Letter-writing turned out to be a nostalgic, romanticizing practice. And the simple, unelaborate writing style was no longer of importance to the correspondents.

De-Ritualization of Correspondence and the Introduction of Visual Literacy

After World War II, private letter-writing increasingly became a spontaneous act of passionate and intimate expression. Everybody could do it, and it no longer had the mark of a bourgeois virtue, but instead that of an extraordinary, even artistic, expression of love. Due to this de-ritualization of amatory communication and the refocusing of the love declaration as an explicit and faithful communication, a severe conflict arose. In the de-ritualized context, the former adequate language of love seems to have lost its meaning[20] and—even more troubling—there was no new "unused" way to express love. One solution to this vexing problem was in a combined communication: Some tried to add meaning by including texts of popular love songs. With this new focus on linguistic semantic content, the stereotypical language of love became impossible. At this time, visual literacy was introduced in the everyday love letter—as a form of implicit language critique and as a symptom for lost meaning. Language was no longer the one and only way to say "I love you."

Changes in design and layout gradually altered the ways of presenting the letter (Wiese, 2000). There was no sudden visual turn in private literacy. From the start of the 20th century and even 20 years earlier, one can observe the gradual beginning of visual communication in handwritten letters. One writer inserted small signs—religous crosses or hearts—within the written lines, another designed his letter as a friendship album, still others complemented their writings with sketches as the bourgeoisie used to do in books of sketches (in the 18th and 19th centuries, see Wyss, 2002).

From the 1960s onward there is no engagement correspondence in the ZLA. It is possible that people continued to write letters nonetheless, but that they no longer had the same impact as they did 60 years before. What I note in my love-letter archive is correspondence over a limited time span when couples were separated for work or studies. Correspondence was also used in the case of hidden relationships (with priests, with married people) to stay in touch with each other, and to communicate openly, because at this time the use of the telephone in a family context was observed and could become a family event—and of course it was not possible to enjoy privacy or to hide telephone communication from other members of the family.

But it goes deeper: The letters in Fig. 5.3 and 5.4 show that language was no longer an adequate medium in which to express love. The authors of love letters preferred a synaesthetic expression, featuring color or even a mathematical graph.

13. Juli 1965

Dieses Blau erinnert
mich an Dich:
Seine Tiefe, seine Leucht-
kraft und seine Ausstrahlung
wird es immer behalten.
#

Fig. 5.3. Synaesthetic letter: verbal and visual expression are equivalent and complementary.

Another example is that of a mathematical design, with which the writer constructed a formula of love:

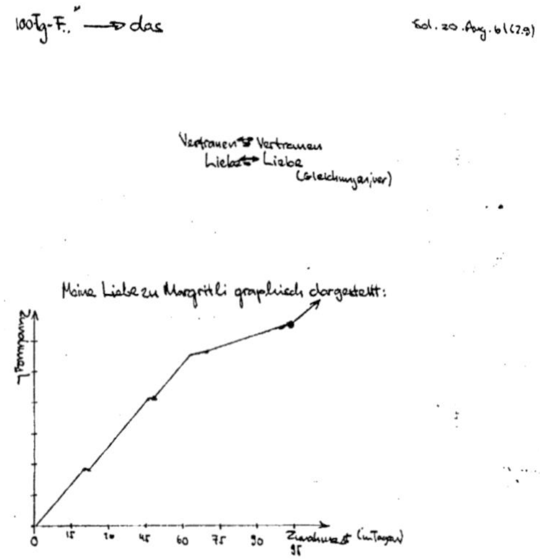

Figure 5.4. Declaration of love in the form of a graph: x = days, y = love.

In the 1970s, many unmarried couples lived together and there was no longer a need to perform certain rites leading up to marriage. The (general) frequency of letter-writing decreased, so much so that the German post office launched the ad "Schreib mal wieder"[21] ('Why not put it in writing?').

New motifs could be observed: Young people and elderly adults wrote not only in Swiss dialect, but also in very long paragraphs, stream-of-consciousness style, about their feelings, about morality, and the authenticity of life. Younger and shyer people tried to get dates by writing love letters in the school setting. Another novelty that marked the 1970s was that children began to write love letters and notes in the classroom (Canaan, 1990; Cherubim, 1981; Wyss, 2003a; Ziegler, 2006); children even began at this time to write short love letters.

Like the letters of married couples in the 19th century and during war times, love letters became more a means to sustain a mutual everyday life. Therefore, the epistolary style of the letters resembled that of a personal diary.

If lovers lived in the same town, they wrote small notes, which they put on each other's beds, on mirrors in baths, or on refrigerators. The letter began shrinking to the size of a note. It seems that writing lost its importance to the discourse of love.

The consequence of this casual treatment of letter-writing was a different performative function. The letters are seen more as unexpected gifts, nice to have as a source of joy, but lacking the existential and relationship-constructive impact the bourgeois bridal letters enjoyed as an introductory ritual.

In the 20th century, the love letter evolved into a whole textual family of written intimate communication and showed new forms and codes of intimacy: erotical nicknames for men and a more variable and colloquial addressing (Wyss, 2000).

The 21st Century—Correspondence on the Internet

The new media will determine a development in the kinds of choices people are able to make with regard to the medium in which they choose to express themselves. The new media can—if the sociocultural contexts of the users allow it—support new forms and uses of communication and new attitudes toward language use and communication. In this case, the new medium changes the social value of the media itself, which will in turn lead to a rearrangement of the media functions and uses (Krotz, 2003).

The invention of media and their adaptation to personal and individual communication creates the possibility of intermedia crossings analogous to the mixture of text types (Wyss, 2003b).

A clear cut came in 1971 with the rise of a new writing technique: e-mail. E-mail was first used by U.S. academics on ARPANET and, nearly 10 years later, by European academics as well, and in the 1990s an increasing number of people in the Western world had access to e-mail facilities in their offices, in Internet cafés, or at home.

With the advent of e-mail communication, the informal and socioemotional communication on Internet increased, as Rice and Love (1987) showed as early as the 1980s. This referred both to the frequency of the messages written as well as to the socioemotional contents of these messages. Moreover, Schönberger (1999), writing about the German situation in the 1990s, noted that people showed certain preferences in the selection of the addressees of their e-mails: The users do not write letters to unknown people, but to friends (90% domestic, 61.3% foreign) or acquaintances (73.6% domestic, 52.9% foreign). This is the quantitative part. For the qualitative part, it seems that the new media offer the user a surplus—a new, exciting, occasionally confusing experience of self with risks of losing the self that were not easy for the early members of the virtual community to handle.

The e-mail constitutes the invention of a new middle-class writing habit in the northern hemisphere. In the 19th century, the alphabetization line distinguished those educated either at home or at school from those with only little education or none at all. Internet literacy divides other groups and milieus, and sometimes the dividing line goes between the groups (as the European project on the virtual society has shown, see http://virtualsociety.sbs.ox.ac.uk/). All milieus (the academic context included) show different specifications of nonusers. For the purposes of this chapter, however, it is obvious that love letters are increasingly written on computers and sent by e-mail.

The letter is the medium that serves to celebrate the lovers' fantasies, for instance the discontinuous "quasi"-mutual remembering of things they have experienced (in real life or online) or even the invention of some actual erotic or sexual fantasies occurring exclusively in, by and through the text.

Every transfer of a text or text type from one medium to another is marked by two leading principles: on the one hand there is the so-called stylistic inertia principle (Bausinger, 1972/1984), which contends, generally speaking, that parts of the text type or characteristics of the genre are conserved throughout the media change; on the other hand, however, there is an innovative force that leads to a media-specific innovation.

"Ordinary" Amatory Correspondence by E-Mail

There are couples living in the same flat or working at the same place who use e-mail to write notes to each other. They also write e-mail love letters or even organize and discuss their vacations using e-mail. This resembles the use of written communication forms that were described earlier. However—and this is a renaissance of 19th-century correspondence practices—there are also many couples who have their first "conversation" or a first "date" on the Internet, then take a liking to one another and begin corresponding. The letters could be flirty at the beginning, but when the two fall in love they initiate a "real" amatory correspondence. The messages no longer constitute an assumed identity to mask or hide the goal of letter-writing: The aim is to get to know each other better, to express very private feelings and to be intimate with somebody else in the real (or virtual) world (see Baker, 2005; Lenz, 2006).

Many people begin a love correspondence by placing a personal advertisement on a dating site and then exchanging letters. In these letters they tell their partner about their everyday life, confide secrets and tell stories about their families. Like the couples in the 19th century, today's couples correspond daily or several times a day. Their linguistic self-representation is a way of forging a relationship (*Beziehungsarbeit*), a way of establishing closeness, trust, and intimacy[22] (Wyss, in press).

These letters are not colloquial, elliptic, and noncoherent. Instead they show a complex and well-organized topic development, are well structured, and use salutation and valedictory forms, as well as unmarked language, correctly, and thus demonstrate greater literacy.

Communicative standards are evolving on the Internet and many surveys show that, if a nonvirtual relationship is desired after corresponding, a first nonvirtual contact occurs with a phone call. More physical reality is created by writing faxes, where the handwriting tells more about a person. Then after some time, a first date establishes a love relationship in real life, or a sudden silence lets the other know that there is no desire to meet in real life (Döring, 2000; Turkle, 1995).

The linguistic construction of a relationship[23] is established with responsive and cooperative performances (on the locutionary and illocutionary levels) or sibling or couple codes (see Langford, 1997; Rowe, 2001, 2007). The letters reveal a great deal of private life, which creates not only closeness; the exchange of personal experiences, opinions, and interests also builds up mutual trust. Furthermore, these letters demonstrate several forms of the construction of intimacy. The most important features are the caring words, the declaration of love and the establishment of mutual fantasies (Wyss, in press).

But the Internet not only transports e-mails from one mail host to another. Aside from textual and lexical Internet culture codes (the e-mail layout with header, body and signature, the abbreviations, smileys, emoticons) and computer-mediated communication adaptation features (the contextualizing of the letters by talking about office problems, the naming of a certain experiential lack due to the virtuality of the context, the problems in de-anonymizing each other; see Wyss, 2003b), there is a more crucial difference between the handwritten correspondence of the 19th century and current e-mail correspondence: the perception of one's own identity as well as that of the other. In the 19th century, the future bride and bridegroom did not know each other in a sense very different from that of the Internet: In the 19th century, although the two writers had not met yet, they knew a lot about each other from their parents, other relatives, and the like. But on the Internet, the ignorance of the other is real. People may have only a nickname or a user ID, which, while stimulating certain fantasies, leaves it unclear whether this cyber identity is desirable and to what extent it correlates with the actual author. This creates not only some crucial psychological problems but also problems on a more linguistic level. First of all, because readers cannot know whose utterances they are reading, they experience something utterly new: an enormous and bewildering lack of context. Such context is the missing fundament for the handling of meaning, although through further communication the communicators do establish their own, strictly literate context from the verbal information they are gathering.

Online Textual Flirting: A New Written Communicative Genre

Virtual reality (VR) is a real or simulated environment that is characterized by its vividness and interactivity. The experience the author has while using this virtual medium is real. The Internet offers a stage for role-playing games with masks and pseudonyms (Beisswenger, 2000, 2001; Turkle, 1995). This creates a perfect environment for flirting. In flirting, it is crucial to be kept guessing about the actual desires of the addressee. On the Internet, where it is unclear who is hidden behind a user ID or a nickname, this situation is the norm. But how does flirting operate through e-mails? By communication with written messages? No glances, no attractive bending of the head, no blushing is visible in the asynchronous texts, but there is still a big secret, a technically created mystery. The dialectic between one's own virtually constructed "desirable" identity and the desired identity increases the thrill of this specific communication.

A flirt could start as a synchronous communication in the chat room or Multi-User Dungeon, after which there might be a first e-mail exchange, without commitments. There are, however, cyber love stories[24] in which intimacy and passion become central. In flirting, it is important to remain in a noncommittal state (Döring, 2000). Flirt letters are playful, volatile, momentary, minimalist verbal meetings without a clear objective (or with a well-camouflaged objective; see Cameron & Kulick, 2003).

> dearest romeolino
> wollte dir nur eine versuesste nacht wuenschen . . . . . .
> angel
> (ZLA 4889)
>
> ('dearest romeolino
> just wanted to wish you a sweetened night . . . . . .
> angel')

Users toy with one another by feigning confusion, they make interesting allusions, ask rhetorical questions, and use erotic nicknames[25] (Danet, 2001). The nicknames turn into self-selected pet names in the duty of attention control.

From: Orophea@xxx.com
To: beramis@xxx com
Subject: AW.: Romeo calling
Date: Tue, 17 Aug 1999 12:40:01 EDT
habe was vergessen . . . .an was sind die philosophen gescheitert?? an der liebe etwa??? [. . .]
kläre mich doch bitte auf . . . [. . .] (ZLA 3975)

('forgot something . . . why did the philosophers fail?? . . . could it have been because of love??? [. . .] enlighten me please . . . [. . .]')

The flirt text becomes more effective, that is to say the sender remains longer in the center of attention, the vaguer the text is. As a produced, ambiguous self, its writer does not want to be an author and instead uses intellectually stimulating quotations to pad his or her e-mail.

So the queston is whether a flirt letter can be called a love letter. Indeed, the flirt letter is the playful counterpart to a love letter. It is a letter written in a romantic context and might be compared to the older letters of a love affair. It does not make much sense to define the love letter according to the possible states of love, which differ not only in terms of historical context, regional and cultural setting, sexual orientation and gender lines, but also from one romantic phase to another. An important differentiation, however, is the performative status of the flirt letter. Flirting by e-mail is not only an exchange of communication but also a way to linguistically construct a relationship. Its linguistic actions are designed to achieve curiosity and interest without a commitment. At every moment of the flirting process, it is unclear whether there will be a further reply. The "lovers" keep a certain emotional distance. Nonetheless, the colloquial style of the letters affects a certain closeness. The colloquial style, also referred to as language of closeness,[26] is a result of conceptual orality in the production of the text (Assmann, 1995; Koch & Oesterreicher, 1985/1994). There are studies of adolescents and teenagers turning to short message services to instigate flirting, so that if they are rejected it is not public or even in physical presence, so it is easier to avoid losing face (Brown, 2002; Thurlow & Brown, 2003).

It can happen—when both partners reply in a speedy rhythm—that this form of e-mail communication becomes almost synchronous, like the communication on a private channel in a chat room. The quick writing results in a noncoherent text that does not follow the syntactic and orthographic rules. Flirting online is a communication by cryptic texts with erotic allusions, sometimes with a (pseudo-)poetic character but also with stereotypical sequences from the traditional Western discourse of love.

habe das gedicht noch nicht gelöscht . . . ich muss sagen, es berührt mich doch . . . . . . aber lieber wurde ich in einem boot dahergleitent bei mondschein und sternenhimmel, in die unendlichkeit oder, in einem meer aus rosen mich

> niederlegen und mich vom blütenduft berauschen lassen und den himmlischen stimmen folgen, die mich in die ewigkeit verlocken. (ZLA 3976)
>
> ('have not yet deleted the poem . . . I must say, it does touch me . . . . . . but I would prefer to float in a boat in the moonlight under starry skies into infinity, or to lie down in a sea of roses and be carried away by the smell of roses and follow the celestial voices, which beckon me to eternity')

The stylistic characterization is ambivalent as well. The letters are expressive texts that stretch dialogical structures. All of these features also belong to orality.

An Intermedial Historical Comparison—Similarities and Differences From the 19th Century to the Present

Not every text type is as rich as the love letter, which allows us to obtain new knowledge and permits a generalization of writing practices. The love letter is advantageous for media-historical sociolinguistic comparative research, not only because it is very frequent, but also because it can be found in different media, social milieus, and historical periods.

As noted earlier, there are three different strands of written love communication, which cannot be classified according to a chronological progression alone: Bridal correspondence is the 19th-century practice of letter-writing by a couple engaged to be married and persists in the 20th century. During this later period, however, letter-writing is mixed up with other written and oral communications and is important for middle-class couples. In this new context correspondence (letter, fax, e-mail) becomes a supplement.

As a third strand the Internet (in particular e-mail and the World Wide Web) creates two new forms of intimate writing: the e-mail correspondence and online flirting. On the one hand there are love letters in which not many adaptational features have changed, whereas on the other hand one finds e-mails that differ in form, content and function very much from e-mail correspondence: flirt e-mails, whose rapid exchange causes them to develop a performative function, which might be called online flirting.

During the 19th century, writing love letters was personal, but not private in today's meaning of the word. The letter was a secretive means of making sure that the relationship during the engagement was lived beyond the family's control. Letter-writing at the beginning of the 20th century was an obligation and a civic virtue. When corresponding became a supplement, the letter could occupy another function: no longer assigned to a distinctive behaviour, it offers the possibility of conversing in a "civilized" way about things that usually cause shame. Today people sit in front of their computers and write love letters during work. They regard love-letter writing as a pleasant distraction.

In other cases, the love letter and/or love note is connected to creativity and having fun. The separation of two people is thus no longer the reason for written communication.

With the differentiation of the love letter in e-mail correspondence and in the context of online flirting, we can observe the effect of intermediality: the Internet transforms a formal oral communicative performance into a written one.

This development represents a very strong new impact on the culture of writing. With the advent of new writing features such as e-mail and text messaging, the culture of writing is currently changing much as visual culture changed with the invention of video, handy cam and Internet cam. The rhythmic structure and quasi-synchronous nature of these new tools have rendered the practice of writing fully performative and thus made of writing a form of experiencing.

Acknowledgments

I would like to thank the Swiss Canton of Solothurn that supported my research and publishing work generously with the grant *Werkjahr des Kantons Solothurn 2008.*

A former version of this chapter was first published in the *Journal of Historical Pragmatics* (John Benjamins Publishers Ltd. Amsterdam, New York).

Notes

1. Bühler (1934/1999) mentioned in his theory of language that all linguistic signification can be divided into three functions: the expressive, the appellative, and the representative text types. He follows a well-known differentiation of literary genres into the poetic (expressive), the epic (representative), and the dramatic (appellative), as formulated in classical genre theory (Plato, Aristotle, and Horace).

2. Sandig (2006) and Brinker (2001), for instance, are exponents of the position that, from a merely holistic point of view, the text function is conceived of as the principal and, as Sandig pointed out, the prototypical feature among other textual dimensions. Other positions focus on the complexity of textual meaning, which cannot be fully described by pragmatic theory, as for instance Heinemann and Viehweger (1991), Adamzik (2000), Fix (2000) and Feilke (2003).

3. The concept of transformation is related to Lévi-Strauss (1958/1959) and indicates the change of status between the empirical textual material and the text as a significative entity.

4. To protect the authors of the letters, I have changed all identifying features such as names and sometimes even locations.

5. Since 1998 I have assembled a corpus of 7,500 love letters written in German between 1850 and the present in Germany and Switzerland by women and men, Christians and Jews, highly educated persons and those with little schooling. Contributors have been

sending me copies of their own letters (notes, e-mails, text messages), letters written by their parents, or by their children, for the scholarly analyses since Winter 1997/1998 (for further details on the archive see www.evawyss.ch).

6. The archive also carries an implicit ideological valorization: The dearth of material extant that makes an investigation of amatory communication possible is due not only to the intimacy of the texts, but also to the fact that archives predominantly collect letters and autobiographical sources from artists, poets and other public personalities in relation to the high value of the bourgeois cultural practice. There are therefore very few authentic documents written by so-called ordinary people (cf. Elspass, 2005). I developed the Zurich Love Letter Archive (ZLA) in order to collect authentic love letters for a linguistic study on love letters. In public and private archives of the 19th and 20th centuries, only few examples of love letters can be found (cf. Joris & Witzig, 1992; Lystra, 1989).

7. The telegram (since 1860), the postcard (since 1870), the telephone call (since 1890), the fax (since 1964), the e-mail (since 1971), instant and text messaging (ICQ, 1996), and SMS or text messaging (since 1992 in Europe and Asia). Practices surrounding text messaging differ in some countries, depending in part on the price structure of the service.

8. The result of the intermedial comparison is often considered to be a method that leads to a theoretical concept of intermediality, one which takes the results of the comparison as a way of revealing the so-called nature of a medium. My own interest in intermediality is not in media essences or media ontologies, but rather in the functions and practices they entail.

9. At the beginning of the 19th century, the so-called bridal letter was on the one hand a legal term referring to the letter written by the bride or the bride's family, in which the dowry was specified, and amounted to a "written record of that brought in by the woman" (DRW, German Dictionary). In the 18th century the term *letter to the bride* was used. This term describes for instance the letters written by Friedrich Schiller to his bride Charlotte von Lengfeld from 1787 to 1790, the letters of Eduard Mörike to Luise Rau (1829–1833), and Theodor Storm's letters to Constanze Esmarch, 1844–46.

10. This fact shows how self selected norm authorities are painting an image that is far away from actual language use, and thus, linguistics should take these differences into consideration.

11. The rules of business writing were taught in school and served as the models most people had access to. Members of the bourgeoisie, by contrast, learned their writing style at home by reproducing or copying model texts such as poetry and letters from canonized writers such as Klopstock, Goethe, or Schiller written in the second third of the 19th century.

12. The visual codes of politeness and distance have been weakening since 20th-century letter-writing, as explained Ehlers (2004).

13. In the ritualized love letters of a bridal correspondence, separation was not the cause for writing. The exchange of letters was arranged in the form of an agreement. The parents determined how many letters the "young" people should or could write. And it was the mother's duty to read them (or to listen to their reading by the son or daughter).

14. Elspass (2005) discovered in some important papers on 19th-century literacy an implicit doubling of the low–high culture bias: when they focus on lower class literacy they focus on submissive speech acts, which obviously determines linguistic style and will distort results on literacy of "ordinary" people (cf. Jakob, 2000; Klenk, 1997; Mattheier, 1990).

15. The three different fields of "Verbürgerlichung" or "bourgeoisification" are the political-juridical, the socioeconomic and the cultural-religious. German social history has begun only recently to take into consideration non-Protestant formations and to focus as well on the differences between the prototypical Protestant (liberal) and Jewish or Catholic bourgeois culture. For the Jewish bourgeoisie, see Lässig (2004) and the Catholic bourgeoisie; see Dowe (2006).

16. The linking of writing culture with bourgeois literacy only—the lack of integration of a broader literal culture—is the reason "standardization" processes have been revisited (cf. Jahrbuch des Instituts für deutsche Sprache, 2005). Elspass criticizes, for instance, the use of the term "standard" in two directions: on the one hand, he condemns the prioritization of written language over other oral-lingusitic expressions, what he calls a "scriptism," while on the other hand he unmasks the term *standard* in his use in Germanistic linguistics as a deontic term, a term that implies a valorization of how something should be: it stands in fact not only for high-culture German literacy in general, but has to date beeen used as a term to *prescribe* how language should be used.

17. For this period the ZLA gathers more single letters, and there are other reasons for writing: first letters (to plan a first date), letters between people in secret relationships, letters as a means of communication in long-distance relationships.

18. Couples also met in earlier times. It depended quite a lot on their status. While working-class people—if they had enough money to have a family—had more possibilities to meet, the bourgeois bride and groom met in highly staged circumstances: the common promenade/walk, the familial invitation for supper/dinner, the proposal visit of the future groom. These ritualized bourgeois meetings were combined with a certain romanticizing and transfiguring experience of nature ("birds singing," "waves rushing"). This makes clear that the context of face-to-face interaction, too, was not only charged symbolically but was also connected to a tendency to idealize and glorify.

19. Some informants (who were engaged at this time) told me about using the phone (they remembered the old-fashioned phones in the corridor of apartments); others told me about writing letters to each other.

20. This conflict is not to be confused with the critique of romanticism established in 20th-century love letters.

21. See Nickisch (1991, p. 69).

22. In this trust-building context, it becomes clear that the demasking of the online identity or the revelation of a real and different identity becomes an admission of a lie which always comes too late and is therefore made (almost) impossible.

23. The construction of relationships is described in sociological literature (Lenz, 2006) as a form of negotiation.

24. Nicola Döring (2003) showed the different dating contexts and types of pick-up situations on the Internet and lists a typology of love or sex relationships based on Internet communication (cf. Lenz, 2006).

25. See Deuel (1996) on virtual sex as a social interaction.

26. Rowe (2001, 2007) finds a code of closeness, a "sibling code," in the e-mail correspondence of two sisters, which is an effect of the rhythm of communication. Langford (2003) finds a code of closeness, which she calls "baby talk," in Valentine's Day messages in print media.

References

Adamzik, K. (2000). Was ist pragmatisch orientierte Textsortenforschung? [What is pragmatically-oriented text type research?]. In K. Adamzik (Ed.), *Textsorten: Reflexionen und Analysen* [Texttypes: Reflexions and analyses] (pp. 91–112). Tübingen: Stauffenburg.

Aries, Ph., & Duby, G. (Eds.). (1994). *A history of private life.* Boston, MA: Harvard University Press.

Assmann, J. (1995). Kulturelle Texte im Spannungsfeld von Mündlichkeit und Schriftlichkeit [Cultural texts in the orality-literacy range]. In A. Poltermann (Ed.), *Literaturkanon—Medienereignis—Kultureller Text* [Literature canon—media phenomenon—cultural text] (pp. 270–292). Berlin: Erich Schmidt Verlag.

Baker, A. J. (2005). *Double click. Romance and commitment among online couples.* Cresskill, NJ: Hampton Press.

Barthes, R. (1977). *Fragments d'un discours amoureux* [Fragments of an amorous discourse]. Paris: Seuil.

Barton, D., & Hall, N. (Eds.). (2000). *Letter writing as a social practice.* Philadelphia: John Benjamins.

Bausinger, H. (1984). *Deutsch für Deutsche. Dialekte, Sprachbarrieren, Sondersprachen* [German for Germans: Dialects, language barriers, special languages]. Frankfurt a.M.: Fischer. (Original work published 1972)

Beisswenger, M. (2000). *Kommunikation in virtuellen Welten: Sprache, Text und Wirklichkeit* [Communication in virtual worlds: Language, text, and reality]. Stuttgart: ibidem–Verlag.

Beisswenger, M. (2001). Das interaktive Lesespiel. Chat–Kommunikation als mediale Inszenierung: Chat-Kommunikation. Sprache, Interaktion, Sozialität und Identität in synchroner computervermittelter Kommunikation [The interactive reading game. Chat communication as medial staging: Chat communication. Language, interaction, sociality and identity in synchronous computer-mediated communication]. In M. Beisswenger (Ed.), *Perspektiven auf ein interdisziplinäres Forschungsfeld* [Perspectives on an interdisciplinary research field] (pp. 79–138). Stuttgart: Ibidem–Verlag.

Bourdieu, P. (1979). *La distinction—critique sociale du jugement* [Distinction: The social critique of judgment]. Paris: Editions de Minuit.

Brinker, K. (2001). *Linguistische Textanalyse. Eine Einführung in Grundbegriffe und Methoden* [Linguistic text analysis: An introduction to basic concepts and methods]. Berlin: Erich Schmidt.

Brown, A. (2002). *The language and communication of SMS: An exploratory study of young adults' text-messaging.* Unpublished bachelor's dissertation, Cardiff University, Cardiff, UK.

Bühler, K. (1999). *Sprachtheorie. Die Darstellungsfunktion der Sprache* [Language theory: The representational function of language]. Stuttgart: Lucius & Lucius. (Original work published 1934)

Butler, J. (1990). *Gender trouble: Feminism and the subversion of identity.* New York: Routledge.

Cameron, D., & Kulick, D. (2003). *Language and sexuality.* Cambridge: Cambridge University Press.

Canaan, J. E. (1990). Passing notes and telling jokes. Gendered strategies among American middle school teenagers. In F. Ginsburg & A. Lowenhaupt Tsing (Eds.), *Uncertain terms. Negotiating gender in American culture* (pp. 215–231). Boston: Beacon Press.

Chartier, R. (1997). *Correspondence: Models of letter–writing from the Middle Ages to the nineteenth century.* Cambridge: Polity Press.

Cherubim, D. (1981). Schülerbriefchen: Neben–Kommunikation [Student notes: Auxiliary communication]. In J. Baurmann, D. Cherubim, & H. Rehbock (Eds.). *Beobachtungen und Analysen zum nichtoffiziellen Schülerverhalten innerhalb und außerhalb des Unterrichts* [Observations and analyses of unofficial student exchanges inside and outside of the classroom] (pp. 107–168). Braunschweig: Westermann.

Danet, B. (2001). *Cyberpl@y. Communicating online.* Oxford, New York: Berg.

Döring, N. (2000). Romantische Beziehungen im Netz [Romantic relation on the net]. In C. Thimm (Ed.), *Soziales im Netz* [Social life on the net] (pp. 39–70). Wiesbaden/ Opladen: Westdeutscher Verlag.

Döring, N. (2003). Internet–Liebe: Zur technischen Mediatisierung intimer Kommunikation [Internet love: On the technical mediation of intimate communication]. In J-.R. Höflich & J. Gebhardt (Eds.),*Vermittlungskulturen im Wandel. Brief—E-Mail—SMS* [Changes in medial cultures: Letters, email, and SMS]. Frankfurt a.M., New York, Wien: Peter Lang Verlag.

Dowe, C. (2006). *Auch Bildungsbürger. Katholische Studierende und Akademiker im Kaiserreich* [Even educational citizens: Catholic students and academics in the empire]. Göttingen: Vandenhoeck & Ruprecht.

Durkheim, É. (1967). *Les règles de la méthode sociologique* [The laws of sociological methodology]. Paris: Les Presses universitaires de France. 16th edition (Collection: Bibliothèque de philosophie contemporaine; original work published 1894)

Eckert, P. (2000). *Linguistic variation as social practice: The linguistic construction of identity in Belton High.* Oxford: Blackwell.

Ehlers, K.-H. (2004). Raumverhalten auf dem Papier. Der Untergang eines komplexen Zeichensystems dargestellt an Briefstellern des 19. und 20. Jahrhunderts [Spacial behavior on paper: The decline of a complex sign system as represented in letter writers of the 19th and 20th centuries]. *Zeitschrift für germanistische Linguistik, 32,* 1–31.

Elspass, St. (2005). Standardisierung des Deutschen. Ansichten aus der neueren Sprachgeschichte 'von unten' [The standardisation of German: Views from recent historical linguistics 'from below']. In L. M. Eichinger & W. Kallmeyer (Eds.), *Standardvariation: Wie viel Variation verträgt die deutsche Sprache?* [Variation in the standard: How much variation is the German language undergoing?] (pp. 63–99). Berlin, New York: de Gruyter.

Ettl, S. (1984). *Anleitungen zur schriftlichen Kommunikation. Briefsteller von 1880 bis 1980* [Navigating written communication: Letter writers from 1880–1980]. Tübingen: Niemeyer.

Feilke, H. (2003). Oberfläche, Kontext und Lexikon [Surface, context, and lexicon.]. *Referat auf der GAL–Tagung 26.9.2003.* [Proceedings from the GAL Conference Sept. 26th, 2003]. Tübingen.

Fix, U. (2000). Das Rätsel. Bestand und Wandel einer Textsorte. Oder: Warum sich die Textlinguistik als Querschnittsdisziplin verstehen kann [The puzzle: The condition and change in one text type or Why text linguistics can be regarded as interdisciplinary].

In I. Barz, U. Fix, M. Schröder, & G. Schuppner (Eds.), *Sprachgeschichte als Textsortengeschichte* [Language history as text type history] (pp. 183–210). Frankfurt am Main: Peter Lang.

Goody, J. (1977). *The domestication of the savage mind.* Cambridge: Cambridge University Press.

Grosse, S. et al. (1989). *»Denn das Schreiben gehört nicht zu meiner täglichen Beschäftigung«. Der Alltag kleiner Leute in Bittschriften, Briefen und Berichten aus dem 19. Jahrhundert* ["For writing is not among my daily activities." Everyday life of ordinary people in petitions, letters, and reports from the 19th century]. Bonn: Dietz.

Häcki Buhofer, A. (1985). *Schriftlichkeit im Alltag. Theoretische und empirische Aspekte—am Beispiel eines Schweizer Industriebetriebs* [Everyday writing: Theoretical and empirical aspects—Case study of a Swiss firm]. Bern: Lang.

Heinemann, W., & Viehweger, D. (1991). *Textlinguistik. Eine Einführung* [Text linguistics: An introduction]. Tübingen: Niemeyer.

Heyne, M. (1892). *Deutsches Wörterbuch 2. H–O* [German dictionary 2. H–O]. Leipzig: S. Hirzel.

Jakob, K. (2000). Zwischen Alltag und Institution. Antrags- und Beschwerdeschreiben am Ende des 19. Jahrhunderts [Between the mundane and the institution: Writing for the purpose of petitioning and complaining at the end of the 19th century]. In A. Häcki Buhofer (Ed.), *Vom Umgang mit sprachlicher Variation. Soziolinguistik, Dialektologie, Methoden und Wissenschaftsgeschichte* [On dealing with linguistic variation: Sociolinguistics, dialectology, methods, and scientific history] (pp. 299–310). Tübingen, Basel: Francke.

Joris, E., & Witzig, H. (1992). *Brave Frauen, aufmüpfige Weiber. Wie sich die Industrialisierung auf Alltag und Lebenszusammenhänge von Frauen auswirkte (1820–1940)* [Good 'women' and insubordinate 'wenches': How industrialization affected the everyday life and meaning of life for women]. Zürich: Chronos.

Klenk, M. (1997). *Sprache im Kontext sozialer Lebenswelt. Eine Untersuchung zur Arbeiterschriftsprache im 19. Jahrhundert* [Language in the context of social life: An investigation of workers' writing in the 19th century]. Tübingen: Niemeyer.

Koch, P., & Oesterreicher, W. (1985). Sprache der Nähe—Sprache der Distanz. Mündlichkeit und Schriftlichkeit im Spannungsfeld von Sprachtheorie und Sprachgebrauch [Language of proximity, language of distance: Orality and literacy in the range of linguistic theory and language use]. *Romanistisches Jahrbuch, 36,* 15–43.

Koch, P., & Oesterreicher, W. 1997. Schriftlichkeit und Sprache [Literacy and language]. In H. Günther & O. Ludwig (Eds.), *Schrift und Schriftlichkeit. Ein interdisziplinäres Handbuch internationaler Forschung (HSK 1. Halbband)* [Writing and literacy: An interdisciplinary manual of international research (HSK v. 1 of 2)] (pp. 587–604). Berlin, New York: de Gruyter.

Krotz, F. (2003). Kommunikation im Zeitalter des Internet [Communication in the age of the internet]. In J. R. Höflich & J. Gebhardt (Eds.),*Vermittlungskulturen im Wandel. Brief—E-mail—SMS* [Changes in medial cultures: Letters, email, and SMS] (pp. 21–38). Frankfurt a.M., New York, Wien: Peter Lang.

Langford, W. (1997). "Bunnikins, I love you snugly in your Warren." Voices from subterranean cultures of love. In K. Harvey & C. Shalom (Eds.), *Language and desire. Encoding sex, romance and intimacy* (pp. 170–185). London: Routledge.

Lässig, S. (2004). *Jüdische Wege ins Bürgertum. Kulturelles Kapital und sozialer Aufstieg* [Jewish means into society: Cultural capital and social advancement]. Göttingen: Vandenhoeck & Ruprecht.

Lenz, K. (2006). *Soziologie der Zweierbeziehung. Eine Einführung* [Sociology of pairs]. Opladen/Wiesbaden: Westdeutscher Verlag.

Lévi-Strauss, C. (1958/1959). *Strukturale Anthropologie I und II* [Structural anthropology 1 and 2]. Frankfurt a. M.: Suhrkamp.

Lystra, K. (1989). *Searching the heart. Women, men, and romantic love in nineteenth-century America.* New York, Oxford: Oxford University Press.

Mattheier, K. J. (1987). Formale und funktionale Aspekte der Syntax von Arbeiterschriftsprache im 19. Jahrhundert [Formal and functional aspects of the syntax of workers' written language in the 19th century]. In A. Betten (Ed.), *Neuere Forschungen zur historischen Syntax des Deutschen* [Recent research on historical German syntax] (pp. 286–299). Tübingen: Niemeyer.

Messerli, A. (2002). *Lesen und Schreiben 1700 bis 1900. Untersuchung zur Durchsetzung der Literalität in der Schweiz* [Reading and writing from 1700 to 1900: Investigation into the spread of literacy in Switzerland]. Tübingen: Niemeyer.

Nickisch, R. M. G. (1991). *Brief* [The letter]. Stuttgart: Metzler.

Ong, W. J. (1982). *Orality & literacy. The technologizing of the word.* London, New York: Routledge.

Polenz, P. von. (1999). *Deutsche Sprachgeschichte vom Spätmittelalter bis zur Gegenwart. Band III: 19. und 20. Jahrhundert* [History of the German language from the late middle ages to the present. V. 3: 19th and 20th century]. Berlin, New York: Walter de Gruyter.

Rice, R. E., & Love, G. (1987). Electronic emotion. Socioemotional content in a computer–mediated communication network. *Communication Research, 14*(1), 85–108.

Rowe, C. (2001). *Genesis and evolution of an email sibling code.* Unpublished doctoral diss., University of North Carolina, Chapel Hill.

Rowe, C. (2007). Building "code": Development, maintenance, and change in a private language. *American Speech, 82*(3), 235–261.

Sandig, B. (2006). *Stilistik der deutschen Gegenwartssprache* [Stylistics of modern German]. Berlin, New York: de Gruyter.

Schikorsky, I. (1990). *Private Schriftlichkeit im 19. Jahrhundert. Untersuchungen zur Geschichte des alltäglichen Sprachverhaltens kleiner Leute* [Private writing in the 19th century: Investigations in the history of daily speech of private citizens]. Tübingen: Niemeyer.

Schikorsky, I. (2001). "Dein bis in den Tod"—Zur Sprache der Liebe unter den Bedingungen des Krieges ['Yours to the death': On the language of love during wartime]. In S. J. Schierholz (Ed.), *Die deutsche Sprache in der Gegenwart* [The German language in the present day] (pp. 69–79). Frankfurt a.M./New York/Wien: Peter Lang Verlag.

Schönberger, K. (1999). Neue Online–KommunikationspartnerInnen? Qualitative und quantitative Annäherungen [New online communication partners? Qualitative and quantitative approaches]. In U.-D. Reips (Ed.), *Aktuelle Online–Forschung. Trends, Techniken, Ergebnisse.* [Current Internet science. Trends, techniques, results]. http://www.gor.de/gor99/tband99/pdfs/q_z/schoenberger.pdf [03-22-2007]

Schultz, L. M. (2000). Letter-writing instruction in 19th century schools in the United States. In D. Barton & N. Hall (Eds.), *Letter writing as a social practice* (pp. 109–130). Amsterdam/Philadelphia, J. Benjamins Publ.

Schulze, G. (1995). *Die Erlebnisgesellschaft. Kultursoziologie der Gegenwart* [The experience society: Cultural sociology in the present age]. Frankfurt am Main, New York: Campus.

Schwender, C. (in press). Letters between home and the front—Expressions of love in World War II fieldpost letters. In E. L. Wyss (Ed.), *Transformations of passion. The mediatisation of intimacy in letters, postcards, TV, internet and mobile phone.*

Thurlow, C., & Brown, A. (2003). Generation txt? The sociolinguistics of young people's text-messaging. *Discourse Analysis Online, 1(1).* Retrieved December 6, 2007, from http://extra.shu.ac.uk/daol/articles/v1/n1/a3/thurlow2002003-paper.html

Turkle, S. (1995). *Life on the screen. Identity in the age of the Internet.* New York: Simon & Schuster.

Vincent, G. (1995). Eine Geschichte des Geheimen [A history of the secret life]. In Ph. Ariès & G. Duby (Eds.), *Geschichte des privaten Lebens. 5. Band: Vom Ersten Weltkrieg zur Gegenwart* [History of private life. V. 5: From World War I to the present] (pp. 153–344). Frankfurt am Main: Fischer.

Wiese, I. (2000). Entwicklungslinien der Präsentationsweise medizinisch–wissenschaftlicher Zeitschriftenartikel [Development lines in the means of presentation of medical scientific journal articles]. In I. Barz, U. Fix, M. Schröder, & G. Schuppner (Eds.), *Sprachgeschichte als Textsortengeschichte* [Historical linguistics as text linguistics] (pp. 125–131). Frankfurt am Main: Peter Lang.

Wittgenstein, L. (2001). *Philosophical investigations.* Oxford: Blackwell Publishing. (Original work published in 1953)

Wyss, E. L. (2000). Intimität und Geschlecht. Zur Syntax und Pragmatik der Anrede im Liebesbrief des 20. Jahrhunderts [Intimacy and gender: On the syntax and pragmatics of address in the loveletter of the 20th century], *Bulletin VALS/ASLA, 72,* 187–210.

Wyss, E. L. (2002). Fragmente einer Sprachgeschichte des Liebesbriefs des 20. Jahrhunderts im Spannungsfeld von Sprachgeschichte, Geschichte der Kommunikation und Mediengeschichte [Fragments of a linguistic history of the loveletter of the 20th century in the range of historical linguistics, history of communication, and media history]. In U. Schmitz & E. L. Wyss (Eds.), *Briefkultur des 20. Jahrhunderts* [Letter writing culture of the 20th century] (pp. 57–92).Osnabrücker Beiträge zur Sprachtheorie 64.

Wyss, E L. (2003a). Liebesbriefe von Kindern, Jugendlichen und Erwachsenen. Eine Textsorte im lebenszeitlichen Wandel [Loveletters of children, youth, and adults: A synchronic text type change]. In A-H Buhofer (Ed.), *Spracherwerb und Lebensalter* [Language aquisition and age] (pp. 71–86). Basel: Schwabe.

Wyss, E. L. (2003b). Metamorphosen des Liebesbriefs im Internet [Metamorphoses of the loveletter on the internet]. In J. R. Höflich, & J. Gebhardt (Eds.), *Vermittlungskulturen im Wandel. Brief—E-Mail—SMS* [Changes in medial cultures: Letters, email, and SMS] (pp. 199–231). Frankfurt a.M., New York, Wien: Peter Lang.

Wyss, E. L. (in press). Communicating, constructing and performing intimacy. In E. L. Wyss (Ed.), *Transformations of passion. The mediatisation of intimacy in letters, postcards, TV, internet and mobile phone.*

Ziegler, E. (2006). Identitätskonstruktion und Beziehungsarbeit in bayerischen Schülerzetteln [Identity construction and relational work in Bavarian classroom notes]. In C. Dürscheid & J. Spitzmüller (Eds.), *Perspektiven der Jugendsprachforschung* (pp. 165–183). Frankfurt/M.: Peter Lang.

Chapter 6

From Postcard to E-Card

Tradition and Change in the Smallest Spaces

HAJO DIEKMANNSHENKE

University of Koblenz

Abstract

When the correspondence card was introduced in Austria and Germany in 1869 and 1870, respectively, it was initially met with skepticism rather than acceptance. Intended as a cheaper and faster alternative to the traditional letter, the correspondence card achieved its breakthrough during the German–French War in 1870/1871. After the transformation of the correspondence card—which was at first solely text-based—into the picture postcard, the "wide world" could be delivered into people's homes via postal mail. Political and social events were captured, communicated, and even collected. Although linguistically, the postcard adhered to the conventions of the traditional letter for a considerable time, a unique postcard style began to emerge slowly. Communication by means of postcards can be regarded as a comparatively nonbinding act, which primarily fulfills the function of keeping up social relations similar to much mobile phone, short message service (SMS), and instant messaging (IM) communication, especially by teenagers—managing the network instead of instrumental communication content. It is in particular the specific relationship between picture and individually created text that serves as the "communicative carrier." Especially in the case of the holiday postcard, attempts can be observed to evade the accordance between cliché pictures and stereotypical text structures by composing individualized, often humorously intended texts. Many postcards now do this themselves—providing an ironic picture and message. A growing tendency toward such rather "creative" compositions can be noticed in the course of time. More recently, it has been primarily the e-card that—although having the traditional postcard as a "role model"—has accelerated change in the "communication form" of postcard through innovative forms of layout and composition.

E-Cards—The End of the Brightly Colored Paper Cards From All Over the World?

Today, e-mail has taken the place of the letter in many areas: People use e-mail to arrange dates and appointments, convey information, tell friends and acquaintances about daily events, debate questions, send out greetings, and much more. Although in the past, it was the letter that was written and sent for these purposes, e-mail and telephone calls have now assumed these social tasks. Although the letter as a means of communication has undergone several functional changes, the letter itself has not disappeared, defying all prophecies of doom.

A similar change in communication can be observed in the case of postcard communication. Meanwhile, the e-card has challenged the traditional postcard, but the postcard is even less likely to be replaced by its digital counterpart than the letter is by e-mail. It is true that a high percentage of mail-order requests today no longer use mail-order postcards and the traditional postal way, but are directly placed by the Internet; likewise, notifications are usually sent electronically. Nevertheless, postcards are still enjoying immense popularity, particularly in the private sphere: 63 million vacation postcards[1] processed alone in 2005 by the International Post Centre (IPZ) in Frankfurt, Germany, speak a clear language. Although we can assert significant changes from letter to e-mail communication, this claim cannot be held when comparing the e-card with the postcard. Here, the postcard model is still prevailing, with new forms, such as video cards and similar phenomena, emerging only slowly. So far, only little innovation is to be found, particularly in language usage and style. Additionally, e-cards are obviously not yet considered appropriate for all kinds of specific occasions. For example, the traditional postcard is likely to continue to be reserved for expressions of sympathy (although usually enclosed in an envelope). At this point, condolences are rarely conveyed in the form of e-cards. The deep-rootedness of the e-card in the postcard tradition makes it seem reasonable to take a look at the development of postcard communication and then, considering this background, to further discuss tradition and variation.

The Idea of Heinrich von Stephan

Following the British historian Eric Hobsbawm (2004), we may refer to the first half of the 19th century as an "age of superlatives." The Industrial Revolution was running at full steam, new inventions were changing people's living and working conditions, particularly in Europe and North America; the cities were growing and, with them, the forced and voluntary mobility of large numbers of people. This mobility was significantly enhanced by the development of the railway system, which substantially altered the time-and-space structures of modern societies especially in the second half of the 19th century (Schivelbusch, 2000).

However, not only were people and freight now able to overcome huge distances compared to previous conditions: communication saw a marked acceleration, too. The transport of letters was no longer performed by the traditional stagecoach,[2] but came to be taken on by train on newly constructed rail tracks, which were less susceptible to weather conditions than the old stagecoach roads. The immigrants who were pouring into the industrial centers were now able to maintain contacts more easily to their relatives and friends living far away. So what would be more obvious than also thinking about faster and more efficient ways of communication compared to the traditional letter?

When in 1865 the German Postmaster General Heinrich von Stephan began propagating the idea of a faster and less expensive alternative to the letter (Täubrich, 1996), was not aware of the fact that he was actually inventing a new way of communication, which until today naturally belongs to the everyday forms of written expression. Nowadays, almost everyone has already made use of this form without thinking about it. The academic study of postcards and their diverse realizations, the changes brought about by mass production, and the way postcards are linguistically arranged by their users has so far occupied a rather marginal place in linguistics,[3] which could possibly be explained by the apparent banality of what is communicated on postcards. On closer look, however, the postcard presents itself as a very informative form of communication that reveals a lot about mentalities and expectations on a massive scale, about linguistic strategies concerning conventions and the infringement thereof, as well as about the complex communicative relation between image and text.

Despite the pioneering of the German postmaster general, in 1869, Austria was the first country to introduce the so-called correspondence card, followed by the North German Union in 1870. The most obvious advantage of the postcard compared to the letter is its significantly lesser postage cost: The postage of a correspondence card was only half the price of a regular letter. Also in the same year Great Britain started circulating its first postcards, and in 1873 the United States followed suit:[4] Although initially the bourgeoisie showed skepticism toward this kind of public communication, worried that the servants could read what was in fact reserved for master and mistress, soon the postcard became widely accepted by all social classes. The widespread use of the new form of communication during the German–French War from 1870 to 1871 convinced even the critics of the postcard. In the first 5 months of the war alone, 10 million so-called field postcards were sent home from the front (Kaufmann, 1985). Because of its ease of usability, postcard communication was not only popular with the soldiers who often had little experience in writing, but it also offered considerable advantage to censorship: The contents on a postcard could be examined much more easily than the often voluminous war letters sent in envelopes (Täubrich, 1996).

Fig. 6.1. Correspondence card of the North German Union.

The original correspondence card was purely designed for written content, containing a space for the recipient's address on one side plus some sort of "instructions for use" (Fig. 6.1) for correct application, which indicates that the early users were not yet familiar with using these cards, similar to the early instructions about the use of the telephone. In this respect, Rule 6 of the instructions seems to be of particular interest: "The sender is not obliged to identify himself or herself by name"—which appears to be unthinkable in traditional letter communication. The blank reverse side of the postcard, in contrast, is reserved for the individual message. Soon after, this blank space came to be filled with the first drawings, vignettes, and illustrations (first on an individual basis, later machine-made), thus becoming the predecessor of the picture postcards. Although picture postcards were already admitted in 1885 (Schuler, 1999), it was only on February 1, 1905 that "messages are officially allowed to be posted next to the address, a reform that is epitomised by the vertical separating line, which until today has been a characteristic feature" (Hedinger, 1992). Consequently, this date can be regarded as the birth of the picture postcard we know today, where one side is completely reserved to the picture whereas the actual text message has to be placed on the left of the reverse side.

The Whole World in Small Format

In the 19th century, the advent of photography in particular influenced mass communication considerably. A seemingly authentic (i.e., "true") picture of the world and its major events could now be communicated to a large group of people. For a long time, however, only a small number of people were privileged to own cameras and photographs. Only after the end of World War I did photography became accessible to a larger sections of the population (Jäger, 2000). Until then, it was the postcard that sought to satisfy the "hunger for images" by initially using graphical devices that were later succeeded by photographs.

The popularity of picture postcards can be largely attributed to the fact that they do not only represent an inexpensive and fast means of communication, but also enable people to acquire pictures on virtually every topic and occasion. Thus, from the beginning, people not only sent out picture postcards, but also indulged in collecting them, often in specially designed albums. Some cards were only written in order to enlarge the addressee's collection, or to become eventually incorporated into the sender's own collection, as for example a card from 1903 sent out by Roosevelt with the following comment: "My post card collection wants filling up" (Phillips, 2000). The postcard collectors' lust for images can be understood as the expression of a widespread collecting passion. Especially the collectible pictures that the meat extract producer Liebig supposedly had been distributing together with Liebig products since 1875, made this form of collecting popular all over the world (Jussen, 2002).[5]

If one compares the depictions on those small pictures measuring 110 × 70 mm with the contemporary postcard motifs of that time, one can notice a concordance as to the clichéd production of exoticism. The multicolor image of the big, wide world, although bound to the contemporary stereotypes and prejudices, was suddenly visualized on a massive scale (Figs. 6.2 and 6.3).

Fig. 6.2. Collectible picture by Liebig (1887–1888), series 110 ["Negroe"].

Fig. 6.3. Postcard from 1920 (Phillips, 2000).

Although the two illustrations in Figs. 6.2 and 6.3 may be regarded as displaying rather harmless foreigner stereotypes, politics soon begin to exploit the postcard for propaganda purposes, especially in World War I where visual propaganda was purported by the means of postcards (Diekmannshenke, 2006). Particularly in the German context, this kind of propaganda postcard can be considered to represent the most important predecessor of the modern political poster (Diekmannshenke, 2006; Fig. 6.4).

Fig. 6.4. German war postcard and first political poster (1916).

All countries involved in World War I produced postcards depicting the enemy by means of negative stereotypical imagery for propaganda purposes and for an ideological strengthening of the home front. Here, one could already find those concepts of the enemy that would dominate the enemy propaganda subsequently: the "primitive and dirty" Russian, the "yellow danger" from Japan, the Prussian "beast," but also positive self-images such as Uncle Sam or Germania and Marianne as symbols for the United States, Germany, and France, respectively. At the same time, some selected postcards were published as early versions of the modern political poster (Lebeck & Schütte, 1980; Linhart, 2005).

Postcard Communication: Social Action in the Wake of Letter Communication

The act of writing and sending out postcards represents a form of communication that manifests itself above all in text types expressing social action. In turn, these text types can be differentiated according to their specific occasion, which constitutes the decisive criterion for text-type classification. In this respect, postcard communication closely follows in the tradition of private letter communication by adopting the letter's essential textual components, such as indication of place and date, salutation, main part, and complimentary close plus signature. The implementation of these elements, however, is not as strictly enforced, and in addition, the main part is extremely compact due to limitations of space on a postcard. Whereas a minor proportion of text types to be found on postcards (e.g., order forms or polling cards) implement speech acts such as informing, demanding information, ordering, or others, the majority of the remaining text types deal with social acts. These social acts are multiply interlinked with further actions by what German linguistics calls INDEM relations. INDEM relations are multistrand acts. Performing one act, one simultaneously realizes another act. For instance, by signing a visitor's book and praising the host, one fosters the social relations of both. Compared with the private letter, these social acts are rather noncommittal because they normally do not require any immediate response but serve the purpose of establishing contact, respectively the renewal therof, as well as the maintenance of existing social relations.

> *Hi* [first name]*!*
> Na, wie geht's? Ich dachte, so langsam könnte ich mal wieder was von mir hören lassen, gell?!!
>
> "Hi [First name], How are you? I thought it was about time to get in touch again!!" (private picture postcard)
>
> *Hallo* [first name]*!*
> *Freust Du Dich von uns zu hören?*
> *Nachträglich alles Gute zum Geburtstag von den Trierern*
> *M. & C.*

"Hallo [First name], Are you happy to hear from us? Happy belated birthday from your friends from Trier M. & C." (private picture postcard)

By contrast, private letter communication manifests itself as an exchange of letters, where a written action requires an according response. If that reaction does not take place within an adequate period of time, the entailed consequences can either include a posterior explanation or apology on the part of the respondent or leads to an inquiry on the part of the first writer. Also, a lack of response may possibly result in disturbance or even discontinuation of the relationship.

This does not apply to postcard communication; if it does at all, however, it happens in a moderate fashion, for example, if the expected holiday postcard is written and sent out with great delay:

Hallo [first name]*!*
Viele Grüße aus dem „wunderschönen Norden"! Die Karte bekommst du zwar erst, wenn ich längst wieder da bin, aber das macht ja nichts. [. . .] Aber ich sage ja immer: Hauptsache überhaupt eine Karte bekommen. Also freu dich über diese wunderschöne Karte!
Bis Montag, Dienstag, Mittwoch, . . . T.E.

"Hallo [First name], Best wishes from the "beautiful North"! I know you'll get the card only when I'm back, but nevermind. [. . .] But at least you got a card. So enjoy this nice card. See you again Monday, Tuesday, Wednesday, . . . T.E. (private picture postcard)

Hi [nicknames]*!*
Heute ist der letzte Tag und wir werden zwar früher ankommen als diese Karte, aber das hindert uns nicht daran euch zu schreiben.

"Hi [nicknames], Today is the last day. We'll probably be home earlier than this card, but that does not keep us from writing to you." (private picture postcard)

Thus, writing and sending a postcard can actually be likened to an act of friendship (Szépe, 1979). Despite the postcard's noncommittal nature, its very handwrittenness constitutes a major binding element of communicative action, finding particular expression in the obligatory signature, which is present on virtually every postcard. The generally low degree of commitment also shows on postcards that are composed by one person but signed by several people (Schuler, 1999). These points are related to current research in mobile phone/SMS/IM research on the use of such media to continuously manage social networks and a sense of ongoing connectedness.

The expectation about receiving a postcard at certain occasions is considerably high and already shows in the early days of postcard communication, for instance

Fig. 6.5. Holiday postcard (1900; Phillips, 2000).

when a writer in around 1900, notes) at the end of the card (taking a humorous stance; Fig. 6.5), "I am well and enjoying myself and follow the fashion, you see, in sending Ansichtskarte [German for *picture postcard*]."

The Interplay of Image and Text

The early picture postcards provided little space for the writer's individual messages because image and text had to share the space on one side, as illustrated by the following example (Fig. 6.6).

Fig. 6.6. Early picture postcard with text.

Fig. 6.7. Early picture postcard.

Many unsent postcards display an image side that at first sight might surprise the viewer with blank spaces extending into or framing the picture; these blank spaces are reserved for the text message (Fig. 6.7).

Until German postcards underwent official rearrangement in 1905, the communicative carrier was dominated by the image. The figures on postcard production and on the sale of picture postcards give impressive evidence of the communicative importance of the early picture postcards. In 1899, factories throughout the German Empire produced 88 million cards, and only 14 years later, in 1913, the German Reichspost transported 22 billion picture postcards, corresponding to an average per capita consumption of 30 postcards (Kaufmann, 1985). According to an official count in 1900, picture postcards made up 46% of the total number of postcards delivered (Gold, 1999: 14).

By inserting a text either into a blank space below the postcard picture or directly into the very picture, the postcard writer creates a qualitatively new texture, in which image and text interact for the recipient, even if the text itself represents nothing more than a simple greeting (Fig. 6.6). Whereas the example given here may be representing a comparably unspectacular "communicative vehicle," others provide more insight. During the era of the Empire, anti-Semitism was widespread among the German population, which was also reflected in certain patterns in contemporary postcard production (Gold & Heuberger, 1999). Sending a postcard represents an intentional communicative act that gives information about the spread of such attitudes very similar to "bumper stickers" on the back of automobiles, and signature files in e-mail (Campell-Kibler, 2003). Particularly, the picture motif is selected with the addressee in mind (Schuler, 1999). Those who send out such postcards can usually count on a shared attitude among sender and recipient, which allows us to draw conclusions about certain mentalities. Such postcards document the vast distribution of those kinds of mentalities, which—during the age of the Empire (1871–1914)—was even reinforced through the use of postcards. This has been demonstrated by Otto May (1998) in his study of the education of subject people in Prussia and by Frank Bajohr (2003) for the widespread phenomenon of the so-called "Spa-Antisemitism."

Sometimes the image serves to reinforce and clarify the written message on the postcard. Although the sample text here is obviously anti-Semitic, it receives its specific meaning only in combination with the picture motif (Fig. 6.8).

> *Geehrter Herr Pfarrer! Da ich mich aus dem Judengewimmel der Stadt in obiges Lokal geflüchtet habe, um in Ruhe ein Pilsner zu schlürfen, gedenke ich Ihrer & sende die besten Grüße*

> "Dear Father, Having sought refuge from the swarms of Jews in the city to the above shown restaurant where I can peacefully drink a Pils, I am remembering you and sending you my best wishes."

Here, the deontic component (Hermanns, 1986) of what is communicated becomes utterly obvious. The writer does not simply convey his personal praise of the Jew-free atmosphere of the particular hotel, but by means of the postcard's picture motif he also implicitly promotes a certain action against Jews in order to achieve that kind of atmosphere in other places, too. Thus, what is communicated implicitly via the picture exceeds the importance of the actual postcard text, which serves the explicit manufacturing of consent between the communication partners.

Fig. 6.8. Anti-Semitic postcard (1887; Stöckle, 1982).

Like many other kinds of speech acts, postcard communication typically features a certain complexity of interrelated actions, which is denoted as 'INDEM-Relation' in German linguistics (Harras, 1983; Heringer, 1974). What distinguishes postcard communication from other speech acts is the fact that postcards not only encompass the performance of purely illocutionary acts, but these acts also interact with those conveyed by the corresponding picture. In the example given in Fig. 6.7, a social relationship is maintained by saying hello via postcard. Through the act of writing this message on this particular postcard, the writer at the same time assumes a mutual agreement with the recipient on a certain attitude toward Jews.

This kind of communicative act receives special significance when it comes to humorous postcards, which represent a large proportion among postcards (Diekmannshenke & Grünewald, 2005). Such humorous cards already occurred in the early days of the picture postcard. In these cases, too, the focus lies on the social activity, because making someone smile or laugh constitutes social interaction and helps to maintain a social relationship. Additionally, the writer takes the chance to present him or herself as a humorous person, which will probably enhance his or her prestige.

The Dissolution of the Norms of Letter Communication

Along with the massive emergence of postcards goes the development of a new style of correspondence, which, because of the public nature of the message, is characterized by "allusion, abbreviation, incompleteness, fragmentation" (Wietek, 1977). Typical features of postcard style are the frequent omission of the subject pronoun ("Sind gut angekommen" ["arrived safely"]; also Fig. 6.11) compare to telegraphs,[6] and IM/SMS or online chat,[7] resulting in elliptical sentences and various colloquialisms. In contrast to the telegram, however, where the price depends on the number of words sent, one reason for this specific style in the postcard lies in the limited space provided on a postcard. Additionally, the postcard's success is to a great extent based on its being a communicative alternative to the stylistically strict norms of the letter, offering less experienced writers a relieving release from stylistic constraints (Polenz, 1999). Illustrating the positive impact of the postcard, Kaufmann (1985) pointed out that

> the postcard ideally met the needs for standardised short messages, requiring little time, reflection, and little experience in dealing with handwriting and syntax. While the letter used to require a certain degree of individual style and articulacy, in fact, even a certain amount of time and leisure, the limited space of the card offered writers the possibility to confine themselves to abbreviations, standard or set phrases and clichés without losing face due to a lack of expressive capability. Writing letters was mainly a vehicle of expression for the educated. The card encouraged everyone, even people without much experience or leisure time, to contact each other through writing. (403)

It would be unwise, however, to pejoratively describe the postcard as "the letter for the ordinary people"—as has occurred at times in the past. After the bourgeoisie had overcome their initial skepticsm, they soon started to make use of the new form of communication because they were taken with the innovative possibilities thereof—without giving up letter communication, however.

Looking at early postcards, one can see that in many cases they do not deal at all with assertive speech acts (i.e., the conveyance of information[8]), but that postcard communication—as already shown—centers around social activities: "Groß ist das Bild, der Raum ist klein, drum kann mein Gruß auch kurz nur sein. Frl. Gr. Wilhelm" ("Big the picture, small the space, therefore my greeting can only be a short one. Miss Gr. Wilhelm") (Postcard from 1905; cited in Grosse, Grimberg, Hölscher, & Karweick, 1989:101). Even cards that seem to serve the purpose of pure information conveyance, such as the following example, in fact convey complex actions, in this case attending one's social duties by informing the loved ones.

> *Lieber Freund Albert. Mein Vater ist Dienstag in Oberhausen gestorben, wird nach hier überführt und Morgen Freitag 11Uhr von hier (Bahnhof) aus Beerdigt. Mit frl Gruß Ang. Weber.*
>
> "Dear friend Albert. My father died tuesday in Oberhausen and will be transported here. The funeral takes place from here (train station) tomorrow, friday, at 11 o'clock. Best wishes, Ang. Weber." (Postcard from 1907; cited in Grosse et al., 1989:102)

Just as with other innovative forms of communication, the users do not at all adhere to the intentions of the inventor, but adapt the medium to their own communicative needs and to the particular communicative occasion (Diekmannshenke, 1999, 2000).[9]

For a long time, the tradition of letter-writing determined the way people dealt with the new means of communication.[10] The text components that were obligatory for the letter also dominated the appearance of the postcard text, although in an abbreviated form. The quest for appropriate phrasing and the mixing of different styles of writing can be taken as a sign of a certain lack of confidence in dealing with rules and conventions that were just in the process of evolving—a phenomenon that can be observed whenever new forms of communication and new text types emerge (Eichhoff-Cyrus, 2000). It is true that—also because of the limitations of space—the new way of communication offers a relief from many constraints posed by letter communication and enables the writers to a relatively carefree handling of German spelling norms: "Endlich bekomst eine Karte von mier hoffendlig gefält sie Dier Gruß an Alle H.Hund" ("Finally, you get a card from me, I hope you like it. Greetings to all. H. Hund.") (Postcard from 1906; cited in Grosse et al., 1989:103).[11] Sometimes, however, the tense relation between strict conventions regarding text types and individual self-liberation thereof becomes evident on a single postcard:[12]

L.G. Lieblich wie auf Blumenwegen Möchst Du durch das Leben ziehn. Und des Himmels rechter Segen Möge Dir zur Seite Stehn. Sei herzlich gegrüßt von Die. Gretchen Teile Dir eben mit das wir einen kleinen Jungen am 3 d.M. bekommen haben Alles gesund nur unsere Maria hat die Lungenentzündung Gruß an Mutter u Schwester

"L.G. May you move through your life as on flowered paths, and may Heaven's blessing always be on your side. Warm regards from Die. Gretchen. I'm just informing you that we got a little baby boy on the third of the month. Everybody's in good health except for our Maria who has got pneumonia. Regards to Mother and Sister." (Postcard from 1900; cited in Grosse et al., 1989:102)

In this example, the essential message follows the stylized greeting rhyme, appearing almost inappropriate in this combination. Also, the sample shows the stylistic inconsistency in the writer's use of language: The last sentence can be described as an "afterthought" pertaining to conceptional orality (Dürscheid, 2003; Koch & Oesterreicher, 1985), whereas the actual greeting follows a (rhymed) specimen text that is classified as belonging to conceptional literacy.

At the same time, one can observe the emergence of new conventions regardless of the linguistic competence of the particular writer.[13] For example, the author of the following postcard (Fig. 6.9) not only has the financial means to have an indvidual greetings card made for him using a photograph—photograph postcards were very popular around the turn of the century (Maas, 1975)—but he also uses a refined, elaborated language. Despite these signs of originality, the text adheres to the same established standards that we still encounter on modern holiday postcards.

Fig. 6.9. Private picture postcard (1912).

Altona 31/8.12
Meine l. Berg's
Erst jetzt bin ich dazu gekommen, Bilder von unserer Wyker Tour anzufertigen und so senden wir Ihnen zum Andenken an die schönen Stunden in Wyk diese Karte, damit Sie Wyk und uns nicht ganz vergessen. Viele Grüße
Von Hermann u Selma Degenhardt

"My dear Bergs, Now that I have finally got round to produce pictures from our trip to Wyk, we're sending you this card to remind you of the great time we had there, so that you don't forget about us and Wyk. Best wishes from Hermann and Selma Degenhardt." (Jäger, 2000:54)

Here again, we can find features that are still typical of our contemporary postcards. People generally expect others to send such holiday postcards, even if the writers[14] perceive this activity as a nuisance (Schuler, 1999). As the expectations are mutual, the social commitment seems to be binding to such an extent that people sometimes send out postcards even if the vacation is already over (illustrated by the indication of the place of permanent residence "Altona," which is a part of Hamburg). In the present example, however, the belated sending of the postcard may be justified by the writers' decisions to send their friends an individual, prestigious photo-postcard instead of a mass-produced postcard.

Although the concept of "vacation" during the first decades of the 20th century is in no way comparable to modern mass tourism, as this activity then used to be reserved mainly for the bourgeoisie, there already existed the tense relation between cliché and individualism, which is an idiosyncratic feature of holiday postcards. On the one hand, writers make use of the mass product that provides a stereotypical depiction of vacation in text and image, but in turn they also attempt to pen a creative text message, the originality of which being somehow doubtful, as it still follows certain conventions. Whereas in the early days of postcard communication humorous illustrations sought to establish originality, recent times have seen writers increasingly striving for creativity in writing (Fig. 6.10).

Fig. 6.10. Private holiday postcard.

The picture itself does not necessarily need to be interpreted as a humorous motif as such, but it is the text message on the back of the postcard that adds a humorous note to it:

> *Hallo, Ihr Lieben!*
> *Wir machen hier zwar nicht Ferien auf dem Bauernhof, aber die Jersey-Kuh ist hier zusammen mit der German Occupation eines der touristisch meistverarbeiteten Objekte. Ansonsten: Alles prima! Herzliche Urlaubsgrüße*
> *G. & S.*
>
> "Hello, our dear friends, We're not spending our vacation on a farm, but next to German occupation, the Jersey cow is one of the most popular tourist objects here. Otherwise, everything fine! Best wishes from our vacation, G.& S."

Even early guides of etiquette illustrate the difficulties of postcard communication by underlining which purposes postcards may be employed for, and for which aims postcards are considered an inappropriate means of communication:

> The postcard serves the disposal of all unnecessary ballast of writing; it demands neither a splendid form of address, nor a detailed declaration of honour, nor long-winded statements. It allows and necessitates short, simple and factual communication—postcards are used for: agreements, sociable meetings, previously forgotten postscripts, short business orders or requests; amicable or sociable messages without secretive background, orders, memories etc. (Kronoff, 1925, cited in Diers, 1987:25)

The fact that picture postcards were enjoying huge popularity was even acknowledged by the lady who authored an early guide on (postcard) etiquette:

> A popular variant of the postcard, especially for lethargic mail carriers, is the picture postcard. It already conveys through the picture what the sender would have to express in words, leaving only little space for short messages. (Kronoff, 1925, cited in Diers, 1987:25)

The explicit declarations of constraints for the use of postcard, however, provide more interesting insight:

> It is not appropriate to use a postcard in order to make a request or to give thanks to somebody; in both cases this would be inconsiderate. Only the letter is suitable for such occasions. It is illegal to send bills and payment requests on a postcard to debtors or businesspeople. Also, it is a matter of sensitiveness not to use insulting forms of address or messages, remarks of suspicion, reproaches, admonitions etc. on an open postcard; this would not only constitute an unforgivable indiscretion towards the recipient, but also an exposure of the sender himself. People with integrity admonish or reprimand on a personal basis and without witnesses, in order to respect the person admonished. Likewise,

> a written expression of disapproval must be concealed in a closed envelope. A postcard may not be directed to people in a higher position, such as superiors, patrons or other persons in authority. In addition, postcards are not suitged for contacting elder relatives, old friends or distant acquaintances; these all demand and deserve communication by letter and do not need to give up this right in any way. (Kronoff, 1925, cited in Diers, 1987:25f)

Compared with the letter, postcard communication requires much less commitment: A reply message is neither required nor expected by the sender of the postcard. Therefore, for private relations as for contact with authorities, institutions, and organizations, the exchange of letters is much more common, whereas "postcard exchanges" are comparably rare.[15] Here, unlike with letters, a special occasion on the one hand and certain expectation on the other are the driving factors for composing and sending postcards.

Linguistic Patterns of Postcard Communication

Over time, every form of communication and every type of text generates a set of specific paradigms of action and stereotypical phrasing patterns (Yates & Orlikowski, 1992). These are co-decisive for the recipients' individual expectations, which, like all linguistic phenomena, are subject to change. Of course, this is also true for postcard communication. Whereas the front sides carrying the pictures represent items of mass consumption that compete for the favor of potential consumers by adopting certain trends (but also reflect massly disseminated attitudes and opinions; mostly in an affirmative, but at times also in a critical way), the text messages on the reverse side express the writer's individual style and also reflect the everyday usage of writing and style conventions.

Basically, we can distinguish two writing strategies, which do not mutually exclude each other but sometimes even occur in "mixed forms." One of them consists of the strict observance and compliance of the relevant conventions. Especially holiday and picture postcards, but also postcards used for certain occasions: for birthdays and other anniversaries, death, holidays, and others) adhere to this pattern (Fig. 6.11).

Fig. 6.11. Private holiday postcard (1960).

Liebe Familie Kropp!
Aus unserem Urlaub recht herzliche Grüsse. Ihre Juliane [illegible] *mit Gatten*

"Dear Kropp Family! Greetings from our vacation. Yours, Juliane [illegible] and husband"

Stereotypical phrasing corresponds with the sterotypes of imagery that are valid at certain times; they illustrate the notion that the aim of maintaining social bonds alone is a sufficient reason for writing and sending out holiday postcards to loved ones and colleagues (Cantauw-Groschek, 1993). Sometimes, this very obligation is claimed by postcard, as in the following example:

Liebe [first name], *wo bleibt Deine Karte von Deinem Urlaub? Du wolltest dich doch melden?*

"Dear [first name], where's your vacation postcard? Didn't you say that you were going to get in touch?" (private postcard)

The following example further illustrates the phenomenon that until today holiday postcards predominantly have been featuring stereotypical phrases; the picture motif chosen clearly evokes the clichés of a vacation in Venice but does not at all relate to the text message (Fig. 6.12). In this case, it is also striking that the writer renounces an explicit form of address, which underlines the noncommittal nature of the communicative act.

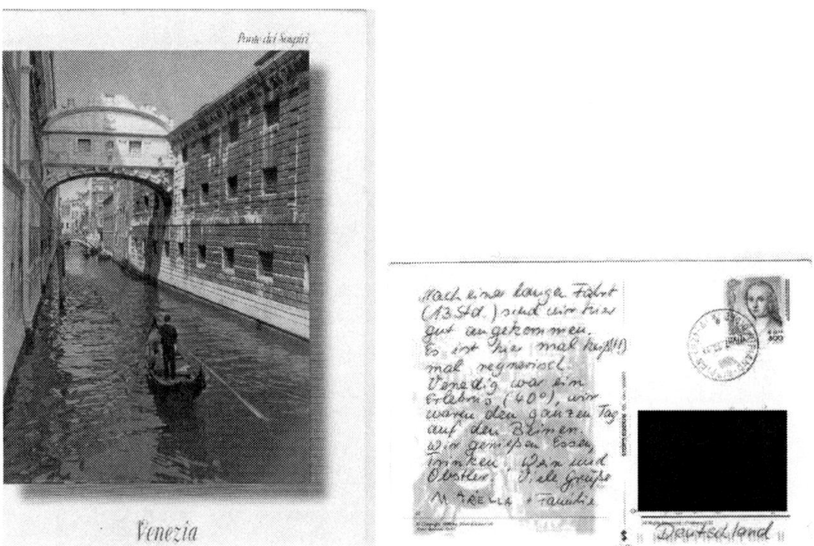

Fig. 6.12. Private holiday postcard from Venice (2000).

Nach einer langen Fahrt (13 Std.) sind wir hier gut angekommen. Es ist hier mal heiß(!!) mal regnerisch. Venedig war ein Erlebnis (40°), wir waren den ganzen Tag auf den Beinen. Wir genießen Essen, Trinken, Wein und Obstler. Viele Grüße Mariella + Familie

"After a 13-hour journey we arrived here safely. Here, sometimes it's hot (!!), sometimes rainy. Venice was quite an experience (40°C), we were on our feet all day. We're enjoying food and drink, wine and fruit schnapps. Best wishes, Mariella + family."

Here, the information content is very small. Instead, the following expectations are being confirmed: that despite a long journey the sender has arrived safely, that Venecian summers are hot (although rain is also possible sometimes), that exploring an unknown city demands a lot of exercise, one enjoys the culinary pleasures—in short, that the vacation has been a noteworthy undertaking: "Venice was an experience." It may well be one of the traids of postcards to allow very dense information to be produced and transmitted in a small and easy-to-handle medium.

The historical and cultural standardization of many picture motifs (especially of vacation cards) provides the recipient with important clues concerning the interpretation of the postcard. This is also true for the overall composition of the card and the choice of a particular typography for preprinted text (especially on condolence cards). In a similar way, the choice of the particular form of address represents a significant indicator for a social relationship (Ermert, 1979).

The lesser degree of commitment also leads to a change in the constellation of the people involved. Whereas letters are usually directed from one person to another, postcards are often addressed to several people: to an entire family, to people sharing rooms, to a group of colleagues, to fellow club members, and so on. Otherwise, collective addressing usually only occurs in special types of letters, such as circulars or open letters. In addition to multiple addressees, postcards can also have collective senders (see, e.g., Fig. 6.12) such as families, school classes, sports clubs.

Additionally, the postcard's noncommittal nature enables the use of certain linguistic diatypes without entailing any loss of image. If we look at the early days of postcard communication practices, we notice that such diatypes (ever-recurring text samples) did not yet exist from the beginning. As the examples just mentioned demonstrate, communicative patterns were often adopted from established letter communication. In many cases, especially the first of several consecutive communicative acts on the postcard results from such historical paradigms (Yates, 1989a, 1989b; Yates & Orlikowski, 1992).

The changes of social and societal activities since the second half of the 19th century are also reflected in postcard communication. Around the turn of the 20th century, postcards were frequently bought and sent out in order to enrich personal postcard collections (Grosse et al., 1989); this practice spawned the industrial production of postcard albums. This pronounced interest in collecting has diminished considerably until today, but has never completely disappeared. In the wake of mass tourism, sending holiday postcards has become one of the most dominating activities. Whereas in the early days of the postcard, the medium was used by travellers to convey a picture of a far and unknown world to the ones at home, this attitude has changed meanwhile, such that the prestigious value of a picture postcard sent from destinations like Nepal, South Africa, or Peru may be higher than of a postcard from Bavaria, Austria, or the Netherlands (at least from a German or middle European point of view). The most elitist and exclusive travel destinations, however, are either not suited for sending picture postcards (e.g., the Himalayan mountains, touristically undeveloped isles in the South Pacific, or the Amazonas jungles), or they are preferrably documented in a different way (particularly by semiauthentic photographs, advertising by exclusive hotels or illegally exported or expensive souvenirs).

In fact, text messages on holiday postcards are nowadays defined by a linguistic compositional pattern that could be described as a "holiday frame." Schuler (1999) stated that the most frequent topic dealt with in postcard texts is travelling and everything that is related to it.

> Next to a description of the holiday destination and a modal and temporal, partly even causal illustration of the ongoing travel, the evaluation of the journey plays a large part. Besides, descriptions of the destination's weather, accommodation and food seem to represent obligatory information. Here, retrospective and anticipatory evaluations alternate with illustrations and and judgements concerning the actual situation when writing the postcard. Present activities plus plans for the future take up as much space as does information concerning the writer's own state of feeling. (47)

In this respect, postcards can communicate the conception of a successful vacation, as propagated by modern touristic advertising in ads, spots, and especially in holiday and travel catalogues. This standard presented in touristic media is sent home in form of a fulfilled expectation, with the picture on the postcard generally supporting this expectation.[16]

> *Hallo* [first name]*!*
> *Wir sitzen momentan in einer Hotelbar und genießen unseren Urlaub. Hotel und Essen sehr gut, Wetter ok für T-Shirt und Shorts, Wasser warm genug zum Schwimmen und Strand so leer, daß immer ein paar Liegen frei sind.*
> *Also, ein toller Urlaub! Viele Grüße*
> *F. und N.*[17]

"Hallo [first name], At the moment we're sitting at the hotel bar and are enjoying our vacation. Hotel and food very good, weather fine for t-shirts

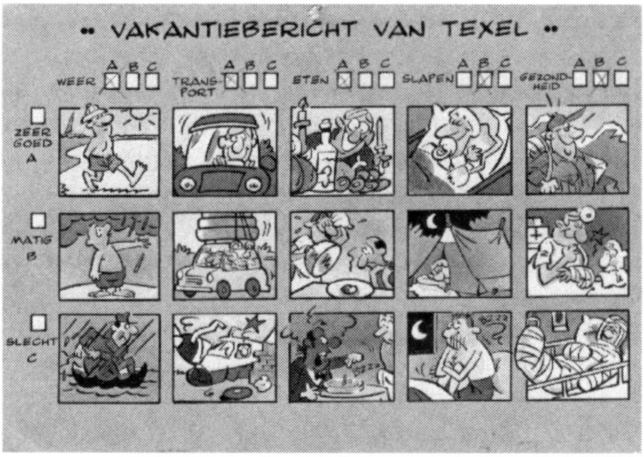

Fig. 6.13. Playing on holiday clichés.

and shorts, water warm enough for swimming and beach so empty that we always find free sunloungers. Thus, a great vacation. Best wishes, F. and N." (private picture postcard)

Although there is often a certain correspondence between the stereotypes of image and text, other examples clearly show the writer's attempt to escape these clichés, either by employing linguistic creativity (Fig. 6.10) or by selecting a picture motif that is perceived as humorous, and by adding witty comments to it (Fig. 6.13).

Consistency and Change in the Style of Communication

The roughly 140 years of postcard communication have seen phenomena of change, but also of surprising consistency. The latter pertain to certain expectations about when postcards are (supposed to be) composed and about which subjects are commonly featured on postcards. Particularly postcards that are composed on the occasion of travel and vacation prove to be relatively resistant to change. With the beginning of mass tourism, traveling has lost its exclusivity, becoming a somewhat ordinary activity. Nevertheless, tourists continuously tend to select picture motifs that show touristic sights or that are perceived as "typical" and combine the image with text patterns featuring standardized subjects and, frequently, clichés. Although today Internet cafés can be found in many vacation resorts, the vacation e-card merely represents an additional complement to the traditional picture postcard, as e-cards usually lack the vacation spot's multicolored image clichés that are displayed on the omnipresent postcard racks.

Despite the significant technical progress in postcard production over the last decades, the topics conveyed have remained relatively static. As most postcards are dominated by social communicative acts, shifts of societal norms and values are also reflected in the postcard texts. Still, examples where taboos are infringed on or abolished can rather be found in the picture motifs.[18] Significant changes can be observed particularly in the area of formal conventions concerning form of address and complimentary closings. Here, too, postcards reflect societal conventions. Whereas postcard communication initially was strongly influenced by rules of letter writing, this impact diminished over time. What Werner Besch (1996) stated concerning the change of salutational conventions in the German language is also reflected in the form of address on postcards. The salutation "Sehr geehrte Damen und Herren" ("Dear sirs" / "Dear Sir or madam") has been subsequently replaced by "Liebe/r" ("Dear"), which today is often substituted by a simple "Hallo." At the same time, the variety of salutations has considerably increased, going along with a striving for a certain linguistic originality. An example of this creativity, some writers tend to borrow salutations or complimentary closes from the language of their holiday destination, whereas the main part of the message is still written in their mother tongue:

> *Sa watdee khrap [first name]*
> (meaning: Hallo [first name]) (private picture postcard)
>
> *Buenos Tardes Lovely [first name]* (private picture postcard)
>
> *¡Holá [first name]!*
> *Que tal? Muy bien, gracias. na, was heißt das wohl? [. . .]*
> *Hasta luego K.*
>
> "¡Holá [First name]! Que tal? Muy bien, gracias.—Guess what that means? [. . .] Hasta luego K." (private picture postcard)

As Schuler (1999) showed, the forms of address are generally kept very short and do not show many variations. However, younger postcard writers tend to use a number of variations that can be characterised as youth-specific: for instance, there is an increasing number of salutation forms like "Hallo," "Hallöle," "Hi," "Hey," "Na," or generalised German regionalisms like "Moin," respectively "Moin, moin," which all typically occur in Internet communication (Diekmannshenke, 1999, 2000; Runkehl, Schlobinski, & Siever, 1998). Thus, it is not surprising that such kinds of salutations are also typical features of e-cards. In many cases, these forms of address have even come to replace the traditional "Liebe(r)" ["Dear"] (Schuler, 1999). Postcards and also e-cards that do not contain any direct form of address can often be found in rather formal social relationships (e.g., between colleagues) or in cases where the recipient is not personnally specified. In fact, relevant style guides have long deemed it "appropriate" for postcard communication to do without salutation. Roughly 20% of the writers surveyed by Schuler (1999) completely renounce the salutation, as also shown on the postcard on Fig. 6.12, for instance: the name given in the destination address field of the postcard replaces the actual address, thus leaving more room for the (body) text.

Nowadays, the style of the postcard text is kept rather colloquial, which corresponds to the noncommittal and fleeting nature of the occasion for writing a postcard. Often, the writer does not care about logical connections, strict cohesion, and stylistic standards, which, however, can also be claimed for the early postcards and, for SMS/IM and even e-mail messages.[19] What has changed, however, is the style of writing. Postcards of our days tend to contain a lot of everyday speech and sometimes youth language (i.e., developments and tendencies in everyday speech are in turn reflected in postcard messages).

> *Hier kommt die Karte aus England! Letzte Woche waren wir in London und wollten Docs kaufen, aber es war so wahnsinnig viel los, es ist echt unvorstellbar. Wir sind bestimmt 5 oder 6 mal die Canaby-Street [sic] auf und ab gelaufen, und in jedem Geschäft waren die innerhalb von ein paar Minuten ausverkauft. Wir probieren's nächste Woche also noch mal, aber die haben alle gesagt, daß die Größen 5 + 6 so schwer zu kriegen sind, weil die fast jederhat [sic] . . .*
> *Na ja, ich probier's noch mal. o.k?! Wir sehen uns dann in der Schule,*
> *Ciao M.*

> "Here's the card from England! Last week we went to London to buy Docs, but it was unbelievably crowded. We must have we must have walked up and down Canaby [sic] Street at least 5 or 6 times, and in every store they were sold out within a few minutes. We're going to give it another try next week, but everybody tells us that it's hard to get sizes 5 + 6 because almost everybody wears these sizes. Well, I'll keep trying, ok? See you again at school,
> Ciao M." (private picture postcard)

For example, especially younger writers use a rather casual tone, which is defined by a high degree of conceptional orality (Koch & Oesterreicher, 1985) and correspondingly shows many features of emotionally determined language use, for example, using smileys, exclamations, capitals, signatures, and so on (Schuler, 1999; see Figs. 6.13 and 6.14). In fact, young people's postcard language is characterized on the one hand by a mix between the features mentioned and a continuous adherence to conventions of written expression on the other, which especially shows in the relatively strict observance of orthographic norms. Thus, we notice that the casual dealing with certain rules of written expression, which can be observed in certain domains of internet communication (especially in publicly accessible and nonmoderated chats) has so far not entered postcard communication stylistics, although occasionally postcards feature graphic devices such as smileys that are familiar from internet communication (Fig. 6.14).

Fig. 6.14. Private holiday postcard.

Besides, especially young people tend to use acronyms and inflexives such as *lol* (*laughing out loud*) and *g* (*grin*) in postcard communication, presumably also influenced by internet communication.

E-Cards

Postcards that are sent electronically, generally referred to by the term *e-cards*, are nowadays offered in large numbers, providing alternatives to the old postcard for diverse occasions (anniversaries, invitations, or simply "quick greetings"). In some respects, the e-card has already taken the postcard's place. Research on the internet reveals a surprisingly broad spectrum of possibilities for users to send out these "virtual postcards." For instance, e-cards free of charge are offered by car dealers, dancing schools, political parties, private individuals, museums, artists, bookshops, mail-order companies, basketball clubs, newspapers, travel agencies, companies, publishing houses, Internet providers, universities, and even by railway stations (e.g., the so-called Hundertwasser Railway Station in Uelzen, Germany). There are even a number of Web sites that are exclusively dedicated to free e-cards, earning revenue by ads placed there (Fig. 6.15).

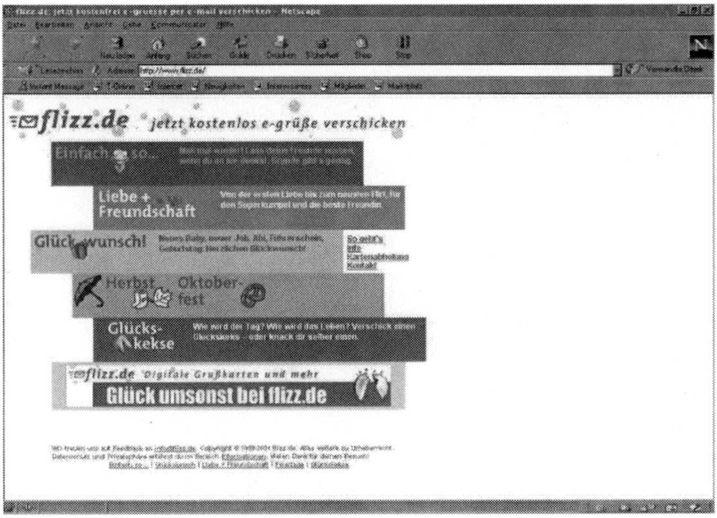

Fig. 6.15. E-card provider on the Internet.

It is true that e-cards can be printed out and downloaded onto the computer. However, if this option is waived by the user, they only possess a "material status" for as long as they are stored on the particular provider's server. Usually they are removed from the server after a relatively short period of time. Therefore, some users tend to print out the e-cards they receive on a color printer and stick the pages together afterward—thus reproducing the traditional postcard format (Fig. 6.16).[20] The fact that some users try to artificially rebuild typical postcard features may also indicate that traditional expectations concerning text types continue to work in the new form of communication.

Most of the traditional postcard's properties are also valid for its electronical counterpart. The sender selects the picture motif with regard to the addressee; the text produced is closely connected with the particular image content, and a remarkable emphasis is placed on the act of *greeting* as a communicative means of maintaining a social relationship. Nevertheless, we can detect several important differences compared to the traditional postcard. In contrast to the "classic case," the recipient is automatically notified by e-mail that there is a postcard on the provider's server, ready to be obtained by a mouse click.[21] Some years ago, this kind of notification e-mail used to inform the recipient that there was a postcard on the server waiting for download, although without specifying the sender. Today these notifications include more information, i.e., stereotypical formulations such as "XY has sent you an e-card," combined with a link to the particular Web site.

Fig. 6.16. "Front" and "back" sides of a private e-card.

In the case of the electronical postcard, the particular service provider is interposed between the sender and the recipient, which is comparable to the regular postal mail parcel or registered mail notifications from the German mail-service Deutsche Post AG announcing that a certain item is to be obtained from a local post office.

The partial or total abolition of the traditional postcard format, respectively of its genre-specific double-sidedness, constitutes another important difference. Still, in many cases the postcard format is at least virtually maintained by simulating the traditional form, respectively by presenting the two virtual "sides" horizontally or vertically next to each other on the screen.[22]

Apart from this, the interplay between front and back sides gets completely lost or has to be simulated. Most e-card services display only one side (Fig. 6.17). In fact, this corresponds to the traditional picture postcard's image side, with the actual text message either placed within the image (following the original fashion of the early picture postcards, Figs. 6.5–6.7) or displayed as a separate text block next to the image.

Fig. 6.17. Private e-card.

Basically, the correlation of image and text is maintained, but it is also subject to important alterations due spacial interaction. For example, the writer in some cases can personally decide on the spacial arrangement of individual image elements, font color, font size, and other creative enhancements like music and background "wallpaper." Thus, the design of the image side is left to the "primary sender" to a greater degree, similar to early postcard use.[23] The simulative character of this kind of e-card also becomes visible where the text message itself is limited in size—although in many cases this limitation is a result of technical conditions, for example, limited space for the text.

In Internet communication, the handwrittenness that is integral to the traditional postcard is substituted by keyboard writing in its manifold specifications. Because of the various multimedia designing devices such as acoustic elements or moving images, e-cards are sometimes reminiscent of earlier greetings telegrams, which until today made use of these very elements, although limited in amount by technical constraints; besides, traditional telegrams, like e-cards, are not composed in handwriting, due to the way they are transmitted. In contrast to the postcard, the telegram has almost completely disappeared as a text type. In many respects, the text message (SMS) has assumed the telegram's original function (see Bergs, chap. 3, this volume), whereas e-cards can be said to have taken the place of the traditional greetings telegram. Postcards that partly contain "playful elements" (Diekmannshenke & Grünewald, 2005) still exist as do ones with acoustic features (e.g., sounding a birthday song on opening). Those variants, however, are usually sent in envelopes. Owing to the fact that it is not possible to set down the signature by handwriting, senders and recipients of e-cards regard these as less personal than their "traditional relatives." They are also generally less appreciated, to some extent because they are more troublesome to view, relying on a certain amount of bandwidth, in some cases certain software, and so on (see later on multimedia aspects of e-cards).

The transmission via Internet entails a further dissolution of text type-constitutive properties. Thus, the address and the postage stamp have disappeared from the actual card space, or they are represented as simulations (Fig. 6.18). Whereas the traditional card represents a final material product which is absolutely identical for both sender and recipient, this is no longer true for the e-card. The generation process of the e-card encompasses details in specifically designed slots (e.g., information on sender, addressee, and the text message), in short, classical components of the postcard. These details, however, usually do not appear on the final card when it is displayed on the recipient's computer screen. This is why the virtual preview assumes an important function when it comes to composing the e-card, because the final product is not identical with the sender's entries. On the part of the sender, this often results in text revision or alteration of card components, processes that are virtually nonexistent in classical postcard creation. Similar to other forms of communication on the internet (e-mail, guestbook, discussion board, chat), this kind of text production represents an entry into an electronic form that can be altered and revised at any time (Diekmannshenke, 1999).

Linguistic changes in relation to the language used on traditional postcards, however, are found to a minor extent only. In fact, writers of e-cards can be regarded to perform the same speech acts as writers of postcards, and even the linguistic devices employed do not contain any specific adaptations to e-card communication. Comparing the texts on postcards and e-cards with each other, we cannot determine any differences (yet). Possibly, e-cards and postcards differ with respect to the scope of topics found within the body text, as well as the way in which these topics are dealt with. For while postcards have taken on a more public character, e-cards at least give the impression of being more private in nature, similar to e-mails (see Fig. 6.18). Still, one can notice a certain influence of Internet-based communication, especially on young e-card authors. Emoticons are used as naturally as other graphical devices (e.g. *gg*, Fig. 6.18). At the same time, however, young people are still sticking to the textual components of salutation, complimentary close and signature typical of the traditional postcard. E-cards tend to be used especially for spontaneous short messages or greetings, which is something they have in common with e-mails, though the composition and design of e-cards is often more personal, in particular due to the imagery. Also it is possible to create one's own e-cards.[24]

Additionally, these options are employed in the multimodal composition of e-cards. Visual animations, video clips, and acoustic elements become more and more present on e-cards, taking over elements that have traditionally been applied to other forms of communication, such as special postcards (especially greetings cards) and telegrams. Here, postcard writers who strive for creativity are offered

Fig. 6.18. Private e-card.

a multitude of opportunities. Because the scope for the design of e-cards goes beyond previous possibilities, it is legitimate to call the e-card a multimodal communication type (see Jones, chap. 1, this volume, on other types of multimodal communication in newer media).

Conclusion

The postcard can be regarded as an artefact of a mobile society, where an initial mobility of workers in the late 19th and early 20th centuries has been replaced by a mobility of mass tourism in the second half of the 20th century. People continue to use postcards to send out their noncommittal greetings from afar to the ones at home, even though the telephone call, preferably using the mobile phone, and the phenomenon of text messaging have meanwhile assumed some of these tasks. What telephone calls and text messaging lack, however, is the image, which is an essential element of the picture postcard and, thus, an integral part of the communicative vehicle. The development of the e-card, of machines making it possible to create individual picture postcards, and of multimedia messages, shows that the trend is still going strong. In retrospective, however, it becomes evident that language change, and particularly the change of pragmatics with regard to forms of address and complimentary close, influences the use of language in postcard communication. On the other hand, we can also notice a remarkable consistency. The traditions of letter communication still play a major role, although Internet communication (especially as an expression of youth language usage) meanwhile has left its traces on postcard and e-card communication, too. Thus, some people tend to become a little nostalgic, like Matthias Stolz (2006) who remarked that "it [the picture postcard; HJD] used to be what we call today a note of absence. Being more illustrative than the notice "I'm not at home from the 13th till the 24th of September" displayed in the e-mail client, it used to include a promise at the same time: 'When I get back, I will have a lot to tell you' " (65). It is hoped that we will continue to find postcards in our mailboxes like the one in Fig. 6.19.

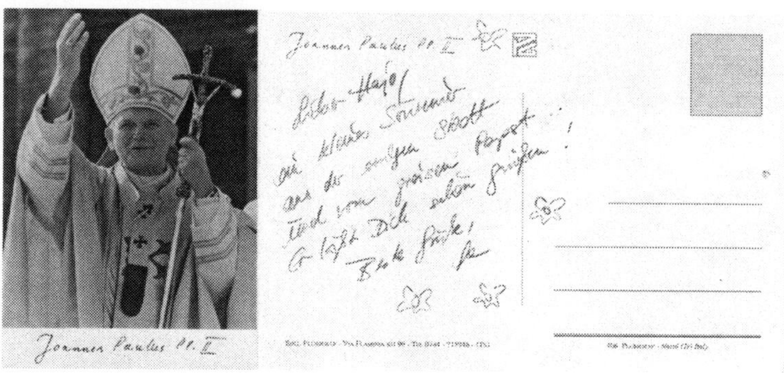

Fig. 6.19. Private postcard.

Lieber Hajo, ein kleines Souvenier aus der engen Stadt und vom greisen Papst . . . Er läßt dich schön grüßen! Beste Grüße, fee

"Dear Hajo, here's a small souvenir from the cramped Eternal City and from the grey Pope—he's sending you his best regards! Best wishes, fee"

Acknowledgments

I would like to thank Moni Reif, Charley Rowe, and Ron Rice for their insightful comments.

Notes

1. http://www.dpwn.de/dpwn?xmlFile=1003696&lang=de_DE (07-23-2007). It is true that the amount of postcards has significantly diminished compared to 30 years ago or even to the beginning of the 20th century. However, this is in no way a sign of an imminent demise of the postcard, as discussed here.

2. For the development of the modern postal system in the United States, the Pony Express as well as the Wells, Fargo & Co. stagecoach service played a central role.

3. A concise research overview can be found in Diekmannshenke (2002).

4. In 1861 already, a predecessor of the postcard as we know it today was published in the United States, though still without postage print (Till, 1994/2003).

5. Even in the 1860s, the chocolate company Stollwerck and the French department store *Au bon Marché* were using these pictures for their advertising; however, it is Liebig who set the standards for these means of mass advertising (Jüssen, 2002). These pictures are similar to the cartoon sheets distributed with bubble gum sheets (reference from Ron Rice). Also art works were distributed on postcards (Till, 1994/2003)—a common way to collect art images.

6. See Bergs (chap. 3), and Rowe (chap. 4).

7. See Quan-Haase on IM (chap. 2) and Bergs on SMS (chap. 3), this volume.

8. This may be slightly different with cards sent home from war: In those cases, the information conveyed receives another, higher value. But even in this case, social activities, especially the maintenance of the relationship endangered by the war, play a major, if not central, role.

9. For media genre evolution see Yates and Orlikowski (1992). Parallels to the development of postcard communication are to be found in the transformation of internal corporate communications in the United States from the end of the 19th century to today. The emergence of formal communication styles favours the control of communication networks (Yates, 1989a, 1989b).

10. Interestingly, mainly academic studies on letter communication take the postcard into consideration, which more or less demotes the postcard to a letter in short-form disguise (e.g., Beyrer & Täubrich, 1996; Grosse et al., 1989; Wyss & Schmitz, 2002).

11. The same reservations people have against internet communication (concerning a degeneration of language) were also applied to postcard communication in the past.

12. See also Rowe (chap. 4, this volume), on stylistic hybridity.

13. See Millar (chap. 11, this volume), for a description of the phenomenon for e-mail.

14. Also do many receivers—consider the annual ritual of Christmas card mailings.

15. The postcard exchange between Diers and Rieckhoff (1987) is an exception, which can be regarded as aesthetic experiment, playing with postcards' form of communication and particularly with its material variety. In the same way, the many postcards that author Jurek Becker wrote to his friend, actor Manfred Krug and his wife (Krug, 1997) as well as to his own son Johnny (Trunk, 2004), represent a communicative sequence and, in the first case, have without a doubt caused relevant responses from the recipients' part. Still, Becker's postcard series do not form a postcard exchange, as they largely lack a mutually constructed narrative.

16. Such diatypical formulation patterns are to be found time and again, be it "Gruß und Kuß" (Love and kisses) on old postcards, "Meld Dich mal wieder" ("Get in touch soon") or "Wetter gut, Essen gut" ("Weather fine, food fine/o.k,"). Corresponding expressions in English might be for example "Having a great time" or "Wish you were here."

17. "Hallo [first name], At the moment we're sitting at the hotel bar and are enjoying our vacation. Hotel and food very good, weather fine for t-shirts and shorts, water warm enough for swimming and beach so empty that we always find free sunloungers. Thus, a great vacation. Best wishes, F. and N."

18. In an impressive way, this is demonstrated by Phillips' collection spanning the entire 20th century, as it clearly shows permanence and change. However, it must be said that in the early days of the postcard, there even existed a number of 'explosive motifs' depicting erotic or pornographic content. Such postcards, though, were rather collected than sent.

19. See Bergs on SMS, Quan-Haase on IM, Rowe on e-mail, and Millar on e-mail (all this volume).

20. For instance, a student presented me her collection of e-cards in this very form. Thanks to Kerstin Luxem and to Eva Barz, who has also allowed me to use her e-cards.

21. Meanwhile, several providers of e-cards (e.g., the basketball club Telekom Baskets Bonn give senders the choice between recipient's notification as described above and direct delivery by e-mail, with the latter being similar to the delivery of postcards in envelopes, or to notification of a telegram.

22. However, it must not be neglected that this in fact happens in a virtual way only. The e-card is displayed within the browser window, and consequently its size depends on the size of the particular computer screen, on its resolution, on the browser software installed, and so on. Together with the other picture elements on the computer screen, it forms a visual unit.

23. By the way, providers of postcard machines take a similar approach by allowing users to create individual postcards from a choice of picture motifs or instant photos. The Multimedia Messaging Service represents a further variant.

24. For example by Siemens_artLab: http://www.artlab.at/galerie_aktuell_detail.php?id=11&PHPSESSID=05b951fbf91a1ad562f7737aaea373dd#; 08-17-2007.

References

Bajohr, F. (2003). *"Unser Hotel ist judenfrei." Bäder-Antisemitismus im 19. und 20. Jahrhundert* ["Our Hotel is Free from Jews." Spa antisemitism in the 19th and 20th century]. Frankfurt/Main: Fischer.

Besch, W. (1996). *Duzen, Siezen, Titulieren. Zur Anrede im Deutschen heute und gestern* [Address in German yesterday and today]. Göttingen: Vandenhoeck & Ruprecht.

Beyrer, K., & Täubrich, H.-C. (Eds.). (1996). *Der Brief. Eine Kulturgeschichte der schriftlichen Kommunikation* [The Letter. A cultural history of written communication]. Heidelberg: Edition Braus.

Campell-Kibler, K. (2003). Gender muffled: Hackers and masculinity. Paper presented at IPrA [The International Pragmatics Conference], Toronto.

Cantauw-Groschek, C. (1993). "Liebe Kolleginnen und Kollegen . . ." Urlaubspostkarten an die Arbeitsstelle. ["Dear Colleagues . . ." Holiday postcards to the workplace.] In D. Kramer & R. Lutz (Eds.), *Kultur-Tourismus—Tourismus-Kultur* [Culture—tourism—culture of tourism] (pp. 143–169). Münster & Hamburg: Lit.

Diekmannshenke, H. (1999). Elektronische Gästebücher. Wiederbelebung und Strukturwandel einer alten Textsorte [Electronic guest books. Revival and structural change of an older text-type]. *Zeitschrift für Angewandte Linguistik (ZfAL), 31*, 49–75.

Diekmannshenke, H. (2000). Die Spur des Internetflaneurs. Elektronische Gästebücher als neue Kommunikationsform [Traces of the Internet-flaneur. Electronic guest books as a new form of communication]. In C. Thimm (Ed.), *Soziales im Netz. Sprache, Beziehungen und Kommunikationskulturen im Internet* [Social action in the Internet. Language, relationships, and cultures of communication in the internet] (pp. 131–155). Opladen & Wiesbaden: Westdeutscher Verlag.

Diekmannshenke, H. (2002). "und meld' dich mal wieder!" Kommunizieren mittels Postkarte ["Write again soon!": Postcard communication]. In E. L. Wyss & U. Schmitz (Eds.), *Briefkommunikation im 20. Jahrhundert* [Correspondence in the 20th century] (= Osnabrücker Beiträge zur Sprachtheorie, vol. 64) (pp. 93–124). Duisburg: Gilles & Francke.

Diekmannshenke, H. (2006). Polit-Postkarten: Propaganda, Wahlwerbung, politische Kommunikation [Political postcards. Propaganda, politicial advertising, political communication]. In H. Girnth & C. Spieß (Eds.), *Strategien politischer Kommunikation. Pragmatische Analysen* [Strategies in political communication] (pp. 97–120). Berlin: E. Schmidt.

Diekmannshenke, H., & Grünewald, D. (Eds.). (2005). *Lachen à la carte. Europäische Humorpostkarten aus dem 19. und 20. Jahrhundert. Die Sammlung Streitenfeld* [Laughing à la carte. European humor postcards from the 19th and 20th century. The Streitenfeld collection]. Alzey: Stadt Alzey.

Diers, M. (1987). Raum für Mitteilungen. Nachrichten aus einem Kartenhaus [Space for communication. Messages from a house of cards]. In M. Diers & J. Rieckhoff, *PostkARTen* [Postcards] (pp. 24–34). Heidelberg: R.v. Decker.

Dürscheid, C. (2003). Medienkommunikation im Kontinuum von Mündlichkeit und Schriftlichkeit. Theoretische und empirische Probleme [Media communication in the continuum of orality and literacy. Theoretical and empirical problems]. *Zeitschrift für Angewandte Linguistik (ZfAL), 38*, 37–56.

Eichhoff-Cyrus, K. M. (2000). Vom Briefsteller zur Netikette: Textsorten gestern und heute [From the letter-writer to netikette: Text types yesterday and today]. In K. M. Eichhoff-Cyrus & R. Hoberg (Eds.), *Die deutsche Sprache zur Jahrhundertwende. Sprachkultur oder Sprachverfall?* [The German Language at the turn of the century. Culture of language or language decay?] (pp. 53–62). Mannheim: Duden.

Ermert, K. (1979). *Briefsorten. Untersuchungen zur Theorie und Empirie der Textklassifikation* [Letter genres. Research on theory and empiricism of text classification]. Tübingen: Niemeyer.

Gold, H. (1999). Stimmungsbilder. Die Postkarte als Medium des (frühen) Antisemitismus [Ambience. The postcard as a media of early anti-Semitism]. In H. Gold & G. Heuberger (Eds.), *Abgestempelt. Judenfeindliche Postkarten. Auf der Grundlage der Sammlung Wolfgang Haney* [Labeled. Anti-Semitic postcards. Based on the Wolfgang Haney collection] [exhibition catalogue] (pp. 13–19). Heidelberg: Umschau Braus.

Gold, H., & Heuberger, G. (Eds.). (1999). *Abgestempelt. Judenfeindliche Postkarten. Auf der Grundlage der Sammlung Wolfgang Haney* [Labeled. Anti-Semitic postcards. The Haney collection] [exhibition catalogue]. Heidelberg: Umschau Braus.

Grosse, S., Grimberg, M., Hölscher, T., & Karweick, J. (1989). *"Denn das Schreiben gehört nicht zu meiner täglichen Beschäftigung." Der Alltag kleiner Leute in Bittschriften, Briefen und Berichten aus dem 19. Jahrhundert. Ein Lesebuch* ["For writing is not among my daily activities." Everyday life of ordinary people in petitions, letters, and reports from the 19th century]. Bonn: J. H. W. Dietz Nachf.

Harras, G. (1983). *Handlungssprache und Sprechhandlung* [Business speech and the business of speech acts]. Berlin & New York: Göschen.

Hedinger, B. (Ed.). (1992). *Die Künstlerpostkarte. Von den Anfängen bis zur Gegenwart* [Postcards from artists. From their beginning to the present] [exhibition catalogue]. München: Prestel.

Heringer, H.-J. (1974). *Praktische Semantik* [Applied semantics]. Stuttgart: Klett.

Hermanns, F. (1986). Appellfunktion und Wörterbuch. Ein lexikographischer Versuch [Message and dictionary. A lexicographical attempt]. In H. E. Wiegand (Ed.), *Studien zur neuhochdeutschen Lexikographie* [Studies in New High German lexicography] (vol. VI.1, pp. 151–182). Hildesheim & Zürich & New York: Olms.

Hobsbawm, E. (2004). *Europäische Revolutionen. 1798 bis 1848* [European revolutions. 1798–1848]. Köln: Parkland. (Original work published 1966 as: The Age of Revolution)

Jäger, J. (2000). *Photographie: Bilder der Neuzeit. Einführung in die Historische Bildforschung* [Photography: Pictures in modern times. Introduction to the research on historical pictures]. Tübingen: Ed. diskord.

Jussen, B. (Ed.). (2002). Liebig's Sammelbilder. Vollständige Ausgabe der Serien 1 bis 1138 [Liebig's Trading Cards: Complete edition of series 1-1138]. Berlin: The Yorck Project [CD-ROM].

Kaufmann, G. (1985). Die Postkarte im Spiegel der Kultur und Gesellschaft. [The postcard in the mirror of culture and society.] In R. Lebeck & G. Kaufmann (Eds.), *Viele Grüße . . . Eine Kulturgeschichte der Postkarte* [Best regards . . . A cultural history of the postcards] (pp. 399–437). Dortmund: Harenberg.

Koch, P., & Oesterreicher, W. (1985). Sprache der Nähe—Sprache der Distanz. Mündlichkeit und Schriftlichkeit im Spannungsfeld von Sprachtheorie und Sprachgeschichte

[Language of closeness—Language of distance. Orality and literacy in the range of language theory and language history]. *Romanistisches Jahrbuch, 36*, 15–43.

Kronoff, F. von [i.e. Frida Hummel]. (1925). Lebensart. Ein Wegweiser des feinen Taktes für Haus und Welt [Way of life. A guide to good behavior for home and away] (6th ed.). Reutlingen: Enßlin & Laiblin.

Krug, M. (Ed.). (1997). *Jurek Beckers Neuigkeiten an Manfred Krug & Otti* [Jurek Becker's news to Manfred Krug & Otti]. Düsseldorf & München: Econ.

Lebeck, R., & Schütte, M. (Eds.). (1980). Propagandapostkarten I. 80 Bildpostkarten aus den Jahren 1898–1929 [Propaganda Postcards I. 80 picture postcards from the years 1898 to 1929]. Dortmund: Harenberg.

Linhart, S. (2005). *"Niedliche Japaner" oder Gelbe Gefahr? Westliche Kriegspostkarten 1900–1945* ["Dainty Japanese" or Yellow Scare? Western war-postcards from 1900–1945]. Münster & Wien: Lit.

Maas, E. (Ed.). (1975). *Das Photoalbum 1858–1918. Eine Dokumentation zur Kultur- und Sozialgeschichte. Ausstellung im Münchner Stadtmuseum 26. März–15. Juni 1975* [The photo album from 1858 to 1918. A documentation of cultural and social history. Exhibition in the Munich Municipal museum 26 Mar.–15. Jun. 1975] [exhibition catalogue]. München: Lipp.

May, O. (1998). *Deutsch sein heißt treu sein. Ansichtskarten als Spiegel von Mentalität und Untertanenerziehung in der Wilhelminischen Ära (1888–1918)* [Being German means being loyal. Picture postcards as reflection of mentality and education in the age of Wilhelm (1888–1918)]. Hildesheim: Lax.

Phillips, T. (2000). *The Postcard century. 2000 cards and their messages.* London: Thames & Hudson.

Polenz, P. von (1999). *Deutsche Sprachgeschichte vom Spätmittelalter bis zur Gegenwart. vol. 3: 19. und 20. Jahrhundert* [History of the German language from the late middle ages to the present. Vol. 3: 19th and 20th centuries]. Berlin, New York: de Gruyter.

Runkehl, J., Schlobinski, P., & Siever, T. (1998). *Sprache und Kommunikation im Internet Überblick und Analysen.* [Language and communication on the Internet. Overview and analyses]. Opladen: Westdeutscher Verlag.

Schivelbusch, W. (2000). *Geschichte der Eisenbahnreise. Zur Industrialisierung von Raum und Zeit im 19. Jahrhundert* [History of rail travel. The industrialisation of space and time in the 19th century]. Frankfurt/Main: Fischer. [first 1977]

Schuler, C. (1999). *Kommunikationsmittel 'Postkarte'—Form, Funktion und Gebrauch* [Postcard communication. Form, function, and use] (master's thesis). Typoscript. Bonn.

Stolz, M. (2006). Wetter gut! Die Ansichtskarte stirbt aus. Ein Abschiedsgruß [Weather is fine. The picture postcard is dying. A farewell]. *DIE ZEIT 7*(9), 65.

Szépe, H. (1979). Überlegungen zum Text der Ansichtspostkarte im Zeitalter des Massentourismus [Research on the text of the picture postcard in the century of mass tourism]. *Sprache im technischen Zeitalter, 72*, 314–316.

Täubrich, H.-C. (1996). Alles auf eine Karte. Correspondenz-Postkarte statt Briefkorrespondenz? [Everything onto one card. Correspondence postcard instead of correspondence by letter?] In K. Beyrer & H.-C. Täubrich (Eds.), *Der Brief. Eine Kulturgeschichte der schriftlichen Kommunikation* [The letter. A cultural history of written communication] [exhibition catalogue] (pp. 112–119). Heidelberg: Edition Braus.

Till, W. (1994/2003). *Alte Postkarten* [Early postcards]. München: Weltbild.

Trunk, T. (Ed.). (2004). *Lieber Johnny. Jurek Beckers Postkarten an seinen Sohn Jonathan* [Dear Johnny. Jurek Becker's postcards and letters to his son Jonathan]. Berlin: Ullstein.

Wietek, G. (1977). *Gemalte Künstlerpost. Karten und Briefe deutscher Künstler aus dem 20. Jahrhundert* [Painted artist-postcards. Postcards and letters from German artists of the 20th century]. München: Thiemig.

Wyss, E. L., & Schmitz, U. (Eds.). (2002). *Briefkommunikation im 20. Jahrhundert* [Correspondence in the 20th century] (= Osnabrücker Beiträge zur Sprachtheorie, vol. 64). Duisburg: Gilles & Francke.

Yates, J. (1989a). *Control through communication: The rise of system in American management.* Baltimore: Johns Hopkins University Press.

Yates, J. (1989b). The emergence of the memo as a managerial genre. *Management Communication Quarterly, 2*(4), 485–510.

Yates, J., & Orlikowski, W. J. (1992). Genres of organizational communication: A structurational approach to studying communication and media. *Academy of Management Review, 17*(2), 299–326.

Chapter 7

History Now

Media Development and Textual Genesis of Wikipedia

CHRISTIAN KOHL

University of Koblenz-Landau

WOLF-ANDREAS LIEBERT

University of Koblenz-Landau

THOMAS METTEN

University of Koblenz-Landau

Abstract

This chapter argues that Wikipedia is a user-driven knowledge system and hence is able to incorporate controversies and different points of view. By an analysis of the discussion-article-relations of the main-article of the philosophy portal, we try to make explicit the internal management of collaborative writing processes and the way in which the process of article-discussion-interactions in Wikipedia leads to the incorporation of controversies and different points of view.

The analysis of the main-article of the philosophy portal focuses on the period from July to August 2005 because this was a period of increasing collaboration. By extracting basic patterns of article-discussion-relations we expect to make clear which specific medial functions of Wikipedia were at work. These functions have an enormous impact on the process of collaborative writing because the process of writing and thinking about writing itself becomes transparent as a whole. Collaborative writing in Wikipedia thus becomes a process discussing different discourse positions and incorporating multiple aspects by rewriting and rearranging the article. These interchanges of articles and discussions along with the dynamic development are what distinguishes Wikipedia articles from classic encyclopedia articles.

At the moment, the most prominent example of Wiki-systems is the Internet encyclopedia Wikipedia[1]. If one looks at the Wikipedia as an encyclopedia, it is part of the long history of encyclopedias. According to Mittelstraß (1967) the *Encyclopédie* of Diderot and d'Alembert can be called the first modern encyclopedia. Two characteristic features that contrasted this encyclopedia with others were the formation of a group of authors (in contrast to the single author of earlier encyclopedias; see Elia, chap. 13, this volume) and the claim not only to present existing knowledge, but to work educationally (Haß-Zumkehr, 2001; Volpers, 2002). Subsequently, the Brockhaus, a type of encyclopedia that firmly addressed an (educated) amateur audience, was developed in Germany (Volpers, 2002).[2] Wikipedia differs from the *Encyclopédie* and the *Brockhaus* in three essential points:

1. The principle of *collaboration* becomes decisive: Although groups of authors and publishers are involved[3] with the *Encyclopédie* or the *Brockhaus*, there still is a conventional editorial staff process: An author is responsible for an article, the only intervention in this article can occur through the editorial staff and the publishers. In the case of Wikipedia, however, there are always several authors equally responsible for an article, there is no editorial staff, no publishers, and also no formal, fixed editorial processes.

2. In case of Wikipedia *everybody is able to be part of the authorship.* The authors are mostly anonymous and not necessarily experts. This differs from the *Encyclopédie* or the *Brockhaus.* Furthermore, the authors are not specifically selected and invited by an editorial staff, but become active on their own impulse.

3. There is *no preselection* of dictionary entries, given taxonomy, article size, deadline *or similar restrictions* of more classic editorial processes.

As Stickfort (2002) noted, the Internet medium has "added . . . a new dimension to the former traditional forms of information collection and information spreading" (271). Besides the generally perceived new possibilities of searching (and finding) as well as the distribution of information, the World Wide Web (WWW) opens new possibilities for the construction of knowledge. Wikis are tools for Internet-based knowledge construction that, in comparison to classical processes of knowledge construction, as for example the writing of an article or a monograph, partly allows radical changes. Initially, Wikipedia served as a prominent example of Wiki-systems. Nevertheless, as long as Wikipedia is viewed in the tradition of encyclopedias, determining features, by which Wikis are usually marked, do not stand out clearly. Only with a representation of the media qualities does their real potential become clear.

Hence, this chapter discusses the historical development and qualities of Wikis. Some basic considerations of the dynamic media are made from which the meaning of the historicity of every text becomes evident. Hence, this chapter opens two historical dimensions: the history of the Wiki technology as a media history, and the history of the origin of text as a textual history. The dimension of the textual history arises from the possibilities of the medium, and therefore from this follows the media history.

Media History and Basic Principles

Indeed, Wikis have become popular only since about 2003, and Wiki projects are often seen as an example of "web 2.0." Nevertheless, Wikis already existed in 1994, thus since the beginning of the WWW. "Wiki" is the designation for a type of software, similar to "word processing" or "spreadsheet," which can stand for a number of different programs. The basic principles of this software are called "Wiki principles." However, the concrete applications are often called "Wiki," for example in phrases like "in our Wiki." However, in this case "Wiki" does not refer to the software type, but to the concrete application. What is meant is "in our Wiki of the lecture XY" or "our film Wiki." Another complication arises from the fact that a large number of different software products of the type "Wiki" exists, which vary in their functional circumference. The following section briefly explains what Wikis actually are, which ideas and principles are behind the technology, and the varied application possibilities.

What is a Wiki?

A Wiki is a Web application that enables collaborative work on the Internet (Kleinz, 2003; Leuf & Cunningham, 2001; Möller, 2002). The user should be able to participate without previous knowledge and technical hurdles. Cooperation and common compiling and handling of texts are in the foreground.[4] This principle already has its reflection in the name: "Wiki" comes from the Hawaiin language and means "fast" or "informally" (Leuf & Cunningham, 2001).

The first "Wiki" was used in 1994 by Ward Cunningham for the "Portland Pattern Repository,"[5] therefore, Cunningham is often said to be the "Wiki inventor." According to Leuf and Cunningham (2001) a Wiki is a "freely expandable collection of interlinked Web 'pages,' a hypertext system for storing and modifying information—a database where each page is easily editable by any user with a forms-capable Web browser client" (14). Wiki-software runs on a Web server and enables every user to publish, change, and link texts on the Internet. The user does not need any HTML knowledge or special software. Text is entered as rough text and a conventional browser without auxiliary modules is sufficient. Furthermore, all contributions can be read and worked on by all participants. This way, a large number of spatially distributed individuals can work collaboratively on (hyper-) texts via the Internet. The low technical requirements as well as the simple usage make it possible for "computer laymen" to publish and work on texts on the Internet and to provide hypertexts. Finally, the technology is platform-independent: This means that it can be used with any operating system that has a browser—the only condition is access to the Internet.

On the basis of Cunningham's "Primal-Wiki," meanwhile, many variants of this software exist.[6] On the one hand, these differ by the programming language in which they were written,[7] and on the other hand by their functions and restrictions. This is how for example TWiki[8] developed, a piece of Wiki software that was programmed for operation in enterprises and therefore has a user administration. These different derivatives are also called "Wiki-clones."[9] Above all, they vary in the functional range, which concerns user friendliness and user administration. Some versions additionally have a Plugin architecture, so that additional functionality can be added according to desire. Furthermore, the "configurableness" of the user surface is different from Wiki to Wiki: The spectrum ranges from rigid systems in which changes are not possible without modifications of the source code of the software, to systems that can be based on drafts and can easily be adapted to one's own needs.

On the basis of a collaboration platform for programmers to exchange software design samples, Wikis are meantime used in the most diverse areas of application: internally as a tool for documentation; for organization; and for transfer of knowledge[10]; for the documentation of software[11]; in E-Learning[12]; collecting recipes, restaurant criticisms, or reports on journies[13]; to the creation of textbooks, dictionaries, or encyclopedias.[14]

Wiki Principles

Despite the heterogeneity of the software and the large differences concerning the operational area and functionality, two core elements can be identified that are constitutive for Wikis: simplicity and openness.[15]

Simplicity

The demand for "simplicity" becomes clear with different aspects of the software. The principle particularly becomes clear with the conditions for the usage of the software as well as the requirements of hardware and software. Changing and providing contents takes place in the browser, either in a so-called WYSIWYG editor, or rather text-oriented with the usage of the so-called Wiki syntax. Besides the simple input and formatting of texts, similar easy productions of links are in the foreground; finally Wikis are, above all, tools for the collaborative creation of hypertext networks.[16] The user does not have to have any technical knowledge and does not have to know the programming language in order to be able to cooperate in a Wiki project.

So far, well-known Wiki surfaces are quite text laden, having none too few graphic elements; the emphasis of the production of Wikis is so far clearly on content, not on the design. So far no usability studies or similar investigations concerning operability or the acceptance of different Wiki surfaces exist.

A further crucial factor for the acceptance of Wikis are the low entrance hurdles: The technical facilities necessary for the operation of a Wiki are low on both the client and on the server side, the user only needs a conventional browser, the operator can exclusively use free software and does not need any special know-how for the installation and usage. By now there are providers where Wikis can be rented or used free of charge, similarly to blogs or e-mail services, so called "Wiki farms."[17] Now, even Web content management-systems and learning management-systems frequently offer Wikis besides chat and blogs.[18]

Openness

Next to simplicity, openness is the second characteristic of Wikis. Publicly accessible Wikis hardly have any barriers—in principle all users have the same right to write and read. This means that each visitor can work on pages or create new ones. However, by now, many Wikis have a more or less sophisticated user administration, so that registration may be necessary, for example certain articles on Wikipedia are only processible by signed users or by administrators. Furthermore, with an appropriate server configuration (e.g., a firewall), it is easy to ensure that a Wiki is accessible only to a certain user group, for example coworkers at a particular company. Some Wikis, for instance TWiki, enable the creation of user groups and the synchronization of persons and groups with already existing listings, such as Microsoft Active Directory.

The principle of openness grants many rights and liberties to the Wiki participants, however, at the same time it obligates them to respectful handling of the contributions of others. The "missing" write protection for contents of a Wiki does not inevitably lead to chaotic contents because for each person who deletes contributions or writes nonsense,[19] there are several others who cancel or improve this change again. Cifollili (2003) also pointed out that destroying a contribution requires more expenditure than repairing it. So far there are no comprehensive studies about this behavior of usage, so that this estimation is based only on empirical values (Hennicken & Zahiri, 2003; Leuf & Cunningham, 2001). However, Wikipedia supplies examples of the effectiveness of the Wiki principle: Viegas et al. (2004) have proven that 50% of all examined cases of vandalism on

Wikipedia were repaired within 2 or 3 minutes.[20] Dworschak (2004) and Viegas et al. (2004) see this openness as a fundamental difference to other communication forms on the Internet and claim the fact that everyone can change anything to be the high "ability to reach a consensus" of Wikis. A further consequence of the openness of Wikis is the dissolution of the distinction of author and recipient: The abolition of this separation of roles is frequently stated as characteristic for Internet communication; however, only Wiki systems enable these two roles to be implemented at any time by any user. One can even go further and note that it is a constitutive characteristic of Wikis that all participants can be recipient and author at the same time: Otherwise the Wiki would not function and would appear to be a "normal" Web site. Because of this extension, not only the user's possibilities change, but the requirements also increase. Instead of a unidirectional communication in the form of bare reception, a multidirectional communication takes place: the exchange of knowledge by cooperative work with several other users. A shift away from passive reading toward more interaction and activity takes place. Critical thinking, handling different points of view, the reflection of one's own position concerning other opinions, teamwork, and the cooperative writing of texts must be learned and practiced.

Legal Aspects: OpenSource and OpenContent

The contents of a Wiki are commonly compiled by the users. Therefore, the legal question of which license the contributions use is of great importance. Many Wikis put an emphasis on a so-called "OpenContent" license. OpenContent is an analogical formation to "OpenSource."[21] The term *OpenSource* originates from the software development and describes a license form that deviates from the classical "copyright" by assigning certain rights of usage to the users, instead of limiting them. It concerns the right for handling, the passing on, and the duplication of the source code of programs. Known software projects that follow this model of development are the operating system Linux, the office-package OpenOffice, the Web server Apache, and the browser suite Mozilla.[22] A further characteristic of OpenSource licenses is that they are so-called "virus licenses." This means that products derived from them must have the same license as the output product. Thus, free entrance and free distribution are secured durably; this not only concerns the output product, but also all conversions derived from it.

Similar to the conditions of the software source code, OpenContent stands for the same license model in reference to "knowledge" or "contents."[23] The most prominent OpenContent licenses are the "license for free contents"[24] and the "Creative Commons license"[25] for the German right as well as the "GFDL"[26] for U.S.-American right. Different variants of these licenses exist. The rights of usage, passing on, and subsequent handling can be specified in detail.[27] Wiki contents, however, do not have to use such a license: Although in most Wiki projects the author uses an OpenContent license, any Wiki user can feel free to use a license preferred for the user's own project.

For scientists, there is an additional question of how the quasi-anonymous authorship in Wikis (e.g., Wikipedia) will be evaluated by consultants and employers in the future. Right now contributions in a Wiki are not suitable for addition to a publication list. For this reason, an important stimulus for writing is void (i.e., the production of appropriate reputation, or bourdieuesque "capital").

Collaborative Media and Text History

The participatory possibilities on the Internet are increasing since the spreading of "Web 2.0" uses.[28] In light of the dissolution of a clear role assignment of author and reader as well as the extended technical possibilities of Wiki systems, texts on the Internet no longer appear static, but rather as dynamic forms of writing. Interactivity describes a new dimension of social interaction, which, also by means of the Wiki software, was established as specific writing practice. Texts that develop in Wiki systems are therefore not singular historical events, but possess a temporal structure. Therefore, they must be understood and analyzed as writing processes and less as final works. This results in the initially outlined second perspective: Apart from the history of the development of the medium itself, the medium's own temporalness can be described, which contributes to the possibilities of communication.

Landow (2003) repeatedly stressed that the majority of the web pages on the Internet are still arranged as "nonbook books." Web pages, therefore, are not understood as books, but in the long run they are designed like books. In this way, an established and antiquated sample of historical media is transferred to new, historically younger media (see McNeill, chap. 12, this volume). However, electronic documents are—against the opinion of many teachers, technologists and software developers—not books (Landow, 2003). With the paradigm of the book, three aspects in particular are connected that coin the handling of texts: Texts are to be discretely differentiated from other texts, the roles of author and reader are clearly definable, and texts are static. Landow on the other hand, suggested describing specific applications of the Internet as a "collaborative learning environment." Such a perspective offers the possibility of understanding texts in Wiki systems not as static works, but rather as dynamic processes. Therefore, the following paragraph, in consideration of the designated aspects, focuses on theoretical considerations about how a changed understanding of texts in Wiki systems affects the investigations of the texts.

Text Theory 2.0

The multiple cross-linking of Web sites on the Internet led to the assumption that the discrete distinction of Web sites dissolves. We contradict this first aspect: The discrete distinction of texts does not dissolve on the Internet. Instead of the limits of a book, a direct linking of sites does develop. However, regarding the historical development of media representational forms, the linking of web pages only appears to be a gradual increase in relation to other representational forms. Daily papers and encyclopedias are likewise characterized by such features as complex chaining and modular structures. Although web pages are, on the one hand, directly connected with one another, there is, on the other hand, no problem in making a clear demarcation. The switch from the web page of a car manufacturer to one of a telephone solicitor becomes immediately evident on the basis of the design. To this extent, hypertexts enable a strengthened modularity of texts whose single modules exist in different relations to each other and can be directly interconnected (see Jakobs, chap. 14, this volume). However, regarding content, organization and addressing a clear demarcation of t texts takes place.

Although for the recipient, the identity of web pages remains to a large extent protected, the dissolution of authorship and the mixture of the roles of author and recipient have further reaching consequences. With the development of the Internet, considerations developed by among others Roland Barthes and Jacques Derrida earlier, repeatedly have re-arisen: "hypertextuality embodies poststructuralist conceptions of the open text" (see Landow, 2006:2). Barthes (2006) suggested relativizing the relationship among the writer, reader, and observer. Therefore, he understood the text from the point of view of the process of writing and stated: "[t]he theory of the 'text' can only accompany the practice of the writing" (72). The reader no longer just reads what is offered to him or her, but is actively involved in creating the text. In the context of hypertext theory, this for a long time just meant: as the reader selects his or her reading path within the complex hypertext structure, he or she develops an individual way of reception. Such a way of reading is made possible by hypertext systems; books, however, specifiy the procedure of reception to a large extent. Strongly interlaced hypertext systems, therefore, create a labyrinth with innumerable possibilities for reception.

The picture of such a labyrinth, however, only develops by means of a description of the structures of the hypertexts. As a result of the structural dimension of hypertexts, which is described by the metaphor of the net, hypertexts are mostly seen as nonlinear. The nonlinearity of hypertexts designates specific structures, which can only apply to certain forms of hypertext organization, and exist together with numerous others, hierarchical or linear organization forms. In each case, the recipient only perceives a linear succession of web pages, which he or she can only gradually reconstruct. Storrer (2000) prepared such a structural perspective on hypertexts for linguistics, whereby such an analysis completely corresponds to the linguistic tradition, which is strongly affected by structuralist theory. The Internet encyclopedia Wikipedia is often compared with printed encyclopedias because both media not only aim at a comparable representation of contents, but also show a strong chaining structure, therefore showing structural parallels. If one exclusively analyzes the texts and their chaining structure within the hypertext system, such a comparison seems appropriate. The kind of text (encyclopedia entry) as well as handed-down concepts of the encyclopedia determine the understanding of the Wikipedia participants as well as those who examine the Internet encyclopedia from a linguistic point of view. From this point of view it is only about the structural and thematic organization of the texts, and not about processes of writing. The view on the structural dimension of hypertexts, however, does not clarify the crucial characteristics of Wikis. An analysis of hypertext structures is important; however, it cannot clarify the processes of text emergence that are crucial for an understanding of Wikis.

Grube (2003) said it is alluring to argue from the perspective of nonlinearity that hypertexts are deeply embodied in the tradition of older texts. Therefore, he suggested "that the so far leading criterion of nonlinearity to determine hypertexts should be taken back in favor of the criterion of interactivity" (83). The understanding of the texts shifts if the writing as well as the writing process moves toward the focus. It will then become clear Wikipedia's texts, due to the software, establish a changed cultural writing practice that no longer proceeds from an individual author nor the author's monological expression. The result of this work is not the text as a final work. The dynamic dimension of the texts clarifies that its temporal organization is of special importance. Contributions in Wikipedia no longer represent only single texts in a net of further texts (hypertext), but also must be regarded in their history of development.

Therefore, we want to understand texts—as they develop in Wikis and also occur in Wikipedia—as processes, whereas the processes of the emergence are to be analyzed regarding the temporal course of events. Although the static text appears as a unique event in a historical continuum, the dynamic text embodies the history of its emergence by the act of the participant's writing. A contribution in the Internet encyclopedia, therefore, can be understood as an embodiment of the act of writing and processes of interacting, negotiating, reformulating, and making the contribution easily understood. This dimension of text history can be described as the performative dimension, which depends on the media condition and clearly goes beyond a purely structural understanding. Therefore, three dimensions of the investigation become relevant for the analysis of the texts in Wikipedia: the structural dimension, the media dimension, and the performative dimension.

The performative dimension records the temporalness of the text with respect to the history of text emergence. Those aspects are crucial for the understanding of collaborativly written texts.

Such an analysis stresses the eventful happening of the single text and documents its history as one of interaction. The idea of interactivity, therefore, marks a turning point in the participation of the recipients, who not only can select their own reading paths but also can create text by clicks in the reception. The reader in Wiki systems is also an active writer. With the computer, such a dynamic sampling of symbol structures occurred (see Krämer, 2003). Time becomes a dimension of the text with respect to text emergence; the writing opens an action area for the interaction and common text formatting. Wiki systems extend the area of action by changing the monologic writing into dialogic negotiating. That connection of negotiating and communication clarifies that the reader–text interaction is embedded in a communicative happening as well. The user interacts with the text, but does this knowing that the individual passages are to be assigned to specific participants. The result of this is the communicative connection. The writing, understood as an area of operations and perception, is therefore not only described as a medium of representation; but also as presenting an "interrelation of eye, hand, and machine"(Krämer, 2003:52). Although Krämer, with reference to the outwardness of writing, stated that these not only serve an expanded communication, they also serve the monologic solving of problems. Wikis show that writing also serves the dialogic writing of texts. Articles—as they appear in Wikipedia—represent a snapshot within a constantly progressing and unfinished process of writing of numerous authors. The current shape of a text, therefore, only can be understood because of the history of development of the respective contribution. Wiki systems enable such a text analysis because of the detailed archive of versions of texts, which makes each change of individual authors accessible. This way, the history of each contribution is completely documented from the first entry on.

Text Emergence and Text Analysis

Collaborative processes of writing dissolve the central intention of the author. The authors only work on passages and never on the entire text. Therefore, they mostly take large parts of text and supplement them with their own formulations, which can include individual words, sentences,or whole paragraphs. The statements taken over now attain the character of quotations, whereby the keeping does not guarantee that the quoter means the same thing with the rewording like the preceding person. Thus, those characteristics of the writing emerge, which Derrida (2004, and others) named particularly in his early work about the concept of script: The writing must function in absence of author and reader. The text as a unit carries the traces of all authors. At the same time the sign—the writing—is quotable (i.e., it is part of the linguistic signs to be repeatable). "I would like to insist on this possibility: The possibility of taking out and quotational grafting, which is part of the structure of each spoken or written sign [marque] . . . What would a sign be that could not be quoted? And whose origin could not be lost along the way?" (Derrida, 2004:89). The repeatability of the sign and the repetition in the writing opens up the temporal structure of the text as the practice of writing. By writing, the participants in each case react to the preceding contributions of other authors. As writers, they start a "dialogue" with the already expressed information. This thematically oriented discourse stands out within the history of development of the text and is constantly continued. The statements of other participants are quoted and varied. On the one hand, the text appears to the recipient as a unit; on the other hand the archive of versions and the documentation of discussions clarify the interactive character of the text emergence. Although texts in books are static, Wiki systems make dynamic writing processes possible. Therefore, the interlaced structure of hypertexts is not their crucial characteristic, rather, it is the temporal structure of their emergence by the practice of the writing of several authors.

An analysis of the contributions in Wikipedia should not only regard an individual text, but also include further texts with reference to a topic into the analysis (Kohl & Liebert, 2004). Therefore, it is about the contribution itself, the documentation of its emergence, the archive of versions, and the discussion pages belonging to the contribution. Thus, the discussion pages offer the participants the possibility of commentating their intentions and interests as well as explanations about their own understanding of the text. Such a view includes the structures of hypertext. Apart from such a structural collection of Wikipedia contributions, an analysis of the temporal course of text emergence should take place. Therefore, the participation of participants and their interactions must trace an appropriate representation and analysis of collaboratively developed texts. The text can be examined regarding its temporal structure as well as with a view of text-constitutive actions, and the participation of the participants. An analysis should include both dimensions in order to formulate statements about the emergence and quality of texts. A comprehensive analysis that does not consider the forms of interaction of text emergence can hardly make appropriate statements about the entries in Wikipedia. Only an analysis of the action patterns and intentions of the participants can create a complete picture of text emergence if the dialogical principle of text emergence and the mutual reference of the participants are considered at the same time. This is particularly important because the appearance of the text frequently resembles a monologically written text.

Earlier we remarked that the understanding of text types of the participants also plays an important role. Apart from this, a stylistic analysis, which is closely bound to the conceptions of the kind of text, seems to be of special importance. On the basis of the contribution about philosophy in Wikipedia, the complexity of such an analysis is to be clarified in a few notes. Contributions in Wikipedia usually start with individual expressions and entries. The further process can take place in different kinds of ways. Some contributions only consist of few sentences for a long time, others rapidly develop, and after a short time they already offer a comprehensive overview about a topic. The text emergence does not proceed continuously. Phases of intensive revising take turns with phases in which hardly any changes are made for a long time. In short phases of intensive treatment, few participants can take part, whereas during long phases, in which hardly any changes take place, revising can be intended for many different participants. On the one hand, the development of a contribution depends on the interest of the participants and the respective constellation of participants; on the other hand, factors such as currentness of a topic in further discourses have an influence on text development. Processes of cultural change, therefore, also settle in processes of text emergence.

Wikipedia's contribution about the topic "philosophy" meanwhile belongs to the so-called excellent articles. This contribution was made with the first entry on May 4, 2002. Since then and up to December 2, 2006, the contribution was worked on 1,216 times. Therefore, the contribution was changed with an average of 0.73 times a day. Within the first year, however, only 34 changes were made. However, 14 changes were made alone on August 3, 2006. This clarifies that the process of the handling does not proceed continuously. A look at the participants shows that only four people made the changes on August 3. Viewed alone, the frequency of changes or the number of the participants involved does not reveal much. If one now adds the commentating of the contribution on the discussion pages to the analysis, the result is a complex picture of the history of development of the contribution, the participation of the participants as well as their self-understanding and intentions. The text as a thematically coherent unit is only readable in an appropriate way when the different aspects are referred to each other.

Aspects that were relevant for the analysis of static (i.e., printed texts) must be examined in relation to the considerations discussed. Contributions within Wikipedia appear as discourse shapes that developed within a complex process of collaborative development. The texts embody dynamics and actions of the participants in each case one at a time. An appropriate understanding of these texts, however, must regard the current status of a text with view to its prehistory. The text is an historical document, not in so far as it developed at a certain historical time, but in so far as it embodies the complex history of its emergence. Regarding Wikis in a media-historical perspective, therefore, does not only mean regarding the interrelation of different media in the historical process but also to regard the historicity of the medium as well as the developing texts. Therefore, a linguistic discourse history must—as is becomes clear by the example of the Wikis—be supplemented with its own historicity of the texts. Therefore, a text cannot exclusively be seen as a subject-bound relinquishing of an intention at a certain time, but must also be analyzed with view to its own temporal structure.

Altogether, the three levels mentioned here appear to be relevant for the analysis: Apart from a structural analysis, the description of a text regarding the medium is relevant. The performative dimension of the analysis now describes the actions of the text emergence. The structural dimension of a text, the medium of its constitution, as well as the form of interaction are therefore crucial for the analysis of the texts. Although the form of interaction is mostly quasi-automatically attributed to the monological expression, interactive media, as they are present by means of the Wiki software exhibit dialogical essential structures. Therefore, the written text cannot be solely understood as an expression of a participant any longer, but must be analyzed regarding the communicative happening. The result is the construction that the contributions of Wikipedia are written conceptionally and in media writing; however, regarding the interaction form, they are arranged dialogically.

Some examples will clarify this distinction: While the kind of text of the interview documents and differentiates the different expressions of the participants, the kind of text of the encyclopedia entry presupposes that the individual items of the participants are not presented as such. Both kinds of text—interview and encyclopedia entry—however, are mostly monologic expressions. The interview and the encyclopedia entry are in each case written by one participant. The kind of text and the text emergence therefore can be different from each other. Although the text emergence of the contributions in Wikipedia are based on a dialogic principle, the text appears to be written according to a monologic principle. Therefore, an interview can be understood as a monological expression of an editor, which reports on a preceding discussion that is produced in the form of a dialogic structure. An encyclopedia entry in Wikipedia, however, appears as a monological expression; however, it is based on the dialogic act of the participant's writing.

Although discourse theory regenerates the historical references of the texts to each other, regarding the Wikis it becomes clear that texts themselves must be noticed in their temporal structure. The contributions in Wikipedia embody history not because they show an historical event, but because they are documents of a temporal developing process. Although the discourse notices texts as events of individual authors, the demarcation of the text dissolves in relation to the discourse in the Wikis. Contributions in Wikipedia appear as uniform texts, although they represent—often also controversial—discourses.

Interaction and Construction of Knowledge

The historicity as well as the dynamic emergence of the text refer to two aspects regarding the generation of knowledge. On the one hand, such a text no longer serves only mediation of information. The opinions and considerations of different authors flow into a text. The obligation to create a uniform and commonly arranged text promotes a process of negotiating, which intensifies the work with the positions of other participants. In summary, it can be said that knowledge is not obtained in Wikis but designed in the tacklement with the positions of other participants during an interactive and collaborative writing process. The experiences of the participants as well as the writing processes form the basis of knowledge. Krämer (2003) expressed this aspect of the epistemological writing with the following formula: "[n]ot only communication but also cognition" (30). Krämer referred to an aspect that is part of a long tradition and is usually stated in the prominent formulation of Heinrich of Kleist of the gradual producing of thoughts while talking. The script is thus not really meant as a means of communication but rather as outward equipment or technology by which sense can be mediated. The writing itself becomes a specific practice of interactive and collaborative construction of knowledge.

Using Wikipedia as an example, Forte and Bruckman (2006) pointed out that these have qualities that enable their participants to have specific experiences. The authors describe the participants of Wikipedia as a global community of learners, which is merged into a common process of constructing knowledge. As Forte and Bruckman said, Wikipedia is not suitable for learning processes only for the reason that it represents immeasurable and universally available online resources; but Wiki systems create area of action, which make an interactive happening of knowledge construction possible. Thus, the disruption in relation to the principle book as memory and transmitting medium is carried. Although the book aims at a final work, Wikis arrange interactive and dynamic processes of writing. In most interesting applications, the Internet must be understood as dynamic and interactive because this is exactly where the specific potential is. Therefore, writing becomes a specific means of knowledge construction (Forte & Bruckman, 2006).

The basis of a culture of knowledge then, which is not only constituted with media, is fundamentally based on linguistic interaction, formed by the controversial process of text genesis. This way writing becomes a practice of knowledge construction. From this extended epistemological understanding results the picture outlined here. Knowledge develops during a constructional, unfinished process from the individual experience, during a process of dialogic writing in the new media. Texts—as they develop in Wikis—stand in an historical continuum. They also exhibit an internal history, one of their emergence, which is unfinished and constantly continued.

Notes

1. http://www.wikipedia.org (English original) and http://de.wikipedia.org (German version).

2. In order to prevent a retro-romanticizing of these two works, it should not be ignored that the impact to start both the *Encyclopédie* and the *Brockhaus* came from the side of the publisher, that means that economic interests were the cause.

3. According to Anger (2002) about 1,500 scientific authors are involved in *Brockhaus*.

4. This refers to the present state of the art: Theoretically it is just as conceivable to work on video or audio data with an appropriate Wiki software. This, however, means higher demands both for the software developers and the users.

5. Portland Pattern Repository Web site: http://www.c2.com/cgi/wiki?WelcomeVisitors [03.12.2003].

6. Huhmann (2004) and Möller (2003) compared different Wikis. However, it is not always transparent according to which criteria the evaluations were made, partly they seem to be based on personal impressions. Under http://c2.com/cgi/wiki?WikiChoicetree [21.07.2004] is a listing of Wiki software, arranged according to functions. For reasons of space, we do not list the different Wikis with their respective pro and cons.

7. For a more exact representation of the differences, see Leuf and Cunningham (2001).

8. http://www.twiki.org [30.07.2004]

9. See http://c2.com/cgi/wiki?WikiEngines [12.07.2004] for a list of Wiki implementation in various programming languages.

10. See Puls, Bongulielmi, and Henseler (2002), Hof (2004), and Majchrzak, Wagner, and Yates (2006).

11. See for example, the documentation about the Linux-Distribution Knoppix http://www.knoppix.net/wiki/Main_Page [11.11.2006] or about text processing LyX http://wiki.lyx.org/ [11.11.2006].

12. See Dieberger and Guzdial (2003) and Hennicken and Zahiri (2003).

13. See, for example, the restaurant- und Bar guide "DaWiki" http://www.dienstag-abend.de/wiki/index.php/Hauptseite [11.11.2006] or WikiTravel http://wikitravel.org/en/Main_Page [11.11.2006].

14. See, for example, WikiBooks http://wikibooks.org/ [11.11.2006], Wiktionary http://wiktionary.org/ [11.11.2006] or the best known Wiki Wikipedia http://www.wikipedia.org [11.11.2006].

15. See the "Wiki Principles" (http://c2.com/cgi/wiki?WikiPrinciples [12.07.2004]) and the "Wiki Design Principles" (http://c2.com/cgi/wiki?WikiDesignPrinciples [12.07.2004]).

16. For the term *hypertext*, see Storrer (2000), Nielsen (1996), Kuhlen (1991), particularly about problems of coherence and coherency in such text networks see Storrer (1999).

17. For a commercial offerer see http://www.etouch.net [24.11.2006], for a free offerer see http://www.wikihost.org/?lan=de [24.11.2006].

18. See http://joomla.org/ [24.11.2006] and http://moodle.org/ [24.11.2006].

19. Intentionally destroying content within the Wiki communities is also called *vandalism*. This covers the deletion of content, as well as the inserting of unreasonable, insulting content, or including passages that are not in favor of the common goal of the Wikis.

20. However this statement is based on a highly simplified operationalisation of "vandalism": Only such cases of vandalism that are determined automatically are seized; a deep content-wise analysis, as well as subtle vandalism (which goes beyond the use of dirty words), did not take place.

21. See O'Reilly & Associates (1999), Himanen (2001) and Grassmuck (2002).

22. For a detailed representation of the process of development of OpenSource projects see Kollmann (2002) and Grassmuck (2002:233–317), who also gives an overview over the different license models.

23. See: "Von freier Software zu freiem Wissen" (Grassmuck, 2002:394–404).

24. http://www.cec.nrw.de/Lizenzen/uvm-lizenz1.htm [17.07.2004]

25. http://creativecommons.org/license/ [17.07.2004]

26. Gnu Free Documentation License: http://www.gnu.org/copyleft/fdl.html [17.07.2004]

27. About the legitimation of such lisence models see Lessig (2002).
28. About the term Web2.0 see O'Reilly (2005).

References

Anger, E. (2002). Brockhaus Multimedial 2002 Premium auf CD-ROM und DVD-ROM [Electronic version]. In H. Rösch (Ed.), *Enzyklopädie im Wandel: Schmuckstück der Bücherwand, rotierende Scheibe oder Netzangebot* [Encyclopedia evolution: Jewel of the bookshelf, rotating plane or net offer]. Köln: FH Köln, Fachbereich Informationswissenschaft. (Kölner Arbeitspapiere zur Bibliotheks- und Informationswissenschaft;32). Retrieved February 9, 2005, from http://www.fbi.fh-koeln.de/institut/papers/kabi/band.php?key=42.

Barthes, R. (2006). *Das Rauschen der Sprache* [Speech noise]. Frankfurt a.M.: Suhrkamp.

Ciffolilli, A. (2004). Phantom authority, self-selective recruitment and retention of members in virtual communities: The case of Wikipedia. In *First Monday, 12, 8*. Retrieved July 21, 2003, from http://firstmonday.org/issues/issue8_12/ciffolilli/index.html.

Derrida, J. (2004). Signatur Ereignis Kontext [Signature, event, context]. In *Die différance* [The difference] (pp. 68–109). Reclam.

Dieberger, A., & Guzdial, M. (2003). CoWeb—Experiences with collaborative web spaces. In C. Lueg & D. Fisher (Eds.), *From usenet to cowebs: Interacting with social information spaces* (pp. 155–166). London: Springer.

Dworschak, M. (2004). Rapunzel bis Regenzeit. [From 'Rapunzel' to 'Rainy season'] *Der Spiegel, 10*, 174–175.

Forte, A., & Bruckman, A. (2006). From Wikipedia to the classroom: Exploring online publication and learning. *Proceedings of the International Conference of the Learning Sciences* (Vol. 1, pp. 182–188). Bloomington, IN.

Grassmuck, V. (2002). *Freie Software: Zwischen Privat- und Gemeineigentum* [Free software: Between private and public ownership]. Bonn: Bundeszentrale für politische Bildung.

Grube, G. (2005). Autooperative Schrift—und eine Kritik der Hyptertexttheorie [Self-writing, and a critique of hypertext theory]. In G. Grube, W. Kogge, & S. Krämer (Eds.), *Schrift. Kulturtechnik zwischen Auge, Hand und Maschine* [Writing: Cultural technology between eye, hand, and machine] (pp. 81–114). München: Wilhelm Fink.

Haß-Zumkehr, U. (2001). *Deutsche Wörterbücher—Brennpunkt von Sprach- und Kulturgeschichte* [German dictionary: Focus of linguistic and cultural history]. Berlin/New York: de Gruyter.

Hennicken, D. & Zahiri, C. (2003). *Arbeiten im Netz: Erste Erfahrungen mit der Kooperations- und Austauschplattform Wiki in der Architekten und Planer-Ausbildung / Forschungsprojekt Notebook Universität, learning to learn—mobiles Lernen und Forschen. Universität Kassel, Fachbereich 06, Architektur Stadtplanung Landplanung* [Working in the net: First experiences with the Wiki cooperation and exchange platform in the architect and planner education]. *Kassel. Arbeitspapier.* Retrieved July 26, 2004, from http://www.uni-kassel.de/notebook/publikationen/ap_arbeiten_im_netz.pdf.

Himanen, P. (2001). *The hacker ethic: A radical approach to the philosophy of business.* New York: Random House.

Hof, R. D. (2004). Something Wiki this way comes. *BusinessWeek Online*. Retrieved November 11, 2006, from http://www.businessweek.com/magazine/content/04_23/b3886138.htm.

Huhmann, J. (2004). Noch schneller: Welches Wiki für welchen Zweck? [Even faster: Which wiki for which purpose?] *iX, 4*, 74–78.

Kleinz, T. (2003). Schreibrecht für alle! Mit Wikis Sites gemeinschaftlich betreuen [Writing rights for all! Collaborative coaching with wiki sites]. *c't, 2*, 176f.

Kohl, C., & Liebert, W.-A. (2004). Selbstorganisation der Wissenschaftsvermittlung: Quellentransparenz, Kontroversität und Qualitätssicherung in der Internetenzyklopädie Wikipedia [Self-organisation of the scientific sharing: Source transparency, controversy, and quality assurance in the Wikipedia internet encyclopedia]. *Fachsprache, 26*(3–4), 133–147.

Kuhlen, R. (1991). *Hypertext: Ein nicht-lineares Medium zwischen Buch und Wissensbank* [Hypertext: A nonlinear medium between book and knowledge base]. Berlin: Springer.

Krämer, S. (2005). "Operationsraum Schrift": Über einen Perspektivenwechsel in der Betrachtung der Schrift [Operation Write: On a change in perspective in how writing is regarded]. In G. Grube, W. Kogge, & S. Krämer (Eds.), *Schrift. Kulturtechnik zwischen Auge, Hand und Maschine* [Writing: Cultural technology between eye, hand, and machine] (pp. 23–57). München: Wilhelm Fink.

Landow, G. P. (2003). The paradigm is more important than the purchase. Educational innovation and hyptertext theory. In G. Liestøl, A. Morrison, & T. Rasmussen (Eds.), *Digital media revisited. Theoretical and conceptual innovation in digital domains*. Cambridge: MIT Press.

Landow, G. P. (2006). *Hypertext 3.0. Critical theory and new media in an era of globalization*. Baltimore: John Hopkins University Press.

Leuf, B., & Cunningham, W. (2001). *The Wiki way: Collaboration and sharing on the Internet*. Boston: Addison-Wesley.

Majchrzak, A., Wagner, C., & Yates, D. (2006). Corporate Wiki users: Results of a survey. *Proceedings of the 2006 International Symposium on Wikis* (pp. 99–104). New York: ACM Press.

Mittelstraß, J. (1967). Bildung und wissenschaftliche Enzyklopädien in historischer und wissenssoziologischer Betrachtung [Education and scientific encyclopedias from the standpoint of history and sociology]. *Die wissenschaftliche Redaktion, 4*, 81–104.

Möller, E. (2002). Das Web, wie es hätte sein sollen [The Web as it should have been]. *Telepolis 30.01.2002*. Retrieved October 10, 2003, from http://www.heise.de/tp/deutsch/inhalt/te/11585/2.html.

Möller, E. (2003). Schreibwerkstätten: Fünf Wiki-Engines im Vergleich [Writing workshops: Comparison of five wiki engines]. *c't. 25*, 202–205.

O'Reilly, T. (2005). *What is Web 2.0?: Design patterns and business models for the next generation of software*. Retrieved December 12, 2006, from http://www.oreillynet.com/pub/a/oreilly/tim/news/2005/09/30/what-is-web-20.html.

O'Reilly & Associates. (1999). *Open Source: kurz & gut* [Open source: Short and sweet]. Köln: Author.

Puls, C., Bongulielmi, L., & Henseler, P. (2002). *Leitfaden für den Aufbau einer unternehmensinternen Wissensbasis mit Hilfe von Wiki* [Guide to the construction of a project-internal knowledge base with the help of a wiki]. Zürich: ETH, Zentrum für Produktentwicklung. Retrieved October 29, 2003, from http://e-collection.ethbib.ethz.ch/show?type=bericht\&nr=217.

Stickfort, B. (2002). Das Internet als enzyklopädische Utopie [The internet as encyclopedic utopia]. In I. Tomkowiak (Ed.), *Populäre Enzyklopädien: Von der Auswahl, Ordnung und Vermittlung des Wissens* [Popular encyclopedias: On the choice, arrangement, and dissemination of knowledge] (pp. 271–295). Zürich: Chronos.

Storrer, A. (2000). Was ist 'hyper' am Hypertext? [What is hyper about hypertext?] In W. Kallmeyer (Ed.), *Sprache und neue Medien* [Language and new media] (pp. 223–249). Berlin: de Gruyter.

Viegas et al. (2004). Studying cooperation and conflict between authors with history flow visualizations [Electronic Version]. *Proceedings of the 2004 conference on Human factors in computing systems* (pp. 575–582). New York: ACM Press. Retrieved July 28, 2004, from http://web.media.mit.edu/~fviegas/papers/history_flow.pdf.

Volpers, H. (2002*).* Idee und Begriff der Enzyklopädie im Wandel der Zeit [Ideas and concepts of the encyclopedia in the course of time]. In H. Rösch (Ed.), *Enzyklopädie im Wandel: Schmuckstück der Bücherwand, rotierende Scheibe oder Netzangebot* [Encyclopedia evolution: Jewel of the bookshelf, rotating plane or net offer]. Köln: FH Köln, Fachbereich Informationswissenschaft. (Kölner Arbeitspapiere zur Bibliotheks- und Informationswissenschaft;32). Retrieved February 9, 2005, from http://www.fbi.fh-koeln.de/institut/papers/kabi/band.php?key=42.

Wikipedia: Geprüfte Versionen. (n.d.) Retrieved February 1, 2007, from http://de.wikipedia.org/w/index.php?title=Wikipedia:Gesichtete_Versionen&oldid=2724.

Wikipedia: Gesichtete Versionen. (n.d.) Retrieved January 7, 2007, from http://de.wikipedia.org/w/index.php?title=Wikipedia:Gepr%C3%BCfte_Versionen&oldid=26114070.

II

Changes in Traditional Mass Media Language and Culture

Chapter 8

Radio Broadcasting

From Secret to Public to Private

JÜRG HÄUSERMANN

University of Tübingen

Abstract

In the 1920s, radiotelephony, a means of person-to-person communication, began to be used for radio broadcasting, a medium of public communication. Radio developed linguistic forms of meeting journalistic and artistic aims in messages that were only acoustic. Presently, the Internet allows virtually anybody to use the medium—even for spreading private messages to small communities.

After more than 80 decades, radio is an effective tool in the production of informative or artistic messages and the medium is broadly open to participation. But it is not clear yet how the "democratic" aspect of Internet radio and the tradition of a fast, reliable, and journalistic medium will mingle.

In the beginning of its history, and until the end of the 20th century, *radio*[1] could be defined as being synonymous with audio broadcasting: the transmission of spoken word and music in order to convey journalistic information and entertainment to the public.[2] In the beginning of the 21st century, radio became the term of generally any kind of transmitted acoustic content, which is being distributed in many different ways: terrestrial broadcasting, cable and satellite radio, Internet streaming, and podcasting. One might even include CDs that are distributed as extras in magazines or newspapers.[3]

At one time, public communication was the framework in which radio communication took place. Broadcasting was produced similarly to the publishing of daily newspapers. Large, publicly controlled organizations were needed, in which the single editor or speaker was only a minor figure. Now, this aspect is vanishing as, through the Internet, anybody can publish audio content—seemingly on their own. Additionally, the enormous amount of organization that is needed for one to spread the message is no longer an obstruction. So, radio communication, which stems from the use of devices in point-to-point communication (meant to be "secret" to outsiders), and which used to be a public medium bringing information and entertainment to everyone, is now open even for private messages concealed on obscure Web sites.

The language of radio is "extremely varied" as Gerhard Leitner (1983) pointed out in his study. Altthough it underwent stylistic changes according to the changing social background of its producers, the form of radio messages can be said to be determined by the following factors:

- Communicating consists of speakers addressing listeners (and not viewers or readers): It is meant to be understood solely by listening and by listening only once (even though program recordings and Internet radio do allow playback, and Internet users may do this as needed).
- The acoustic message has little or no visual support: Although modern radio displays and all Internet radio can provide additional text or even pictures, the success of radio communication depends on its acoustic content.
- The language of radio, although addressed to a dispersed crowd, has many features of person-to-person communication: The radio audience is divided into individuals sitting near the loudspeaker or using headphones. This, plus a refined studio technique, allows a person-to-person style of communication (e.g., in volume, speed, and vocabulary).

When radio was just a journalistic medium, the range of the linguistic styles used by its communicators was limited. Now that the Internet has made it possible for anyone to broadcast, broadcasters do not necessarily make use of traditional journalistic form and content. If we agree to call any audio content that is generally accessible "radio," a specific language of radio is disappearing. This, however, is not true for the official stations nor for the restrictions given by the monologic form of addressing a listener via microphone and loudspeaker.

In this chapter, I outline overall features of the development of radio broadcasting in certain time periods and then point out characteristics of the communicative context before discussing important linguistic features of radio in the past. I focus here on European examples, also taking into consideration certain traits of radio in the United States. Conscious of the fact that the systems of public communication differ from country to country, I illustrate common developments of radio as a tool of linguistic communication.

A Medium Discovered for Public Use: 1920s

Wireless transmission of speech (radiotelephony, as opposed to radiotelegraphy) was invented around 1900 as a means of person-to-person communication. It was initially used for military and commercial purposes. The fact that anyone could listen to its messages was considered to be a handicap. It was only after World War I that industry, looking for customers other than the military, started to use the technology for transmitting messages of information and entertainment to a dispersed audience.

The Precursors: Calling One

For thousands of years, long-distance communication had consisted mainly of sending hard copies of messages from one place to another (e.g., by personal messengers or mail). In order to reach the addressee faster, physical types of signaling were used, mainly for short or intermediate distances: fire, smoke, flags, drums, canons, alphorns, and the like.

The discovery of electromagnetic waves provided a means of communicating that was much faster and covered longer distances. The techniques that were developed on this basis needed an apparatus that converted the message into radiant energy before sending it to be re-converted into sound signals after receiving it.

This development did not go directly from the discovery of electromagnetic waves to radio broadcasting. Precursors of radio communication were wire telegraphy, radiotelegraphy, landline telephony, and radiotelephony. From our modern point of view, there have been detours and culs-de-sac on the way to this successful mass medium. But around 1910, when the technical requirements were already existent, still no one could predict that they eventually would be used to compete with journalistically and economically successful media such as the newspaper.[4]

Telegraphy was never designed to become a general journalistic medium. Although it was soon to be a powerful tool for news agencies (Faulstich, 2006), it only served professionals (e.g., as an intermediate bridge between correspondents and editors), who profited from the secrecy of the wired service. Every text had to be encoded in Morse code before being sent and decoded into normal writing in order to be understood by nonspecialists. The messages contained only what could be expressed in letters and numerals (no music, pictures, etc.). They conveyed no paraverbal information, such as pitch, rhythm, or stress (see also Rowe, chap. 4, this volume).

It would be strange to imagine a medium with these particular features as capable of spreading informative, literary, and entertaining messages to a general audience. Still, the successor of radiotelegraphy—radiotelephony—would be able to do just that. Using normal speech, and being capable of transmitting virtually any sound, it was as apt for specialized and professional as well as general and public purposes.

During World War I, wireless telephony increasingly was used, but never predominantly. Still, until the end of the 1910s, industry had made a great number of inventions that made easy transmitting and receiving of sound possible, but there was no more war whose combating parties would buy them. A way of using those inventions for civil purposes, and thus getting private consumers to buy receiving gear, was to produce programs for everybody and receivers that opened access to them.

Until then, radio had been a medium for only two participants. The fact that others would listen to the exchange of information was considered problematic (and in the war, civilians were forbidden to make use of their radio apparatus[5])—although the fact that others would listen was used (e.g., in navigating). When the Titanic sent its S.O.S. call (in Morse), it benefited from the fact that every ship around would hear and go to help.[6]

But after World War I, industry pushed the idea of public broadcasting in many countries. Other players (military, state, post) opposed it. As is seen here, the controversy was resolved in different ways—from liberal systems of private radios (such as in the United States and France) to state-owned and controlled radios (Germany).

Broadcasting: Calling All

The beginning of broadcasting was somewhat similar in all countries: Individuals interested in the new technology started to use it not only for calling other persons, but also for playing music and reading literature, sometimes reading news from the papers, and the like. Some individuals were connected to industry from the start, as was Dr. Frank Conrad in Pittsburgh, a chief engineer at Westinghouse, who played music from records on the air in order to test certain radio devices. The public feedback he received during 1919 encouraged him to schedule regular music programs. Their success motivated Westinghouse to found its own radio station, KDKA which, in collaboration with the *Pittsburgh Post*, started its transmissions on the night of the presidential election, November 2, 1920. Everything still had an experimental character and the news and music program reached an estimated 1,000 individuals. Four years later, election night was covered by more than 400 stations in the United States, and the listeners are said to have amounted to 20 million.[7]

In other countries, as in the United States, the strongest force propelling the new medium was the electrical industry, interested in selling radio receivers to make up for the decrease in military sales after World War I. In Britain, a number of smaller firms of the electrical industry applied for broadcasting licences. They were requested by the post office to form one company. So the British Broadcasting Company Ltd. (BBC) was founded and started their regular broadcasts in 1922.

In other countries such as the Netherlands and Switzerland, radio clubs that were founded to gather listeners and producers alike applied for broadcasting licences (Schade, 2000; Wubbe, 1963). When amateur radio clubs were politically motivated, the political establishment considered them a threat that should be fought by establishing a state monopoly, as was the case in Germany from 1923 on (Dahl, 1983). In the following years every arising political difficulty was used by the government as a pretext to tighten the laws on broadcasting. By 1932, radio was in fact state-owned and state-controlled, with a structure all ready for the Nazis to take over in 1933 (Lerg, 1980).

The First Electronic Mass Medium: Communicative Framework

By the second half of the 1920s radio was established as a powerful and very fast medium of information, education, and entertainment in many developed countries. Radio broadcasting acquired all the information capacities of the newspaper, surpassing it in some important features (like speed, availability, and the natural rendering of voices and music). The competition with the newspaper is illustrated by the fact that up to 1938 the BBC would not present its first news bulletins before 6 p.m. in order not to endanger the sale of the newspapers (Paulu, 1981).

One difference from printed media was the equality of access. Whereas newspapers had to be bought separately, everybody could listen to all radio programs as soon as a receiver was available. On the other hand, radio lacked the mobility and individuality that made reading desirable on the way to work, in trains, restaurants, and so forth. People had to retire to one room in their home or in a community building, where they would often listen as part of a group. Soon, the mobility of the sender (and from the 1950s on, when the transistor had been introduced, of the listener) was adding to the image of a fast and up-to-date medium, which was able to give the latest news as soon as it was available to the editors.

Neither did radio observe political barriers nor geographical or social borders. It could be used as an instrument of political propaganda. Additionally, the whole society was able to listen to its programs and take advantages from their educational content. Being a spoken medium only, it could address illiterate listeners, and it has been a useful instrument in developing areas even 80 years later. Nowadays, the receiving equipment is relatively inexpensive, which makes it easier to use than television and print.

The First Electronic Mass Medium: Linguistic Features

Since its early beginnings, the producers of radio were closely connected with and partly dependent on politics and its makers. As a consequence, early radio texts reflected these dependencies. They tended to be highly formal and stylized.[8] Additionally, the linguistic features of radio in its first decade were also influenced by the competition with the established journalistic media and by the rhetoric of public speaking. Contributors to the programmes of the early years were often prominent authors, who read their elaborate manuscripts.

Although entertainment was an important function of broadcasting from the beginning, in its informational segments it sounded formal and stylized. People who produced radio often were closely connected to politics or the military. The few recordings left from the early years give the impression that the linguistic and prosodic style was influenced by this connection. Early recordings from different countries reveal highly standardized formulas of announcement, literary style in interviews, and a very distinct paraverbal form (loud, clear articulation, and the stressing of several words in each sentence).

This style can also be noted in connection with the use of language that was current in radiotelephony: Wireless communication in its military or other official uses meant obeying strict rules of pronunciation, syntax, vocabulary, turn-taking signals, and so on. Many of these ways of speaking (e.g., the call-signs introduced by *Hallo!* or *Achtung!,* long pauses between different parts of text, and over-accurate articulation) were adopted by the first radio announcers. According to Wolfgang Hagen (2005), this habit of "calling" determines all speech acts of early radio broadcasting.

Calling only slowly turned into "speaking" (Hagen, 2005). This can be illustrated by early call-signs. The famous opening announcement by Arthur Burrows, program director of radio 2LO, which started the BBC, on November 14, 1922, was: *2LO, Marconi House, London calling . . .* Soon thereafter, another famous announcement, *Govorit Moskva* ('Moscow speaking'), was created—this one already revealing the fact that broadcast speech was different from speech used when trying to reach soldiers or ships by air.[9]

New situational uses and genres were invented along with new technical possibilities. For example, radio introduced *live reporting* at a very early stage. Reporters took their microphones to the streets as soon as it was technically possible,[10] in order to cover parades, funerals, rugby and football matches, and so on. For a long time, radio was the medium that gave people the impression of "being there." Live reporters won their fame by their ability to find the right words simultaneously to what was happening. Talking fast and still formulating precisely, reporters could be heard all over the world, finding lexical variation for sequences of events that were quite alike and adding background information (about the players, the events before and after, the weather, etc.) at times when little was happening. The few recordings that are left are proof of a style that differed from that of manuscript programs—borrowing more from literary than from colloquial speech, and adapting to the nature and social context of the event.[11]

Presenting was another important feature of the new medium. Radio needed to convey spoken meta-communicative messages before and after program segments in order to compensate for the lack of visual organizers (which in television are given by inserts and similar simultaneous information). This has changed only little in recent years, when car audio systems and hi-fi receivers have been able to provide the name of the station, program, and song played on the display of the radio. As far as what European recordings from the 1920s and 1930s tell, presenting sounded official in tone and word choice. In most cases the speakers were anonymous (on the air—not to the fans and readers of the radio magazines.[12] But nevertheless, this kind of talking to people beside and about the scheduled program was a first step toward the personal "presenting" style that is used today. Also, the time-dependent medium radio was supported by the print media. Program journals were founded.

They not only offered listings of dates, times, and titles of programs, but also articles about broadcasting politics or radio genres, such as radio drama and the operetta (Bauer, 1993; Currie, 2001).[13]

The *news* broadcast by radio was more or less identical in content and style to the news read in the papers.[14] The French term *journal parlé* describes well that they derived from printed news. In many cases, the news was written by the same agencies that served the newspapers.[15]

But it was soon common knowledge among radio professionals that the language of news had to be different from the one of print media. Rules especially designed for radio speech (e.g., Räuscher, 1929, 1999) prove that shorter sentences, easier vocabulary and other stylistic features were sought when writing for the spoken medium. Still, radio texts were (and are) in the majority not spoken texts but written texts, adapted to speech (secondary orality). The news syntax and lexicon only slowly lost their written touch,[16] although the need for the medium-specific use of language had been common knowledge for a long time.

Still to this day, the absence of the image is considered to be the most distinctive feature of radio aesthetics, but it has long ceased to be judged a handicap:

> We cannot see its messages, they consist only of noise and silence, and it is from the sole fact of its blindness that all radio's other distinctive qualities—the nature of its language, its jokes, the way in which its audiences use it—ultimately derive. (Crisell, 1986:3)

But as early as 1930, the "blind" medium was already considered a challenge for artistic and journalisic expression:

> What is suitable for radio . . . is the purely acoustic. By "purely acoustic," however, it must not be understood as the optically "inexistent," but the acoustically explicable. Rieths work in Cologne has shown me . . . in what

> a good, convincing and wonderfully suggestive way space can be created in radio drama. . . . [The space radio creates] is dynamic, even more than actual space. That is why sound fills it with noise in a different way than actual space—different, because it is not filled with sounds but created, constructed by them. (Leonhard, 1928:161)

Radio does not imitate the empirical space of existing places. It creates its own space. And so, on all levels of radiophonic form, the new medium soon created, or at least adapted, its own signs and genres.

A Powerful Medium: 1930s to 1950s

Number 1 Entertainment and Information Medium

When the 1920s ended, both structures and programming of broadcasting were well developed in most countries. In the United States, two commercial networks were dominating. Ninety-eight percent of nightly radio broadcasting was done by stations that belonged either to the National Broadcasting Company (NBC) or to the Columbia Broadcasting System (CBS). In Britain, the BBC had turned from a private company to a respected corporation granted by Royal Charter (which made it independent, e.g., from political changes in the government), whose monopoly was not to be challenged until after World War II.

In Germany, the Reichsrundfunkgesellschaft (RRG), a government-controlled roof organization, controlled all 10 regional stations. In the Netherlands, two stations (plus one for overseas services) broadcast the programs of private associations; airtime was portioned out according to the number of their members. In France, 24 stations were active, 14 of which were private enterprises. The biggest of them, Radio France, had been bought by the State in 1933.

Radio was the only electronic mass medium—the only source of information and entertainment that would come into the homes gratis (in the United States) or for a low licence fee (10 shillings per annum in Britain, 24 Marks in Germany, 80 Lire in Italy), or by a moderate membership fee (e.g., in the Netherlands). This helps explain the fast increasing number of listeners in all countries between 1920 and 1940.

Broadcasting was considered an effective *propaganda* medium. The German National Socialist Government founded a Ministry of Propaganda and made the Reichsrundfunkgesellschaft one of its departments. They made extensive use of the Deutscher Kurzwellensender, which had been established in Königs Wusterhausen in 1929 (Lerg, 1980).[17] In the meantime in France, a "poste colonial" had been founded in 1930. The BBBCs Empire Service, directed to British subjects in the colonies, later developed into the BBC World Service, was founded in 1932 (Briggs, 1965). The United States was "the only world power without a government-sponsored international radio service" at the outbreak of World War II.[18]

Even in countries that started their radio system with only little regulation, the economic development and the political threats in the 1930s, or the war, lead toward a system of publicly controlled *monopolies*. Most European countries had one monopolistic broadcasting company. But there are also examples, such as France, where on the eve of World War II 12 private and 20 state-owned stations were still active. All attempts to nationalize radio had failed. Only the armistice of 1940 brought a network of seven regional stations controlled by the Vichy government in the south and German-controlled Radio Paris in the occupied area (Tudesq, 1989).

During the war, radio relied on its role as an up-to-date and universally informed medium—and as entertainment, often a comforting source of music and distraction.

A faithful BBC listener recalls:

> The war started in 1939 when I was nine. You could even understand that there was a thing called the black-out in those days and you confined yourself to certain rooms because they were easier to black out. And so if we did go and listen to the radio in the front room I mean quite often it was listen to it was just the glow of the radio. We just were sitting round and looking at this glow and listen to it.[19]

International services also multiplied, and people in the enemy countries were lured with news and music especially designed to lead them to mistrust their own leaders and their broadcasting. In the long run, people trusted those radio stations from which they heard negative news too. The BBC and Swiss Radio Beromünster were popular in Nazi Germany even when German citizens listening to foreign stations were threatened with the death penalty. The French in occupied France listened to the BBC, in order to hear General de Gaulle speak to them.

Soon after World War II, international radios started to participate in the Cold War. The United States, for example, directed several different stations toward eastern Europe. The Voice of America (1954), RIAS (Radio in the American Sector) Berlin (1946), Radio Free Europe (1950), Radio Liberty (1953) profited by transmitters on European soil such as those in Munich and Berlin. The United States even much later created a flying radio station in order to influence people in enemy areas (as was done in Afghanistan in 2001[20]).

Radio stations also became important targets in local wars and revolutions. Latin American military takeovers started with the seizure of radio stations. The Portuguese revolution of 1974 was triggered from a radio station.

Radio in Its Golden Age: Communicative Framework

Although the economic and political backgrounds of broadcasting differed from country to country, radio content differed as well. But some factors were true for almost all of them, for example, the importance of entertainment which was even realized by the Nazis, who changed their attitude of stern political talk already after the first months (Diller, 1980), as well as the intention to keep up with current events in politics, sports, and so on. For example, in their overseas broadcasts they even played jazz, which was banned in Germany.[21]

News broadcasts in the evening—in most cases, in order not to compete with the press—began to be seen as more and more important as the danger of war became visible. Information features and live reports were becoming part of the daily routine. Election campaigns were also held on the radio, as in France in 1936.

As for *entertainment,* radio developed its strong sides very early and established it firmly during the 1930s. Music, daytime serials, and talk shows soon became steady formats that could be anticipated by the audience as well as by the advertisers. And wherever the representatives of literature and the arts got hold of program space, they tried out new forms of drama, feature, and musicals.

In some countries, like the United States, the commercial direction of broadcasting was more favorable for the development of popular entertainment; in others, like Great Britain, highbrow culture programs were invented and refined. But one main factor united broadcasters everywhere: They now mastered a technique. Anything they could was put in sound, staged, or reported. When in 1938 Orson Welles and Howard Koch brought their famous version of H.G. Wells' *War of the Worlds* on the air, they displayed a ubiquitous medium, capable of transmitting live music from a dance hall, interviews from an observatory, on-the-spot reporting from the middle of nowhere—and switching between these places easily and fast.

Apart from the importance of presenting, a regular daily structure was recognized as an important guideline. The listeners knew their station's daily schedule and returned to certain programs because they were always aired at the same time. Much more than its printed precursors, radio is the medium of reassuring repetition: of drama serials and their protagonists, of news and its readers, of sport matches and their reporter announcer, and of songs and their presenters. And the repetition of words and phrases in everyday life goes along with it. Dry sentences became part of everyday phraseology simply because they were repeated regularly in this glamorous medium. Woody Allen's boy protagonist in *Radio Days* gets hit by his rabbi and his parents because he addresses the rabbi with stereotyped words he knows from *The Lone Ranger*, a Western serial.[22]

The *daytime serial* was born in the second half of the 1920s on commercial American stations, whose concern was to produce an optimal environment for advertising messages. (NBC had created a department especially for this purpose as early as 1926.) They produced serials like *Amos' n' Andy*, a program of dialogue featuring two men who were easily recognizable as African Americans because their talking and jokes were similar to what the listeners knew from the tradition of minstrel shows. Sponsored by Pepsodent, the show was aired twice a day, six times a week (Dunning, 1976; Hilmes, 1997).

Between Make-believe and Authenticity: Linguistic Features

Although it is not easy to give a general overview of the changing radio styles over the years (although Scherz-Schade, 2004, did so for German radio, focusing on the language of the news), one can take a look at some innovative forms, which come together by the fact that in the 1930s radio was an established medium with its own forms of expression. The radio speaker and writer of the 1930s mastered the medium and was proud of it. No longer did producers look for ways to bring genres from the stage or the newspapers to radio. They now knew that they were able to adapt any content to the medium. They had discovered that entirely different skills were needed in the task of reporting for the radio than for the newspaper (Kasack, 1929).

The *tone* was now much less "official" than in the first years, when radio communication was still closely orientated to public speaking—not only in the verbal style, but also in loudness and intonation. The BBC news bulletins of the 1920s were read word by word. In the sentence *Here is the lást néws búlletin for todáy* from May 1926, four words are stressed; in the sentence *And here is a spécial bulletin/read by John Snágge* from June 1944, only two words are stressed, and these two words are combined to two coherent intonation chunks with much more melodic variety.

The news readers' voices were crucial for the reassuring effect of radio news. Since the first years of the war, BBC news readers began introducing themselves by name. Likewise, radio documentaries strongly stressed authenticity. Radio had the advantage, contrary to the press, that politicians and other protagonists of the news could be heard in their own voices, or in a dramatization with a professional speaker rendering their speech.

As there was the radio journalist who freely selected parts of politicians' speeches for his programs, there was the politician adjusting his way of speaking to the new medium. Franklin D. Roosevelt invented the fireside chat, a very personal way of commenting on current issues in talks directly addressed to the American people (Breitinger, 1992). The goal here was also to create authenticity.

A Medium at Your Disposal: 1950s to 1990s

Pirates and Deregulation in Europe

During World War II, radio established its position as the main information and entertainment medium. It was able to keep this position until the beginning of the 1950s when *television* first challenged radio's position in entertainment (and with some delay in up-to-date news as well). This turn can be illustrated by the CBS news show *Hear It Now* being transferred to television in 1951. It was then re-christened to *See It Now* (Thovenon, 1997).

At the same time, radio broadened its range. The FM frequencies allowed more diversity in the regulated European radiospace, and educational content in the United States, where commercial radio had dominated. There, 20 FM channels (FM was not popular at the beginning, because the listeners would have to buy new receivers) had been reserved for noncommercial, educational stations by 1945. And in 1967, the Public Broadcasting Act of 1967 made public radio networks possible with independent, educational, and sometimes experimental content (Hemm, 1997; Widlok, 1992).

In countries with a monopoly, such as Germany, FM was used to diversify their range. Only in the 1980s (1970s for Italy, 1990s for eastern Europe) would frequencies be given to private companies, which concentrated on commercial radio.

The number of private radio stations differs from country to country, according to the way of regulation chosen. The content often sounds much alike, because they target the same group of young urban money-spenders.

In a commercial context, structured programming was refined. Instead of mere daily schedules, radios were given a *format,* a specific, daily repeated acoustic shape. A format is composed of content (music, spoken word, advertising), presenting (trailers, teasers, and other forms of promotion), and the structure of the daily program (type, length, sequence of program elements). There are news formats (all news, all sports, all business, etc.) as well as music formats (adult contemporary, middle of the road, contemporary hit radio, etc.). These very narrow formats are typical for the U.S. market, whereas in Europe full-service formats are common, with information and entertainment and a wider range of music (Keith, 1987; McFarland, 1990).

The Accompanying Medium: Communicative Framework

In Europe in the 1950s radio could, for some years, hold its stand against television. The evening was still a prime-time slot. The big radio dramas, old-fashioned but skillfully produced, assured that listeners would be at home to enjoy variety programs, as well as thrillers by Francis Durbridge.

As for the *information* side of programming, radio was still a fast medium able to spread global, regional, and local news as soon as a reporter was on the spot. But by the end of the century, this also became true for television and any of the Internet information forms. Today, only a small portion of the news has to be gathered on the spot. Most news sources are texts released by political or economic institutions and are distributed to the news desks automatically (Bell, 1991). There is no need to be present at a press conference; correspondents can stay at home and use the Internet.

Meanwhile, *dialogue* became more important than ever—dialogue with singers and other stars in the program, and dialogue with listeners over the phone. This faint hint at the listener's participation was developed when alternative, noncommercial stations were possible (as in Germany and Switzerland from the 1980s on). Now ordinary citizens would be able to make their own programs. From the point of view of the established media, the results would seem unprofessional; for the entirely new subjects brought up by the amateurs, a new, radically different rhetoric might be appropriate (Erziehung zum Stummsein, 1984).

On the other hand, participatory programs have been an important educational tool in developing countries for a long time. In Africa, several forms of rural radio have brought development workers and farmers together for decades, in order to produce programs (often distributed on tape) in which experiences are shared and advice is given (Albert & Tudesq, 1995; Bourgault, 1995; Schmidt, 1989). Thus, societies with a high analphabetic rate and oral traditions of education make use of a new medium of spoken language. Typical programs of rural radio in these areas have farmers speak to farmers, which helps avoid the gap between the program speaker and the listener and reduces the problems of comprehensibility.

The Material Medium: Linguistic Features

Whereas the presenters of the earlier decades had proudly announced their programs, conscious of their public role, in the 1960s they came closer to the listener. Many radio hosts and columnists found intimate tones of expression—as if there were no microphones, magnetic waves, and loudspeakers in between them. In fact, presenting may be the field in which most changes in the use of voice and language could be observed (Scherz-Schade, 2004). The radio journalist or presenter of the late 20th century learned to lead a dialogue—even if having no one to talk to in the studio or on the phone.

And of course he or she was a trained interviewer, producing live and recorded dialogue together with people in the news and experts of all kinds. In this way, radio texts became products of cooperation between journalists and the people they speak about. This made it difficult to draw a sharp distinction between the media and the issues and the persons they reported on. Especially problematic was the closeness of the media and public representatives. In the case of the interview, the journalist adopted the language of his or her sources (such as certain lexemes and their connotations) rather than insisting on a neutral vocabulary (Häusermann, 2005).

The issue of comprehensibility has been central to the discussion about radio language from its very beginnings (see Räuscher, 1999). Since BBC's Staff Training Department in 1942, radio training departments have been founded (Briggs, 1985). They have continuously stressed the special comprehension needs of the acoustic medium. Competition has altered the style, not only in presenting music, but also in the language of news bulletins, weather reports, and the like.

News style has changed in relation to the competition against rival stations and even against other news programs within the same institution: Listeners of BBC's Radio 1's news found the style "chatty," "speedy," "punchy," "more jazzy," "racy," "snappy" und "breathless" (Mautner, 2002).

Additionally, the art of quotation was refined. The recording tape, introduced in radio stations only after the war, allowed making short original sound quotations, leading to their shortest forms that are nowadays called sound bites.

Today, original sound quotes are technically reproduced parts of speech that have been previously recorded. The main features of this kind of quotation are that it is pre-recorded and limited in length (from a few seconds to about 2 minutes), with a distinct beginning and ending (although fade-in and fade-out is possible). For spoken text, the sound bite eliminates the clumsiness of indicating the beginning and end of a quote, which cannot be marked by apostrophes and other signs common to print.[23] And, of course, it adds to the authenticity of the text, being recognizable as the voice of the quoted person. Its function is often reduced to just this aspect, as the context of the original statement is given before or afterward (although, in many cases, not at all). But the sound bite also works only with an experienced listener. He or she must be capable of recognizing (a) the changing of voices, (b) the changing of addressing, (c) a presumed argumentative function of the statement in the orginial context, and (d) its argumentative function in the context of the radio text (Häusermann, 1998).

Digital Radio: A Free-for-all Medium?: 2000s

Radio has profited from digitalization in two ways.

It changed the way the acoustic messages were sent and spread. Digitalizing the program before sending it reduced many typical broadcasting problems, such as static or the limited number of frequencies. To the listener, digital radio has brought (in most cases) better sound quality and a greater range of stations to chose from, although little innovation.

And it made it possible to store radio messages in easily accessible form, which gave the listener the freedom to chose the listening time and reduced the problems of a handicap related to a main feature of traditional radio—the volatility of the medium, which had made it fast and attractive, but which also had obliged the listener to tune in at a precise time.

Thanks to the Internet, no one is confined as a listener anymore. The costly side of broadcasting from the amateur's point of view, the price of the transmitting method, is suddenly reduced to a few euros for both server and provider. The question of applying for a licence and being accepted by an authority seems to be gone forever.

Considering the tradition of radio amateurs, radio on the Internet could, because of decreased cost over the years, become the tool of democracy that radio of the 1920s never became. Considering the experiences with any new medium in history, we are only in a transition stage.[24]

Internet Services: Streaming and Podcast

At first, the Internet did not seem to produce anything new for radio. It just served as a platform for existing radio stations for advertising their programs and displaying some additional reading material. But the arrival of the Real Player in 1995 and similar programs has allowed making acoustic messages accessible via Internet. Although live streaming continuously allows for listening to radio programs from all over the world, and thus using it as the background medium it used to be in the decades before, downloads allow one to grab single audio files and listen to them whenever and however often one wants.

Internet radio is not regulated. "Anyone with access to a Web server and the encoding software could create an Internet radio station and broadcast to a potential international audience" (Anderson, 1998:215). Dedicated Internet stations are therefore more specialized in music and/or information and provide alternatives to the streaming of the established radio stations.

These new opportunities for distributing audio artwork were soon used by artists all over the world. Web sites presenting artwork of the most varied kind were founded. Because of their presentation "on demand," the term *radio* is not quite appropriate to describe these products, but they are clearly akin to avant-garde radio drama of old (Baratsits, 2001).

The impulse for an increasing number of artists using sound and placing their acoustic events on the Internet can be seen twofold: The Internet makes it possible to publish and to contact people involved in the same field with little effort. And effective and cheap audio producing software is available, with which anybody can produce sound at home without relying on the expensive technology of a recording studio. Meanwhile, the spoken word, in a form related to the traditional journalistic radio forms, has gained new importance, supported by a consumer product that was designed for an entirely different purpose.

Around the year 2000, audioblogs became an alternative to written weblogs. Music freaks, but also professional and amateur radio journalists posted their columns in MP3 files on the Internet. There were two ways of listening to them: either by streaming (the audio content being played while it was downloaded) or by downloading first and then playing back the file in a media player. Both ways were not very satisfactory. Streaming for big audiences meant compromise in audio quality, whereas downloading could take several minutes even with comfortable bandwiths (a "click–wait system" in the words of RSS developer Dave Winer[25]). A breakthrough was achieved when blogger Adam Curry (of MTV fame) presented his software in 2004, an enhanced version of an existing web aggregator, which could do the downloading automatically.[26]

The software made it possible to "subscribe" to an audioblog. The application checks the different URLs regularly for updates. In the second half of 2004 an increasing number of amateurs were posting their spoken columns on the Internet. Listeners would transfer the blogs to their MP3 players, which had become popular in the late 1990s. In 2005, Apple added the aggregator feature to its iTunes software, which made it especially easy to retrieve audio blogs and copy them to iPods. A few weeks later, this new kind of "broadcasting" was baptized "podcasting."

Podcasting was adopted very quickly by professional broadcasters and politicians such as Sen. John Edwards (D-N.C.), who started his in March 2005 (Walch & Lafferty, 2006).[27] The most widespread form of podcasting is the spoken blog. Here the author records and digitalizes a spoken text of any type (personal column, journalistic report, political comment, etc.). These texts can be enhanced by editing and mixing with music and other sounds. The product is converted into a file in MP3 or another format of audio compression and posted on a server.

For traditional broadcast stations, it has become normal to offer their daily spoken-word programs as podcasts, so listeners will not depend on tuning in regularly. Whenever the application is opened, the subscriber's computer will immediately look for and download updates of the subscribed podcasts.

Only few podcasts and blogs provide up-to-date information. The person who downloads has an elastic relationship with time. It is possible for a disciplined listener to hear just the messages of the day. But updating can be irregular and, while browsing the Internet, one can stumble across older material as well. Because every blog or podcast may produce feedback and reactions to feedback, there can be a series of texts connected to every message—and what is more, these intertextual complexes change from day to day.

Podcasting: Communicative Framework

In 2004 there were fewer than 1,000 podcasts in the United States (Walch & Lafferty, 2006), and probably not too many more worldwide. Two years later, iTunes listed 65,000 podcasts. The Internet is open to anybody who wants to produce and publicize their own audio content. On the users' side, subscribing to a podcast is very different from tuning into a traditional radio station.

In traditional radio, the listener can choose among several stations on the dial of his or her set. The choice is made by ear. One evaluates small parts of speech or music giving hints of the station's format.

On the other hand, the listener of a podcast opens a podcast client, which leads to a podcast directory. In most cases it will be iTunes leading to the iTunes Music Store. There, one will find a podcast section, which lists 16 search categories (from Arts to TV & Film), but also displays dozens of clickable images linked to highlights, selected by obscure editorial principles. It also offers a listing of the "Top 100," which are mostly from commercial origin.[28]

Thus, with podcasting, the listener's selection is made via directories and advertisements instead of by tuning. And the result of the selection is not a station but one single show. The preselection is made by the Music Store (or any other portal), not by program journals as in the days of terrestrial radio.

Apple's Music Store has become a powerful gatekeeper because on the users' side "most podcast traffic, roughly 75%, is driven through Apple (AAPL) iTunes' list of the Top 100 podcasts" (Holahan, 2006). These top lists are now dominated by big media corporations (e.g., movie producers, the BBC, National Geographic, etc.).[29] Independent podcasters can still be found via iTunes, but they have to be sought with skill (and other portals and directories return better results[30]). According to Walch and Lafferty (2006) "the community accepts that iTunes has been both a blessing and a curse for the independent podcaster" (12). They claim that it is not easy to get one's podcast listed in iTunes and that it is almost impossible to succeed in corresponding with Apple on podcasting matters.

On the other hand, there are hardly any barriers for amateur podcasters, and independent podcaster networks create alternatives to the big media podcasts. Or, as Citizen Shelley, host of Citizen against Lies, claims:

> Podcasts are new media—not under any regulation by the government or the FCC. The government can't touch us, we're not regulated. So we have the same freedom of speech and the same complete lack of outside obligations as two people talking with each other in a room. So podcasters are one of the last bastions of completely free speech broadcasting in this country.[31]

Podcasting has widened the variety of radio content. Although many amateur podcasters try to imitate traditional radio, others develop their own genres by mixing factual information with opinion, combined with music, noises, and sound effects. For this reason, amateur podcasting as well as blogging have been criticized of violating rules of traditional journalism (Wenk, 2006): "The public is now assaulted by news and pretend-news from many directions, thanks to the now infamous 'information highway,' " said former CBS correspondent Eric Engberg, who is concerned about a tendency that bloggers ignore journalistic researching principles:

> One of the verdicts rendered by election night 2004 is that, given their lack of expertise, standards and, yes, humility, the chances of the bloggers replacing mainstream journalism are about as good as the parasite erplacing the dog it fastens on.[32]

But many podcasts have something entirely different in mind, and perhaps criteria of traditional journalism cannot be applied to blogs and podcasts. Although they participate in public discourse, their attitude is propagandistic rather than journalistic, and even if they verbally attack traditional news media they do not strive to replace them.

An unsolved question is: Who are the audiences? Podcasting is available worldwide. It is not directed at social or geographical groups. It is not clear who and where the recipients are, until there is feedback. Then the listeners will possibly form their own group, united by the podcast, and possibly also by its content. It is likely that the impact on public opinion of podcasts remains limited unless it is picked up by and discussed in established journalistic media. Polls say that in 2006 there are relatively few podcast listeners.[33]

Podcasting: Linguistic Features

Right now, it seems that the term *radio* has come to mean any acoustic message that is delivered via an electronic medium—live or recorded, by professional or amateur, whispered, spoken, shouted, sung, instrumental, or even consisting of recorded or computer-made noise.

Linguistically, everything is possible. Never-before-heard dialects or slips of the tongue, or fully improvised speech, are just as likely to be heard as old-fashioned verse or highly elaborated essay style. At least in the first years of podcasting, the variety of production is also great, ranging from simple spoken texts to complex combinations of sound and image. Because most independent podcasts can be considered alternative and independent, it is part of their nature to be experimental with form as well. A survey of the styles of podcasts and related texts would no doubt reflect a great variety of linguistic and radiophonic expression.

But despite this, the context also provides its limits. Any speaking through this medium is a rhetorical act and will be understood by its speaker as a task different from speaking outside of the medium. For example, any text will have its beginning and its end, in order to fit into the rigid structure of a web site. So there will be genres and figures and argumentation that are not so much different from rhetorical forms in general—only varying much more than in traditional radio: In podcasts you can hear untrained voices, insults, pornographic vocabulary, unintelligible paraverbal signals, and many more forms that would not match the standards of an established radio station.

Instead of looking for linguistic change inside the podcasts, one should rather look at the textual and intertextual level. To the listener, a podcast is a single program (of a few minutes up to more than 1 hour) downloaded to an MP3 player. Among other podcasts, pop songs, and audiobooks, it is part of the listener's playlist of the day. So, the difference to traditional radio's programs and formats is that the podcast is put into a new context by the listener. It is not embedded in the stream of music and words typical for traditional radio. In some cases, the podcast is not even presented. It just begins without an introduction and ends with the last word of the announcer. Once the text is downloaded, it is isolated from any other product of the same source. In many cases, however, the podcaster presents him or herself and the podcast in a more or less skillfully produced introduction. Similar announcements are repeated within the podcast as well. Additionally, podcasters place each other's promos—short sound clips, which promote their podcasts—in their texts, in order to create connections between podcasts of the most different kind.

The most crucial difference to conventional radio production is that podcasting produces a complex system on the textual level (and not on the level of programming). The way its elements are interconnected changes from day to day, because every new issue of each podcast can refer to different other podcasts.

Radio Communication Through the Decades

When trying to draw a very general line of the development of radio forms, we come to a path from coming to terms with the medium to mastering the medium to fracturing the medium (using its elements and effects in a deliberate way, using them for new artistic expression). I outline four phases.

Phase 1: Coming to Terms With the Medium

The microphone used in the first years was an instrument to which one had to adjust one's own body posture, voice, articulation, and so on. Speakers on the radio learned to talk to their audience despite the demanding technique. To the radio editor, the crucial question seemed to be how genres of the press, literature, and theater could be rendered by radio in an optimal way. Articles reprinted in Dussel and Lersch (1999) spectulated on the effect of the absence of visual signs. The authors argued that spoken language was perceived in a different way when it was not supported by visual information (Bofinger, 1924).

Phase 2: Mastering the Medium

"It was about the third year of broadcasting—we are now in its eighth—when the conviction mightily spread that literature written for the stage cannot be brought in front of the microphone in its original form," said a witness of the early years of radio, German writer Jochen Klepper (1932).

And indeed, the period of struggling with the new medium was quite short, before it turned to a period of handling it with skill. On a social level this fast change in the attitude toward radio aesthetic is paralleled by the rise in the esteem of the radio producers, which can be traced best in the history of the BBC of 1922 to 1926 (Briggs, 1965). This period is a long period of dramatic live reports and of realistic drama.

Phase 3: Fracturing the Medium

The numerous technical innovations available by the 1950s and 1960s allowed for a great variety in sound engineering. The mixer made it easy to fade-in, fade-out, add effects, and produce new kinds of sound. The recording tape allowed cutting and duplicating, thus creating forms of expression that played with the medium. We are still in this period nowadays, where even the simplest podcasts make use of specially created, self-made trailers. The producers of radio are as assertive as ever. Earlier, the tendency of audio freaks was to produce naturalistic effects, but now the technical effects (editing, fading, filtering, delay, etc.) have their own aesthetic message. Live reporting, for example, is still live reporting, but it has become fashionable to let the medium prove its strength by switching quickly and smoothly from one scene to the other. The final sequences of Saturday afternoon's football games of the German Bundesliga are covered by a pool of announcers who can hear each other. Each of them comments for less than 1 minute, before a colleague in another stadium takes over. They often interrupt each other, and the overall sound impression is one of a long series of dramatic sound bites. On the whole, however, live radio reporting, even of sports events, has lost most of its importance. This is now the domain of television. Of course, some spectators listen to the radio commentary in the stands on headsets, people turn to radio sportcasts while driving. But these are exceptional ways of using the radio, within the framework of its important role as a mobile medium.[34]

Radio, in the sense of providing access to audio content, is available now to both professionals and amateurs alike. It is open to public and private messages, to information and entertainment, to education and acoustic art. Has it become the participative medium once dreamed of by the avant-garde of the 1920s and feared by the political establishment of the 1920s? The question cannot yet be answered. Radio is easy and cheap to access for many, but not for those who cannot afford the hardware needed even for the listener.

In terms of access, traditional radio was only potentially international. How far a station reached depended on the power of the transmitting station and on the actual atmospheric conditions. The radio of the Internet is global. Every message can be spread as far as is possible. Several factors hinder the free flow of information: (a) Search engines, which filter and sort information, are of high importance, making critics conclude, "what is not to be found by Google or Amazon does not exist" (Hartmann, 2006:214); and (b) censorship. After the era of terrestrial broadcasting, when it was even difficult to fight international broadcast content by jammers, in the case of podcasting the Internet can much more easily be controlled and censored (Bambauer et al., 2005; Hartmann, 2006).

Phase 4: Fragmentation of the Public Sphere

The development of the media at the end of the 20th century has been seen as a sign of the end of the civic public sphere for some time. Where there once was one public sphere, a group of parallel media spheres has developed. The Internet is seen as a main factor for their replacement by small communities that are united by one interest only and divided geographically and socially (Breitsameter, 2001; Zimmermann, 1995).

A critical analysis by Denis Mocigemba (2006) can be heard as a podcast. He said:

> So you could argue that theoretically, podcasting can be an interesting means of participation or that it might have some potential, but in practice, there is a huge lack of social and political content in the poddersphere and so that—podcasting as you find it to-day can not at all be called a means of democratization or a means of emancipation. Furthermore you might argue that to participate in the poddersphere, a certain media literacy is required.

Conclusion

Like any journalistic medium, radio puts three players into the public sphere: the actor of the event, the recipient, and the journalistic communicator. Journalism has always been a site of interaction between these three. For a long time in radio history, the communicators and political actors were quite near to each other, often even part of each other, whereas other actors and most receiver groups were kept at a longer distance.

The Internet has enabled forms of podcasting and blogging where the roles of communicators and receivers have become redefined and sometimes reversed. Receivers can become communicators without coming closer to the actors of the political or economic sectors. This can be seen as an opportunity. Blogs have not only conveyed frivolous intimate information about nobodies hunting for fame, they have also brought up issues that were neglected. In some cases they have even brought up stories the established news media otherwise might have ignored. As the credibility of the established media is questioned, the new, more or less independent web publications are gaining ground (Lasica, 2006). In this process, the established media have still been a vital factor, because they picked up and spread information brought up by blogs or podcasts. It remains to be seen what kind of interplay will develop from the many variations of acoustic communication we still call radio.

Notes

1. A word that actually means the transmission of information in waves of any kind—deriving obviously from the prefix of words like radiography, radiotelegraphy, and so on.

2. For another study on spoken traditional mass media, see Luginbuehl on television, this volume.

3. The German childrens‘ magazine *National Geographic World* includes a CD, on which a radio play connected to the cover story can be heard. The French learners‘ magazine *Champs-Elysées* includes an audio CD "packed with interesting news, interviews, and features on topics ranging from politics and current affairs to sports, travel, film, and literature." The Austrian monthly CD *Kompakt Magazine rundgehört* presents oral versions of articles from several print magazines.

4. For a detailed and intelligible introduction to the prehistory of radio see Hagen (2005), Lichty and Topping (1977), or Fahie (1901).

5. American radio amateurs were allowed to make use of their gear until their country joined the war in 1917 (Douglas, 1987). By the time of World War II, things had changed completely: Radio was now an instrument of mass communication and propaganda. People were encouraged to listen—if only to their own country's stations.

6. But at the same time, rescue efforts were delayed due to a lack of regulation and chaos on the air (Douglas, 1998).

7. This estimation (Lichty & Topping, 1975) might be exaggerated, since only 1,250,000 U.S. households owned a radio receiver in that year (Sterling, 1978).

8. It was not even clear whether the principles of the freedom of the press should be applied to broadcasting. The BBC, under its first managing director, John Reith, achieved its freedom by fighting hard and proving to be an independent source of information. German radio, on the other hand, got its news from an agency of which 51% belonged to the Ministry of the Interior and the rest to the national press association, news agencies, and two publishing companies (Dussel, 1999). Beginning in 1933, German radio became a department of the Ministry of Propaganda.

9. Already in 1930, the sentence Govorit Moskva was so popular that the *Periodical Radio slušatel* (The Radio Listener) was renamed *Govorit Moskva.*

10. On July 2, 1921, a heavyweight boxing fight was broadcast live from Hoboken, N.J. on an experimental wavelength of 1600 m. (Lichty & Topping, 1975). "The first outside broadcast [by the BCC] to become an almost universally-known annual feature was the sound of the nightingale in the Surrey woods in 1924." (Briggs, 1981:164).

11. Examples can be heard on *Süddeutscher Rundfunk 1924 bis 1994.* (1994) and *75 Years of the BBC* (1997).

12. It is most difficult to compare presenting in older and newer programs because of several reasons: The archives of the first decades have conserved the presented content

rather than the whole programs. In the early times there is hardly anything comparable to today's mixtures of music and personal talking. Music presenters presented the program, they were not considered part of it. The concept of information, education, and entertainment did no include chatting about the weather. And all the same, it is hard to generalize. As a rule, there is hardly a type of radio show that was not already there in the 1920s.

13. There has been little research on the specific content of program journals. But on the other hand, as MacDonald (1988) said, and Briggs (1981) demonstrated, they have been used extensively as a source for program research.

14. See Spina (this volume) on the question of the written vs. spoken nature of newsreaders' speech on Italian television.

15. In Britain, the BBC was forced to subscribe to the agencies owned by the big newspapers. In Switzerland, the news was delivered by the Swiss News Agency (SDA, Schweizerische Depeschenagentur) until 1966 (Scherrer, 2000). In Finland the Finnish News Agency (STT, Suomen Tietotoimisto) had a monopoly on the radio news until the mid-1960s (Salokangas, 1996).

16. Scherz-Schade (2004) sees the most drastic changes in German news as late as the 1990s.

17. The first shortwave broadcast came from Radio Moscow in 1925.

18. http://www.voanews.com/english/About/the-beginning.cfm (viewed 9/30/2006); see also: Wood (1992) and Heil (2003).

19. From the BBC series "20th Century Vox" on BBC 5 (http://www.bbc.co.uk/education/20cvox/media.shtml, viewed 27. 11. 2001).

20. *Time*, 16. 10. 2001 *Die Welt*, 11. 10. 2001 *Clandestine Radio Watch*, 11. 10. 2001.

21. Bergmeier and Lotz (1996) described "Charlie and his Orchestra," an English-singing German big band, especially recruited by the Ministry of Propaganda, who in their texts lampooned politics in Europe and the United States.

22. "You speak the truth, my faithful Indian companion" said Joe to the rabbi (at 12 minutes on the RCA DVD).

23. Quotes in spoken media are handled in different ways, depending on the language. In English, quotes are indicated by saying the word *quote* before the quoted material, and in German by using various forms of subjunctive ('Konjunktiv I' or, if its form is identical to the indicative, 'Konjunktiv II').

24. To see the Internet as a ubiquitous, cheap means of communication, is typical for the wealthy, industrialized countries, in which, like Germany, every second inhabitant has access to the Internet. The CIA world factbook (https://www.cia.gov/cia/publications/factbook/rankorder/2153rank.html, viewed 9/26/2006) cites 48,722,055 German Internet users in 2005 out of a population of 82,400,000. For Brazil it is 25,900,000 users out of a population of 188,000,000, for Nigeria 1,769,700 users out of a population of 131,000,000.

25. Dave Winer: Payloads for RSS. www.thetwowayweb.com/payloadsforrss, Jan 11, 2001.

26. The application was first called iPodder, later Indiepodder. It is now available under the name of Juice at http://juicereceiver.sourceforge.net.

27. German chancellor Angela Merkel has even her own weekly video podcast: http://www.bundeskanzlerin.de/Webs/BK/DE/Aktuelles/VideoPodcast/video-podcast.html, viewed 9/30/2006 (Wanhoff, 2006).

28. Apple gives most varying explanations as to what their top 25 (or top 100) lists are based on. For three different explanations see Rubens (2006), Walch and Lafferty (2006), and the interview with Georg Albrecht, spokesperson of Apple Germany on www.podcastclub.fresh-info.de, August 9, 2006.

29. The Top 10 Podcasts according to iTunes (UK version) of Week 48, 2006, are The Ricky Gervais Show (Guardian Unlimited), Jimmy Carr Live, Borat's Bits, Russell Brand (BBC 2), The Now Show (BBC 4), The Lost Initiative (sky), BBC Test Match Special, Best of Moyles Enhanced (BBC 1), The Secret Policeman's Ball (Amnesty International), Best of Moyles (BBC 1).

30. Walch and Lafferty (2006) prefer the Yahoo! directory to Apple's. A list of 144 directories (on 12/01/2006) is to be found at www.podcast411.com/page2.html.

31. Podcast "Citizen against Lies" of June 7, 2006 (http://shellyl.libsyn.com/rss)

32. In his article "Blogging as typing, not journalism" on www.cbsnews.com/stories/2004/11/08/opinion Engberg has examples for even Microsoft owned www.slate.com working with unreliable news sources.

33. Only 1% of the Internet users download a podcast on a typical day, according to a study by Pew Internet Project of August 2006.

34. 40 percent of radio listening time is spent outside of the homes, said Klingler and Müller (2005).

References

Albert, P., & Tudesq A.-J. (1995). *Histoire de la radio-télévision.* Paris: Puf.

Anderson, S. D. (1998). Internet radio. In D. G. Godfrey & F. A. Leigh (1998). *Historical dictionary of American radio.* Westport & London: Greenwood Press.

Baratsits, A. (2001). *Die Hybridisierung von Radio und Internet* [The hybridization of radio and the Internet]. Wien: ICCM.

Bauer, T. (1993). *Deutsche Programmpresse 1923–1941: Entstehung, Entwicklung und Kontinuität der Rundfunkezitschriften* [German program press 1923–1941; Birth, development and continuity of program journals]. Rundfunkstudien 6. München: Saur.

Bell, A. (1991). *The language of news media.* Oxford & Cambridge: Blackwell.

Bergmeier, H., & Lotz, R. (1996). Charlie and his orchestra. Ein obskures Kapitel der deutschen Jazzgeschichte [Charlie and his orchestra: An obscure chapter of German jazz history]. In W. Knauer (Ed.), *Jazz in Deutschland.* Darmstädter Beiträge zur Jazzforschung (Vol. 4, pp. 13–48). Hofheim: Wolke.

Berkenbusch, G. (1998). *Hörer beraten Hörer: Gesprächsorganisation und Verfahren der mündlichen Textproduktion; eine vergleichende konversationsanalytische Studie zu spanischen, katalanischen und französischen Radiosendungen mit Hörerbeteiligung* [Listeners give advice to each other: Dialogue organization and oral text production technique: A comparative analysis on Spanish, Catalan, and French phone-in radio programs]. Tübingen: Stauffenburg.

Bofinger, A. (1924). *Grundsätzliches zur Programmgestaltung des Rundfunks* [General thoughts on radio programming]. Süddeutscher Rundfunk, Ausgabe B, I (1924), 171–174. (Reprinted in Dussel & Lersch 1999, 31–36).

Bourgault, L. M. (1995). *Mass media in sub-Saharan Africa.* Bloomington–Indianapolis: Indiana University Press.

Breitinger, E. (1992). *Rundfunk und Hörspiel in den USA 1930–1950* [Radio and radio play in the USA, 1930–1950]. Trier: Wissenschaftlicher Verlag.

Breitsameter, S. (2001). AudioHyperspace. In A. Schwanebeck & M. Ackermann (Eds.), *Radio erobert neue Räume. Hörfunk: global, lokal, virtuell* (pp. 67–83). München: Fischer.

Briggs, A. (1961). *The birth of broadcasting. The history of broadcasting in the United Kingsdom* (Vol. 1). London: Oxford University Press.

Briggs, A. (1965). *The golden age of wireless. The history of broadcasting in the United Kingdom* (Vol. 2). London: Oxford University Press.

Crisell, A. (1986). *Understanding radio.* London: Methuen.

Currie, T. (2001). *The "Radio Times" story.* Tiverton: Kelly.

Dahl, P. (1983). *Radio. Sozialgeschichte des Rundfunks für Sender und Empfänger* [Radio: A social history of broadcasting for broadcasters and listeners]. Reinbek: Rowohlt.

Diller, A. (1980). *Rundfunkpolitik im Dritten Reich* [Broadcasting policy in the Third Reich]. (= Bausch, H., Rundfunk in Deutschland 2). München: dtv.

Douglas, S. J. (1987). *Inventing American broadcasting.* Baltimore & London: Johns Hopkins University Press.

Dussel, K. (1999). *Deutsche Rundfunkgeschichte. Eine Einführung* [A history of German broadcasting]. Konstanz: UVK.

Dussel, K., & Lersch, E. (1999). *Quellen zur Programmgeschichte des deutschen Hörfunks und Fernsehens.* Quellensammlung zur Kulturgeschichte 24. Göttingen & Zürich: Muster-Schmidt.

Erziehung zum Stummsein 1984. Jugend und Medien in der Schweiz. Beobachtungen der Schweizerischen Kommission für Jugendfragen. Bern: EDMZ.

Fahie, J. J. (1901). *A history of wireless telegraphy, 1838–1899.* Edinburgh: Blackwood.

Faulstich, W. (2006). *Mediengeschichte von 1700 bis ins 3. Jahrtausend* [Media history from 1700 to the third millennium]. Göttingen: Vandenhoeck & Ruprecht.

Hagen, W. (2005). *Das Radio. Zur Geschichte und Theorie des Hörfunks—Deutschland/USA* [Radio: History and theory of radio broadcasting—Germany/USA]. München: Fink.

Hartmann, F. (2006). *Globale Medienkultur. Technik, Geschichte, Theorien* [Global media culture: Technology, history, theories]. UTB 2723. Wien: WUV.

Häusermann, J. (1998). *Radio.* Tübingen: Niemeyer.

Häusermann, J. (2005). *Journalistisches Texten* [Journalistic writing]. Konstanz: uvk.

Heil, A. L. Jr. (2003). *Voice of America. A history.* New York: Columbia University Press.

Hemm, T. (1997). Auf der Suche nach dem Wort im amerkanischen Radio [Looking for the spoken word in American radio]. In T. Hemm et al. (Eds.), *Radio Soundscapes USA* (pp. 6–21). MUK 111/112. Siegen: Universität-Gesamhochschule Siegen.

Holahan, C. (2006, November 27). What podcasting revolution? *Business Week* (http://www.businesweek.com/technology/content/nov2006/tc20061127_441486.htm, viewed 7/24/2008).

Kasack, H. (1929). Mikroreportage [Microreport]. Die Sendung 29/1931. Reprinted in Schneider, I. (1983). *Radio-Kultur in der Weimarer Republik. Eine Dokumentation* (pp. 128–131). Deutsche TextBibliothek 2. Tübingen: Narr..

Keith, M. C. (1987). *Radio programming. Consultancy and formatics.* Boston & London: Focal Press.

Klepper, J. (1932). Was unterscheidet das Hörspiel vom Drama? [What is the difference between radio play and drama?]. Volksbühne, Berlin, Nr. 10/1931/1932. Reprinted in Schneider, I. (1983). *Radio-Kultur in der Weimarer Republik. Eine Dokumentation* [Radio culture in Germany 1924–1933: A documentation] (pp. 193–198). Deutsche TextBibliothek 2. Tübingen: Narr.

Klingler, W., & Müller, D. K. (2005). ma 2005 Radio II: Radio behält Leitfunktion [Media analysis 2005. Radio III: Radio is still the leading medium]. *Media Perspektiven* 9/2005, 465–477.

Lasica, J. D. (2006). Blogging as a form of journalism. *AUSC Annenberg Online Journalism Review.* 2002-04-29.

Leitner, G. (1983). The social background of the language of radio. In H. Davis & P. Walton (Eds.), *Language, image, media* (pp. 50–74). Oxford: Basil Blackwell.

Leonhard, R. (1928). Die Situation des Hörspiels. Funk 49/1928. Reprinted in Schneider, I. (1983). *Radio-Kultur in der Weimarer Republik. Eine Dokumentation* [Radio culture in Germany 1924–1933: A documentation] (pp. 158–163). Deutsche TextBibliothek 2. Tübingen: Narr.

Lerg, W. B. (1980). *Rundfunkpolitik in der Weimarer Republik. Rundfunk in Deutschland* [Broadcasting policy in Germany: 1918–1933]. Hg. von Bausch, H, Band 1. München: dtv.

Lichty, L. W., & Topping, M. C. (1975). *American Broadcasting. A course book on the history of radio and television.* New York: Hastings House.

Mautner, G. (2002). Kommunikative Funktionen der Hörfunknachrichten. In J-F. Leonhard (Ed.), *Medienwissenschaft. Ein Handbuch zur Entwicklung der Medien und Kommunikationsformen* (Vol. 3. pp. 2006–2019). Berlin & New York: de Gruyter.

MacDonald, B. (1988). *Broadcasting in the United Kingdom. A guide to information sources.* London & New York: Mansell.

McFarland, D. T. (1990). *Contemporary radio programming strategies.* Hillsdale, Hove, London: Erlbaum.

Mocigemba, D. (2006). *Podcasting. Realisation of an old vision?* (http://oekoradar-live.bwl.uni-hohenheim.de/loudblog/get.php?web=balance-p-2006-03-22-75183.mp4, viewed 6/1/2006)

Räuscher, J. (1929). Stilregeln [Style rules]. In Dradag-Geschäftsbericht 1928/1929, pp. 27–28. In J-F. Leonhard (Ed.), (1997). *Programmgeschichte des Hörfunks in der Weimarer Republik* (pp. 423–622). München: dtv.

Räuscher, J. (1999). Akustischer Radiostil [Acoustic radio style]. Vortrag für die Funkstunde Berlin am 24. Juni 1930. In K. Dussel & E. Lersch (Eds.), *Quellen zur Programmgeschichte des deutschen Hörfunks und Fernsehens. Quellensammlung zur Kulturgeschichte* 24 (pp. 70–74). Göttingen & Zürich: Muster-Schmidt.

Salokangas, R. (1996). The Finnish Broadcasting company and the changing Finnish society, 1949–1966. In R. Endén (Ed.), *Yleisradio 1926–1996. A History of broadcasting in Finland* (pp. 107–228). Helsinki: WSOY.

Schade, E. (2000). Das Scheitern des Lokalrundfunks, 1923–1931 [The failure of local broadcasting, 1923–1931]. In M. T. Drack (Ed.), *Radio und Fernsehen in der Schweiz. Geschichte der Schweizerischen Rundspruchgesellschaft SRG bis 1958* (pp. 25–58). Baden: hier+jetzt.

Scherrer, A. (2000). L'essor envers et contre tout, 1931–1937. In M. T. Drack (Ed.), *Radio und Fernsehen in der Schweiz. Geschichte der Schweizerischen Rundspruchgesellschaft SRG bis 1958* (pp. 283–304). Baden: hier+jetzt.

Scherz-Schade, S. (2004). *Deutsche Radionachrichten: Wandel ihres Sprachgebrauchs 1932–2001* [German radio news: Change in its linguistic usage 1932–2001]. Berlin: Technische Universität dissertation.

Schmidt, K. J. (1989). Die Zimbabwe Broadcasting Corporation (ZBC). *Rundfunk und Fernsehen 37*, 524–531.

Schumacher, R. (1997). Radio als Vermittlung von Gegensätzen: Eine Resüme [Radio as adjustment of contradictions: A summary]. In J-F. Leonhard (Ed.), *Programmgeschichte des Hörfunks in der Weimarer Republik* (pp. 1196–1207). München: dtv.

75 Years of the BBC 1997. A celebration of BBC Radio. London: BBC.

Sterling, C. H. (1978). *Stay tuned. A concise history of American broadcasting*. Belmont, CA: Wadsworth.

Süddeutscher Rundfunk 1924 bis 1994. 1994 Ein Streifzug durch 70 Jahre Radio-Programm [Süddeutscher Rundfunk 1924–1994: An exploration into 70 years of radio programs]. 2 CDs. Stuttgart: Süddeutscher Rundfunk.

Thovenon, G. (1997). *Histoire des médias*. Mémo 66. Paris: Seuil.

Tudesq, A.-J. (1989). La confrontation des radios publiques et des radios privées en province: Les expériences françaises 1925–1939 et 1982–1986. In *Presse, Radio et Histoire, Actes du 113e congrès national des sociétés savantes (Strasbourg, 1988)*, Section d'Histoire Moderne et Contemporaine, Tome 1 (pp. 301–318). Paris: CHTS.

Walch, R., & Lafferty, M. (2006). *Tricks of the podcastings masters*. Indianapolis: Que.

Wenk, H. (2006). Mit eigenem Stil und Zweck. Podcasting entwickelt sich zum Hype nach dem Vorbild der Webblogs. *Menschen machen Medien* 7/8 2006 (55), 31.

Widlok, P. (1992). *Der andere Hörfunk. Community Radios in den USA* [The other radio: Community radio in the USA]. Berlin: Vistas.

Wood, J. (1992). *History of international broadcasting*. London: Peter Peregrinus.

Wubbe, R. H. F. J. (1963). Radio-Omroep [Radio broadcasting]. In *Oosthoeks Encyclopedi*, deel 12 (pp. 386–389). Utrecht: Oosthoek.

Zimmermann, H. (1995). *Konsens der Erschöpften—Radio-Kultur in Deutschland* [Consent of the exhausted—Radio culture in Germany]. In S. Müller-Dohm & K. Neumann-Braun (Ed)., *Kulturinszenierungen* (pp. 213–238). edition suhrkamp 1937. Frankfurt/Main: Suhrkamp.

Chapter 9

Is Television Language a Spoken Variety?

A Corpus-Based Study of Italian Television

STEFANIA SPINA

Università per Stranieri Perugia

Abstract

The language of television is generally considered one of many different forms of spoken language; the emphasis is usually on the oral nature of communication through television rather than its technological and thus communicative, textual, and functional specificity as a broadcasting medium, although research on media and communication has shown that the language of broadcasting strongly differs from traditional orality. This chapter illustrates the specificity of the language of television with reference to one selected linguistic feature—left and right dislocation—which is traditionally viewed as typical of spoken language. Instances of left and right dislocation in an Italian TV corpus are compared from a linguistic point of view with instances of the same feature in both Italian spoken and written corpora. This study's assumption—that television broadcast Italian cannot simply be identified with spoken Italian but needs to be distinguished as an independent variety—is confirmed by data on a range of parameters in the use of dislocations that show striking differences between the two varieties; among these parameters, the high frequency of grammaticalizations and of dislocations of clause, the passives/dislocations ratio, and the types of dislocated pronouns show that TV broadcast language, much more than canonical speech, tends to use ready linguistic segments of colloquial register (dislocations of clause and grammaticalizations), as well as more rigid forms like passives. The data on passives/dislocations ratio synthesize the mixed and heterogeneous nature of the language of television, which shares important features with both spoken and written languages. The theoretical perspective presented here closely connects language changes in media to technological issues.

The goal of this chapter is to highlight the distinctiveness of the language of television compared with speech and writing. The main hypothesis is that the language used on television occupies a specific area that lies between spoken and written language (de Araujo, 2003; Tannen, 1980); the study of dislocations used in the three different varieties is a means to test this hypothesis. The methodology proceeds as follows: Left and right dislocations extracted from a broadcast language corpus, which are described later, are analyzed and then compared with dislocations extracted from two other written and spoken Italian corpora. The theoretical perspective presented here closely connects language changes in media to technological issues.

Starting in the early 1980s, it has been repeatedly pointed out that the traditional bipartition between speech and writing does not satisfactorily describe and classify TV broadcast communication (Sabatini, 1982). The language of television does indeed share some features with the other two varieties, but cannot be exclusively identified with either of them. The ponderous volume on *Gli italiani trasmessi* ("Italian broadcast language"; Accademia della Crusca, 1997) more precisely asserts the functional specificity of broadcast language in general:

> Broadcast communication takes place in significantly different ways compared with the other two historical forms, because it shares features of both, but it lacks both the simultaneous presence of participants in a space-time real situation (typical of face-to-face [FTF] conversation), and the stability and the visual decomposability of the linguistic message (typical of written communication) (Sabatini, 1997:25–26).

Despite this, the linguistic peculiarity of broadcast language is to date often neglected, and the language of television is frequently identified with spoken language. In the field of corpus linguistics, for instance, the language of radio or television programs is usually included in the spoken language sections (e.g., in the *British National Corpus*, in the *London Lund Corpus of Spoken English*, in the *International Corpus of English*, and in the Italian *Lessico di frequenza dell'italiano parlato*). The emphasis is thus mostly set on the phonic nature of the communication, rather than on its textual and communicative features, which instead have significant peculiarities compared to the other varieties; the different nature of the language of television is even clearer compared with radio, because television programs use a mix of different codes in addition to spoken language, which is the primary device that is used in radio communication.

Criticism of the traditional classification of linguistic means of production coexists with the effort of considering the language of television as a diversified system, divided into different genres, more than as a homogeneous entity. Just like any other linguistic variety, the language of television incorporates significant variation, and an in-depth analysis may lead to a genre classification with suitably homogeneous features and goals.

A genre classification is the starting point of the *Corpus di italiano Televisivo* ("Corpus of television broadcast Italian") that has provided the data for this study. The language of television, with its genres and subgenres, is then viewed as a textual parameter (Herring, van Reenen, & Schøsler, 2000), the independent variable that affects the linguistic feature selected in this study as a dependent variable (i.e., left and right dislocations). This choice is due to the fact that dislocations are traditionally considered typical of speech; showing that they are used in different ways and to differing extents in TV language means implicitly establishing its distinctiveness.

The aim of this chapter is thus to show how the use of dislocations varies in a significant way in written, spoken, and broadcast language. The first two sections provide a short presentation of some general properties of TV broadcast language compared with speech and writing. The corpora used for the analysis is described, and another section defines dislocations and research parameters. Later, the chapter describes the use of dislocations in the language of television. The chapter then compares broadcast language with spoken and written data; and the last section provides some final considerations.

Heterogeneity of Television Text

Although broadcast language is definitely an oral event, it belongs to that kind of electronic age orality that Ong (1982), in a well-known study, defined as *secondary orality*; apart from its phonic nature, this new orality—said Ong—shares only a few other features with primary orality: the strong sense of community—television audience (Herbert, 1997) comes to mind—and the concentration on the present and the widespread use of a formulaic language.

Taken for granted its nature of secondary orality, the language of television has at least two unique key features: on the one hand its development in a cultural context permeated by writing and print, and on the other its great disposition to heterogeneity.

From the communicative point of view, the language of television is a deliberate and conscious form of orality, deeply rooted in a culture based on the use of writing and print. This is clearly evident for the most frequent written form among television genres: the news. The work for organizing and setting up a news edition, for example, with all its editorial activities of selection, deletion, and combination of different materials, is also conceptually typical of a universe based on writing: "What distinguishes secondary orality from primary orality is its continuing dependence on the analytical and technical and narrative skills that in turn depend on print. Secondary orality is a displaced orality" (Silverstone, 1991:148).

From a linguistic and textual point of view, TV broadcast language is a hybrid, relying on text typologies with very diversified features that may be defined as *mixed*.[1] Going beyond the macroscopic aspect of the mixing of different codes (the verbal code mixed with images, sounds, music, and written text), this mixture can be observed at different dimensions, namely, *means of production*, *genres*, *structure*, *registers,* and *media*.

Means of Production

Ong has already pointed out how secondary orality is based on a mixture of features, some of them typical of spoken communication, and others connected to the world of writing (Ong, 1982). News bulletins provide a clear example of that, being a spoken variety read through the phonic channel, but also representing a written variety because they are produced in written form by journalists; in other words, they are a kind of communication that is written in order to be read. Accordingly, speech and writing—means of production that are usually clearly distinguished—tend to merge into each other. This happens in various forms of newer mass media, whereby "two universes interact with each other producing a great majority of texts in which an actual cooperation of the two varieties of language is found" (Dardano, 1994:177).

Genre Mixing

Together with the widespread usage of quotation, whereby television programs refer to other television programs, producing a self-referential mechanism, several television genres undergo progressive modifications, acquiring features from other genres, and becoming intertextually hybrid. One effect of this phenomenon is that the widespread trend to dramatize facts and news drives the genre of news toward entertainment, producing heterogeneous textual types that are difficult to identify and define (Fairclough, 1995).

Structure

Additionally, television texts are complex structures composed of different parts: a news program, for example, is a unitary text that is structured in different sections[2] (headlines, current events, reports), which in their turn may contain statements, interviews, and other forms of communication like monologues or dialogues, whose register can be more or less formal or spontaneous. Television texts are the result of the juxtaposition of fragments with variable function and extent. The linguistic consequences of this text fragmentation are evident: for example the marked tendency of news headlines to nominalize, a feature that is much less frequent in other parts of the news (Ennaji, 1995).[3]

Registers

In the language used on television, different styles, with particular regard to formal/informal and public/private styles, coexist and alternate (Fairclough, 1995); this is the result of a continuous tension between public/formal and private/informal dimensions (the actual reality of politics, economy, science, and its representation through the electronic medium, that tends to produce a generalized transfer to the private sphere). News programs, for example, transform their original source—public discourse—into a more private kind of discourse that is prepared to be consumed in domestic settings.

Multimodal Interactivity

Even if physically remote, television audiences tend more and more frequently to participate in television programs, intervening with comments or questions via other media like the telephone, short message service, or e-mail (Skokut, 1993; Sobrero, 1997). This kind of multimodal interaction produces differentiated and more complex communicative solutions, if compared, for example, to FTF conversation (see especially Jones, chap. 1, this volume).

From a linguistic and textual point of view, all these issues make television communication a very diversified system, by nature hybrid and heterogeneous, with characteristics that are not always clearly defined.

Specific Features of TV Languages

Although speech and writing, as several studies have demonstrated (Biber, 1988), cannot be considered homogeneous varieties, the use of the two categories of *typical speech* and *typical writing* may prove useful (Biber, 1988); they indicate respectively the unmarked genre of speech, that is FTF conversation, and the unmarked genre of writing, the informative exposition.

The language of television has basically in common with speech the phonic nature and the possibility of seeing the image of the person who is speaking, and thus of catching the person's gestures and facial expressions; as for writing, on the other hand, writer and reader are in different places, not physically sharing the same situation, and the communication is unidirectional and addressed to a large number of people. From a pragmatic and textual point of view, however, the language used on television has its own individuality, due to the technology used, to the ritual produced, and to the relationship established between speaker and audience; in the following section, three of these features, connected respectively to the speaker, to the listener, and to the context, will be briefly examined.

The Speaker

Television shots, even those taken outside of the studio, require a very complex and intrusive technical apparatus, which unavoidably alters the naturalness and the spontaneity of communication. Individuals speaking in front of a camera, even in an extemporized and ad lib interview, inevitably tend to control their words much more than they would do in a typical FTF conversation. Apart from programs that are based on a modality of communication prepared beforehand (news, but also entertainment programs and talk shows, in which hosts work with a predetermined outline), the language of television can rarely be considered spontaneous, even in live broadcasts, being far more mediated and controlled by the speakers' form of communication (Wachtel, 1993): The fact of being potentially viewed by millions of people, as well as the action of speaking in front of a complex technical equipment, necessarily reduce its spontaneity.

Additionally, there is another television-specific feature: Far more than other kinds of communication, television is based on a complex ritual in which each person, upon appearing in video to millions of people, automatically becomes a "TV personality," with a specific role (e.g., host, assistant, or guest in a talk show). The interaction between the different characters who appear in television programs must inevitably conform to a set of implicit conventions which the ritual of the television event brings with it: "in the ritualization of television the spontaneity and the improvisation typical of spoken language diminish; new rules in the actualization of the linguistic interaction arise for the different roles adopted by the speakers who are involved" (Rossini Favretti, 1988:207). Each genre of television program has its specific script, based on which each person has a specific function and is somehow bound to play a part, connected to this role: The anchor reads the headlines, and then introduces the reports, usually reading or working with a written outline; the host of a talk show greets the audience, presents the subject to be discussed, and introduces the guests, calling on them to speak in turn (Fairclough, 1995; Ilie, 2001). In TV broadcast language, therefore, due to the technological apparatus and to the great tendency to stylize, the features of improvisation and spontaneity that characterize typical speech seem to fade if not completely disappear.

The Listener

It is only too evident that TV programs present a great heterogeneity of the listeners. For any kind of program, in fact, the main receiver of visual information is the audience, who plays therefore a fundamental role. In the course of programs that present communicative situations based on dialogue, like programs that include interviews, speakers have another type of conversational partner, one who is physically present at the same time in the studio or in the place of interview.

People speaking on television have thus at least a dual kind of listener: one who is collective and undifferentiated, represented by the audience (as in writing), and the other present at the same time in the communicative situation (as in speech). As Bell (1991) noted: "Playing to two or three layers of audience, each containing diverse segments, is clearly a difficult language skill" (97). As a result, people speaking on television are in some way forced to take into account both kinds of target reactions, of which one is visible and by their side, the other distant and totally unknown. This double interaction inevitably produces new communication strategies, half way between speech and writing, but in no way entirely identifiable to either of them.

Context

It has already been pointed out that one of the features that TV broadcast language has in common with spoken communication is the possibility to visualize the person who is speaking, and thus to also catch paralinguistic features like gestures and facial expressions, or nonlinguistic features like the setting where the interaction takes place; yet television does not always exploit this opportunity, and for reasons of direction or of television aesthetics, it often turns to other solutions, including passing from the close-up of the speaker to different images, which then become the background of the speaker's voice and strengthen and emphasize the speaker's message. Hence, for example, in the course of broadcast news it is rather common to see a journalist reading a report on bad weather framed for few seconds, followed by images of snow, obstructed roads, and blocked cars (Goodman & Manners, 1997). There are also programs that by nature completely ignore speaker shots and focus exclusively on the represented event; during a live broadcast commentary, the audience listens to the commentator's voice but watches the images of the event that is being transmitted. In such cases, just like in written communication, the audience has no access to paralinguistic and nonlinguistic features of the utterances, although they are oral utterances. All these briefly outlined features show that broadcast communication has its own characteristics, a great complexity and internal diversification; despite some similarities, it is clearly distinct from both speech and writing, and this peculiarity inevitably has significant linguistic consequences.

The Data

The analysis of dislocations has been carried out on a corpus of transcriptions of television programs (main corpus); the data extracted has been compared to the data extracted from other corpora (comparative corpora), specifically from a written Italian corpus that includes journalistic and literary texts produced in the last 15 years, and from a spoken Italian corpus that includes transcriptions of conversations and other more formal FTF-interactions, such as school lessons or university exams. The corpora are described more fully in the sections that follow.

The Main Corpus

The reference corpus for the language of television is the *Corpus di italiano televisivo* (*Cit*), in the process of being compiled, which is composed of transcriptions of television programs divided into five genres (current affairs, entertainment, advertising, sport, and news), for a total of 500,000 words. The data extracted for this chapter are based on a corpus subsection called *CitDemo*, which includes 10 programs, transcribed in electronic form and completely annotated with grammatical categories and other parameters (e.g. lexical aspects such as foreign or technical words and multiword expressions) connected to the programs' genre and structure (Spina, 2000, 2005; see also http://www.sspina.it/cit/cit.htm). The annotation standard is the one adopted by the *Text Encoding Initiative* (http://www.tei-c.org), and more specifically by version P4 of its guidelines; the grammatical annotation is based on a tagset developed at the University of Stuttgart (http://www.ims.uni-stuttgart.de/projekte/corplex/TreeTagger/DecisionTreeTagger.html), partially and manually adapted to the aims of *Cit*. The programs included in *CitDemo*, together with their duration and length in number of words, are listed in Table 9.1.

Table 9.1. Programs Included in *CitDemo*

Genre	Date	Channel	Minutes	Tokens
Current affairs				
Il fatto	1/20/99	RaiUno	10′	743
Il fatto	1/29/99	RaiUno	10′	1,185
Pinocchio	10/1/98	RaiDue	80′	12,703
Entertainment				
Coppie (talk show)	2/3/99	Canale5	60′	8,542
Meteore (variety show)	10/7/99	Italia1	60′	5,669
Advertising				
260 ads	March–June 99	all	90′	10,908
Sport				
Football match (live programs)	3/9/99	RaiUno	45′	6,085
Novantesimo minuto (comment)	1/31/99	RaiUno	30′	5,453
News				
TG1	12/1/98	RaiUno	30′	5,123
TG3	8/25/98	RaiTre	30′	5,220
TOTAL			**445′**	**61,631**

This division into five genres deserves some brief remarks: According to its methodological principles, the *Cit* only includes nonfiction television programs. Starting from the assumption that the main parameter that has guided the genre's attribution is aim (Swales, 1990), advertising represents those programs whose aim is to persuade. As far as the other four genres are concerned (current affairs, entertainment, sport, news), their definition has not always been easy, and the aim criterion has not always been the only one to be used. Current affairs programs, for example, are the ones that analyze in depth the most recent news. Their aim is thus commenting on news, but their distinction from entertainment talk show is not always straightforward (Nacci, 2003). As for sport broadcasts, although they are a kind of entertainment program, they have been classified separately because they describe a live event or comment on it in detail shortly afterward. This consideration, which has important linguistic consequences (Marriott, 1997) linked to their great diffusion in terms of audience, has suggested their collocation in an autonomous category. The five genres can also include subgenres, producing the classification reported in Table 9.2.

Comparative Corpora

The data on dislocations extracted from *Cit* have been compared to data extracted from similarly dimensioned corpora, belonging to the two traditional means of production (i.e., speech and writing).

The written corpus is divided into two sections, one of informative texts (articles issued by four Italian newspapers in 2003), and the other of literary texts (samples taken from five Italian novels of the 1990s). The written corpus is composed of about 40,000 words. The spoken corpus is based on a portion

Table 9.2. Genres and Subgenres Included in the Cit

Genres and Subgenres	Aim
Current affairs	Commenting on news
Entertainment	Entertaining
Quiz shows	
Talk shows	
Variety shows	
Advertising	Persuading
Sports	Describing or commenting on a sporting event
Live programs	
Commentary	
News	Informing

of the texts of the *Lessico di frequenza dell'italiano parlato* (De Mauro, Mancini, Vedovelli, & Voghera, 1993), in its Internet version of the *Banca dati dell'italiano parlato* (http://languageserver.uni-graz.at/badip/badip/home.php); Sections A and C of texts recorded in Rome between 1990 and 1992 have been selected, which include respectively informal FTF conversation (Type A) and more formal interactions like university exams, school tests, union meetings, and so on (Type C), for a total of about 50,000 words.

Definition of Dislocation and Research Parameters

Dislocation, in this chapter, is the marked construction found in many languages including Italian (Sornicola, 2002), in which, with reference to the unmarked subject-verb-direct object order, one sentence constituent is displaced to the left (left dislocation, LD) or to the right (right dislocation, RD).[4] The dislocated constituent, which can be the direct object, the indirect object, or less frequently the sentence subject, is repeated (LD) or anticipated (RD) by a co-referential clitic pronoun:[5] (1a) and (1b) are two examples of LD and RD, with the dislocated element in bold character and the clitic pronoun underlined. These examples are two dislocations of direct object, represented by the noun il giornale 'the newspaper':

(1a) **il giornale** l'ho comprato
'the newspaper (it) I have bought'
(1b) l'ho comprato **il giornale**
'(it) I have bought the newspaper'

From a theoretical point of view, this study pursues the approach of Berruto (1985), according to whom for a satisfactory description of dislocations (particularly LD), most of all in a pragmatic perspective, a new category is needed: the category of speaker's *center of interest*. The notion of center of interest, according to Berruto, is not in contrast with the notions of *given* and *topic*, but it refers to another level of pragmatic analysis: whereas the categories of given/new involve the relationship between utterance and context, and the topic/comment categories refer to the informational and textual organization of utterance, "the center/periphery of interest category involves the peculiar relationship between utterance and the speaker . . . the mechanism operating in left dislocation is the movement of a constituent to the role of center of interest: the dislocated constituent is the speaker's center of interest inside an utterance" (Berruto, 1985:70).

Previous Studies

Literature on dislocations in Italian, starting with the study of Meriggi (1938), is rather rich (most of all for LDs), with a great range of approaches, from the syntactic, particularly in the field of generativism, to the pragmatic, functional, or historical.[6] Apart from the differences in approach, the conclusions are rather consistent regarding the following points:

- Dislocations are a multifunctional construction: Their use is connected to diversified functions that cannot often be allocated to a unique type.
- Dislocations are a construction typical of speech, most of all dialogic, informal, and familiar (RDs), but they are not completely unrelated to writing (particularly LDs). Some define them as *conversational constructs* (Duranti & Ochs, 1979b).
- LDs have the function of highlighting an element of the sentence.
- RDs often have the function of producing a sort of familiarity between speaker and listener, or of marking an afterthought, by adding on the right the dislocated constituent that had been anticipated by a clitic pronoun.
- In some cases, most of all in RDs, the clitic is strongly bound to the verb and shows a high degree of grammaticalization.

Research Parameters

After having collected RD and LD occurrences in the main and the comparative corpora and having computed their frequencies, the analysis of data has been carried out on the basis of the following parameters, which seek to account for the complex and multifunctional nature of dislocations.

Syntactic and Semantic Parameters

FREQUENCY OF FUNCTIONS. Within the occurrences of RD and LD, a distinction has been made among the ones in which the dislocated constituent has the function of direct object (DO), indirect object (IO), partitive (PART), topic-based case (TB), locative (LOC), or subject (SUB), as in the following examples:

(2a) Lei **questa possibilità** non glie<u>la</u> vuole dare (DO)
'She this possibility does not want to give to him'
(2b) Tu **a quello** <u>ci</u> credi? (IO)
'(To that) do you believe it?'
(2c) ma tu **quelle pentole che reclamizzi** te <u>ne</u> rimane qualcuna? (PART)
'those saucepans that you advertise have you left some?'
(2d) Quando <u>ne</u> parlava **di quel periodo** (TB)
'When he spoke of that period'

(2e) ma **il cinema** ci andate o in vacanza? (LOC)
'to the cinema do you go or on holiday?'
(2f) Eccole **le mie Giravoltine** (SUB)
'Here they are my Giravoltine'

Frequency of Categories. The grammatical category of the dislocated elements has been considered, specifically nouns (N), pronouns (PRO), or clauses (CL), as in the following examples:

(3a) Lei **un lavoro** ce l'ha per fortuna (N)
'You luckily have a job'
(3b) **Di questo** lei ne può stare certa (PRO)
'Of this you can be certain'
(3c) Lo sa **quanto si paga la verità**? (CL)
'Do you know how much you pay for the truth?'

N Semantic Features. Dislocated nouns have been divided into [±definite], [± human], abstract, concrete.

Types of Pronouns. Within dislocated pronouns, I, II, III person or neuter pronouns have been distinguished.

Subject Frequency. In the sentences that contain dislocations, occurrences with an explicit subject have been separated from occurrences without an explicit subject; the subject presence is a sign of a double topicalization of the dislocated element and of the subject.

Dislocations/Passives Ratio. Dislocations frequency has been compared to passive frequency: the two constructions, as it has been pointed out (Berruto, 1985; Duranti & Ochs, 1979b), have similar features, but they are considered typical of speech and writing, respectively.

Grammaticalization. Cases of clitic grammaticalization have been considered. The following is an example of clitic grammaticalization within a dislocation:

(4) La palla non ne vuol sapere **di entrare**
'The ball has no intention of entering'

Pragmatic and Textual Parameters
From a pragmatic point of view, data extracted from the corpora have been analyzed in order to locate, in the use of dislocations, the dominant function (sometimes it has been particularly difficult to isolate only one function that is dominant with reference to the others, because of the construction's multifunctionality). The following three distinct functions have been identified:

Informational. The dislocated constituent is the center of interest that the speaker wants to focus on. It can be carried out with a single topicalization (if the subject is not expressed) or with a double topicalization: If the subject appears in the sentence there will be a center of interest (the dislocated element) and a topic (the subject, capitalized in the examples):

(5a) **Questo** l'ho capito solo dopo (one topic)
'This I have understood only afterwards'
(5b) LEI **un lavoro** ce l'ha per fortuna (two topics)
'You luckily have a job'

Expressive. It is used to create a sort of familiarity between speaker and hearer, and embraces the use of idiomatic and/or grammaticalized expressions.

(5c) Lo sai **che faccio**?
'Do you know what I am going to do?'

Control. Control is used by the speaker to further specify, owing to an afterthought, a self-correction, a delay, or a lack of planning, a sentence constituent previously anticipated by a clitic, that is often preceded by a pause (#).

(5d) IO lo trovo un atteggiamento molto conservatore # **quello**
'I find it a very conservative attitude # that one'

Dislocations in the Television Corpus

The total number of occurrences of dislocations in the *CitDemo* corpus is 108, that is 0.2% of the total number of tokens. Of these 108 dislocations, 66 are LDs (61% of the total) and 42 are RDs (39%). The controversial issue of numerical prevalence of LDs (Rossi, 1999) is thus confirmed as far as the language of television is concerned.

Syntactic and Semantic Parameters

Frequency of Functions. Of TV dislocations, 71% concern DO, 14% IO, whereas other syntactic functions have smaller percentages (from 6% to 1%). DO higher topicality is well established in all the previous data on Italian dislocations (Berruto, 1985; Bozzone Costa, 1991; Duranti & Ochs, 1979b, etc.), and the same happens in other languages (Givón, 2001), although this prevalence in the case of television is slightly more prominent (more than two thirds of dislocations concern DO). This great prevalence can also be explained, from a quantitative point of

Table 9.3. Dislocations of Syntactic Functions in the *Cit* Corpus (X 100)

	DO	IO	PART	TB	LOC	SUB
Total	71	14	6	5	3	1
LD	76	9	8	5	3	0
RD	64	21	5	5	2	2

view, with the fact that DO is in general the most frequent case; furthermore it has no morphological (affixes) or syntactic (prepositions) marks, but is signaled only by its position within the sentence. It is therefore a typically word-order sensitive case; it is placed after the verb in the unmarked sentences and in other positions when the speaker wants to raise it to center of interest.

As shown in Table 9.3, the prevalence of DO is stronger in LDs than in RDs, where on the contrary IO dislocations increase to a significant extent; the numerical difference between DO and IO is thus less significant in RDs.

Frequency of Categories. The grammatical category that is more subject to dislocations is the noun, which comprises 51% of the total TV dislocations; pronouns are dislocated in less than one-fourth of the total cases (22%), whereas a significant presence of clauses is found (27%), which are the second grammatical category after N to be dislocated. The last one is an interesting phenomenon, also considering that topicalization is often viewed as concerning only the nominal constituents of clauses ("When whole events or states are made topical, they are almost always nominalized," Givón, 2001).

Table 9.4 shows some important differences between LD and RD: The dominant category is N, in the total just like respectively in LDs and RDs; but in LDs, dislocations of PRO are more frequent than they are in the total, whereas in RDs dislocations of CL are more frequent compared with the total number (36% vs. 27%).

Table 9.4. Dislocations of Grammatical Categories in the *Cit* Corpus (X 100)

	N	PRO	CL
Total	51	22	27
LD	48	30	21
RD	55	10	36

This means that apart from the most frequent case of N, LDs frequently bring a pronoun to center of interest (often a neuter pronoun, as in Examples 6a or 6b, or more rarely a third-person pronoun, as in Example 6c), whereas RDs preferably dislocate clauses (they often are questions), a large part of which are grammaticalized or semi-grammaticalized, like the expressions lo so/sai che + CL or lo vedi che + CL (7a and 7b).

(6a) **questo** non me l'aspettavo
'This I didn't expect'
(6b) **di questo** lei ne può stare certa
'Of this you can be certain'
(6c) **di lui** non me ne frega più nulla
'Of him I don't care anymore'
(7a) non lo so **cosa vuol dire**
'I don't know what it means'
(7b) lo vedi **che è lui**
'You see that it's him'

N SEMANTIC FEATURES. From a semantic point of view the dislocated nouns have a rather marked tendency: 80% are [-human], against 20% of [+human]. Concrete nouns are most of all dislocated (*pentola* 'saucepan,' *palla* 'ball,' *cosa* 'thing,' *lavoro* 'job,' for a 60% total) and abstract nouns that indicate sensations and feelings (*speranza* 'hope,' *ipotesi* 'hypothesis,' *ottimismo* 'optimism'), rather than nouns with human referents. In this case the theory of topic hierarchy (Givón, 1976) is contradicted; it expects in fact a greater topicalization for human and speaker–listener-related referents. Givón's theory is on the contrary confirmed by the other considered parameter, the definiteness of dislocated nouns: 95% of them are in fact definite, and thus conform to the hierarchy.

TYPES OF PRONOUN. Another feature that the data have shown concerns the type of dislocated pronouns: considering both LD and RD, the majority are neuter pronouns, which, together with third person pronouns, cover more than 90% of dislocations of PRO; there are no second person pronouns and only 4% first person pronouns. These data indicate that very rarely, with dislocations in TV languages, the speaker's center of interest is represented by a reference to self or to the listener and, more generally, it is a sign of a low level of interactivity and of a low involvement of participants.

SUBJECT FREQUENCY. It is well known that the subject is not mandatory in the Italian sentence; the same happens in sentences with dislocations, where in many cases the subject is not expressed and it is inferred from verb inflexion. Yet the occurrences of sentences with dislocation that have an explicit subject constitute almost half of the total (about 49%); the sentences with dislocation, thus, that already provide the speaker with the possibility of highlighting the center of interest, in addition exploit another syntactic device to focus on another element that is prominent from the informational point of view, that is, the subject. Resulting constructions have thus a double topic: the dislocated constituent and the subject, both with a strong informational prominence (as in 8).

(8) LEI **un lavoro** ce l'ha per fortuna
'You luckily have a job'

DISLOCATIONS/PASSIVES RATIO. Duranti and Ochs underline the similarities between passives and dislocations: both are used (also) to topicalize a sentence element different from the subject (Duranti & Ochs, 1979b). Dislocations also have three important advantages compared to the passives: They can be applied to any element of the sentence (and not only to DO like passives), the sentence subject remains unchanged, and the resulting constructions may contain two topics. The two devices would be somehow in complementary distribution: Speech contains few passives and a great amount of dislocations, which are less rigid and more dynamic, and thus more suited to a real-time communication, whereas writing presents more or less the opposite situation. Berruto goes more in detail and dwells on a particular case, the generic third-person plural subject dislocation, as in Example 9a. This example is exactly equivalent to a passive sentence without dislocation (Berruto, 1985), thus to Example 9b, in which *la posta*, which is the subject in the passive sentence and DO in the sentence with dislocation, the topic in Example 9b and the center of interest in Example 9a. According to Berruto, the two structures—the active one with dislocation and the passive one without dislocation—are in complementary distribution: Wherever there are many passives, most of all in writing, there are few generic third-person plural subject dislocations, and vice versa:

(9a) **la posta** l'avevano fatta per noi
'the post office they had made it for us'
(9b) **la posta** era stata fatta per noi
'the post office had been made for us'

The *Cit* data are interesting because the two features (passives and dislocations) show in the broadcast language corpus very similar frequencies (dislocations are 108, passives 143;[7] their token ratio is about 0.2%). They have therefore a basically equivalent and not a complementary distribution in the language of television.

GRAMMATICALIZATION. About one third of the TV dislocations (34%) can be considered grammaticalized: The clitic pronominal function partially or totally weakens and the pronoun becomes an integral part of the verb, whose meaning is also changed. Grammaticalized dislocations have particular features with reference to those in which the clitic preserves its grammatical function: A close clitic connection to the verb makes these structures basically fixed expressions. The language of television shows a significant number of grammaticalized dislocations, and resulting verbs confirm this trend to the use of fixed expressions: *volercene, fregarsene, esserci, intendersene, volerne sapere, saperlo* (mostly in the interrogative *Lo sai che . . . ?* 'Do you know that' or in the affermative *Lo so che . . .* 'I know that'; see Bazzanella, 1995), and, with the highest frequency, *averci*. In some cases these expressions are completely idiomatic as in 10.

(10) **qui** non <u>ci</u> piove
'About it, there is no doubt'

Pragmatic and Textual Parameters

INFORMATIONAL FUNCTION. The dislocations functions in the TV corpus are mainly informational (55%): Using this function the speaker aims to highlight the sentence element that the speaker considers the center of interest, and to differentiate it from the sentence periphery (Berruto, 1985). A great deal of dislocations that fit into this category (a little less than half) have a double topic, namely the dislocated element and the subject; this confirms that the emphasis on topic and center of interest is one of the dislocation's most important functions, in particular because it can be doubled between dislocated element and subject.

EXPRESSIVE FUNCTION. A significant percentage of dislocations (43.5%) has, however, an expressive function, aiming to give the utterance a more colloquial register and to set a relationship of familiarity between speaker and listener.

CONTROL FUNCTION. The use of the control function, on the contrary, by which the speaker reconsiders the utterance he or she is pronouncing and adds a further element announced by a clitic, is very rare; this kind of function, strictly connected to planning issues, does not concern the language of television very much, which is far less spontaneous and more planned than face-to-face conversation.

Dislocations in the Different Television Genres

Within the *CitDemo* corpus there is a strong variability in the 108 total occurrences of dislocations, depending on the five genres that have been considered (current affairs, entertainment, advertising, sports, news): entertainment programs alone cover more than half the total number of occurrences, and, including current affairs, exceed 80% (advertising has about 10%, news and sports respectively 5% and 3%). Thus, as may be expected, communicative situations with a dialogical interaction between two or more speakers are the most favorable to the use of dislocations. It is nevertheless a little surprising the very low percentage of dislocations in sports broadcasting, which include, apart from live commentaries that are essentially monologues, programs that comment on events, with interviews and discussions, and thus interaction among different speakers.

Another feature that deserves some attention is the distribution of LD and RD in the two genres of news items and entertainment; news, prepared as a written text, have more RDs (considered typical of speech) than LDs (7% and 5%, respectively); entertainment programs, on the contrary, the most "spoken" television genre, report a prevalence of LDs over RDs (59% vs. 52%). This shows, as seen later, that the generic statement that RDs are almost exclusively a speech feature has no foundation in empirical data.

Among the 10 syntactic and pragmatic parameters that are the object of this research, the most significant for the different television genres are analyzed.

Syntactic Parameters

FREQUENCY OF CATEGORIES. Dislocated grammatical categories have a rather homogeneous frequency across the five different genres, with few striking differences that deserve some remarks: First of all, clauses have in all genres the same or higher frequency than pronouns (Fig. 9.1); it is shown later that this is a specific feature of the language of television compared to speech and writing, and that it indicates a trend toward the use of fixed expressions (many of the dislocations of CL, as shown in Examples 7a and 7b, are fixed expressions) or to a markedly colloquial register, as in Example 11.

(11) **'sti cinque mesi che sto passando** li ho passati molto brutti
'these five months that I am passing I have passed them very badly'

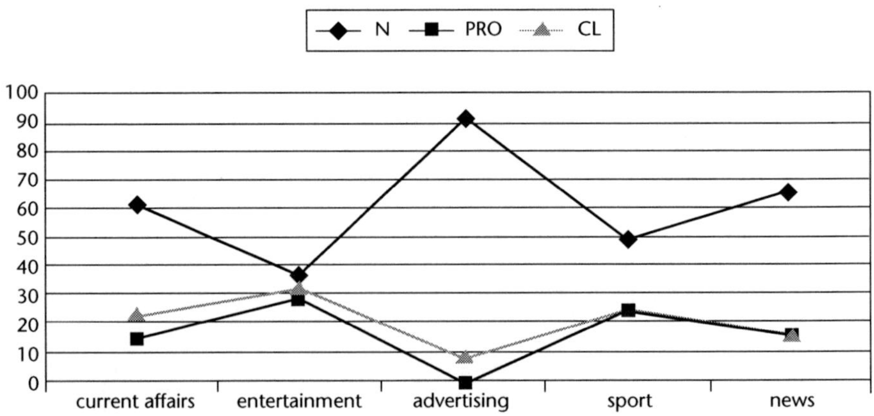

Fig. 9.1. Dislocated grammatical categories in the TV genres.

The most frequently dislocated grammatical category in all genres is N, with higher than 90% peaks in advertising: In this genre there are no dislocated pronouns, but there is a clear tendency to focus on nouns referring directly to the advertised product (Example 12a) or to some of its features (Example 12b).

(12a) **alla maionese** <u>ci</u> pensa CALVE'
'For the mayonnaise cares Calvè'
(12b) **la tariffa** te <u>la</u> fai TU
'the rate you make it'

In both cases, along with the center of interest (the LD noun), the explicit subject is also present, highlighted at the end of the sentence (in Example 12a it is the product brand, in Example 12b the second person pronoun *tu*, so as to stress the advantages of a telephone company that allows the customers to personalize their rates).

N SEMANTIC FEATURES. Only in two of the television genres (entertainment and sports) the majority of dislocated nouns are [+human]; this can be explained with the greater interactivity and involvement of entertainment-based genres; in all other cases concrete nouns are mostly dislocated (100% in advertising) and to a lesser degree abstract nouns.

TYPES OF PRO. The only genre that includes first-person dislocated pronouns is entertainment (Fig. 9.2); in this case as well, however, they are far less frequent than third-person pronouns (there are no second-person pronouns) and neuter pronouns. The latest, in their anaphoric use that refers to something previously mentioned (as in Example 13), are largely preponderant compared with personal pronouns; in the field of dislocations, therefore, pronominal reference to other parts of text prevails over the reference to people involved in the interaction, and thus textuality prevails over involvement and interaction.

(13) perché c' è anche una dignità personale da salvare e **questo** ce <u>lo</u> siamo scordato
'because there is also a personal dignity to preserve and this we have forgotten'

DISLOCATIONS/PASSIVES RATIO. The quantitative equivalence in the use of passives and dislocations has already been highlighted; the genres variability of this parameter is quite high (Fig. 9.3): It ranges from entertainment programs, that have more dislocations than passives; advertising, which displays a perfect equivalence of the two features; current affairs programs, which show small differences; sports broadcasts, that have a much higher frequency of passives; and finally news, which shows the highest passives/dislocations ratio (nine passives every single dislocation).

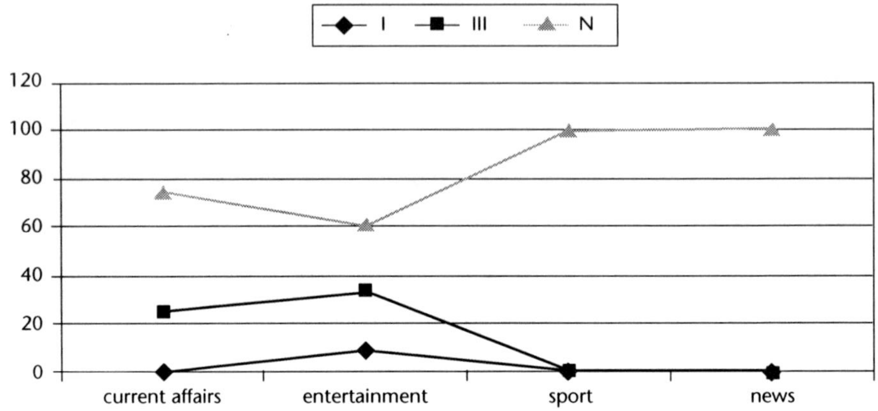

Fig. 9.2. Dislocated pronouns types in the TV genres.

GRAMMATICALIZATION. It has already been remarked that in all programs there is a significant percentage of grammaticalized clitics in the occurrences of dislocations; this feature ranges from 25% to 50% in the different genres (Fig. 9.4); the genre with highest frequency of grammaticalization is news (in 50% of the total, clitic dislocations are grammaticalized). The most "written" genre therefore shows the greatest trend to use grammaticalized and thus fixed expressions.

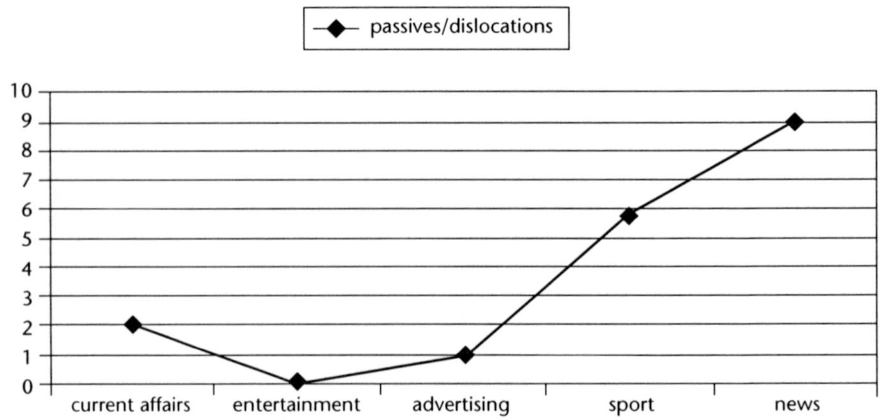

Fig. 9.3. Passives/dislocations ratio in the TV corpus.

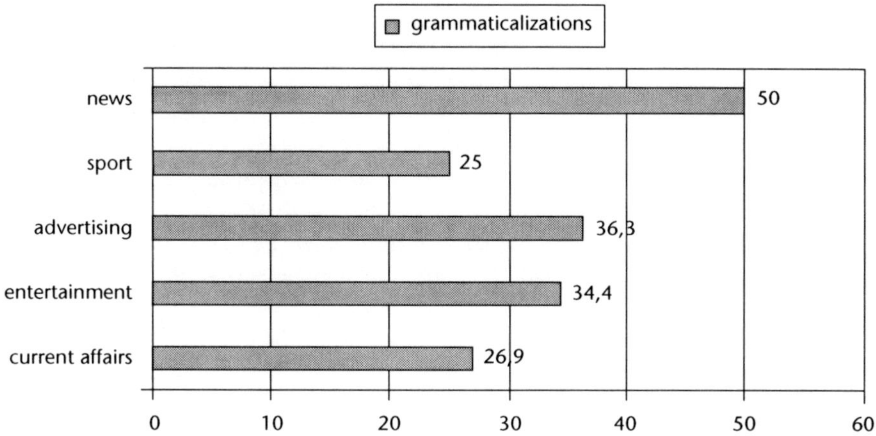

Fig. 9.4. Percentages of dislocations with grammaticalization in the TV genres.

Pragmatic and Textual Parameters

Within pragmatic functions, there is an unexpected result in sports and news genres: In the first, the few instances of dislocations all have an expressive function (there are no cases of informational function). Dislocations are basically linguistic devices that are used by TV commentators in order to make their style more colloquial (Braun, 1998); in the news, on the contrary, the two informational and expressive functions are exactly equivalent (50% each), although news is the most informational genre of all.

Comment on the Data

In the TV corpus, dislocations are 0.2% of the total number of words (every 500 words a dislocation occurs). Passive forms have a very similar frequency; this equivalence is new evidence with reference to the previous literature, which generally assumes that dislocations and passives are in complementary distribution in the spoken language (Duranti & Ochs, 1979b). The language of television uses different devices to highlight particular elements of the sentences: one of them, dislocations, more dynamic and colloquial, characterized by a small segment of information that appears split within the sentence; the other, passives, more rigid and formal, well integrated in a coherent and stable syntax. One may say that these data on passives/dislocations ratio synthesize the mixed and heterogeneous nature of the language of television, which shares important features with both spoken and written languages.

The high frequency of grammaticalizations and dislocations of CL is a sign of a strong tendency to use fixed or even stereotyped expressions, which provide ready packages of linguistic material with an expressive conversation-oriented flavor.

The almost total lack of dislocated first- and second-person pronouns and the predominance of [–human] dislocated nouns show a trend to detachment, low involvement, and low interactivity. Furthermore, this happens with a construction that is purposely used to focus on elements that the speaker considers more important; one could thus draw the conclusion that in the language of television, the speaker's empathy tends to fade, focusing on elements unrelated to the speaker and to the listener.

The dominant pragmatic function of dislocations in TV languages is the informational one, within which there is a strong presence of double topicalized dislocations (the frequency of explicit subject dislocations is globally high); nevertheless the expressive function also plays an important role (little less than half of the dislocations have this dominant function), and this shows a considerable trend to use a conversational register.

Comparison Among Television, Written, and Spoken Corpora

From the quantitative point of view, the first data on the frequency of dislocations in the three corpora of TV, written (WR), and spoken (SP), is a picture of the general differences among the three different varieties; Table 9.5 shows that, in a very symmetric way, the frequencies of TV, normed to 100, are exactly halfway between speech and writing frequencies. The quantitative prevalence of SP over WR is a well-established evidence in research on dislocations (even if they are also present and well-attested in writing); the new evidence is the frequency difference between spoken and television language, and the placement of the latter in an intermediate position between speech and writing.

Differences among the three corpora are also found in LD and RD distribution within the total number of dislocations (Fig. 9.5): The graph shows a progressive increase in the distance between LD and RD, starting with writing, where the

Table 9.5. Frequency of Dislocations in the Three Corpora

	Occurrences	Occurrences Normed to 100
TV	108	0.2
SP	169	0.3
WR	55	0.1

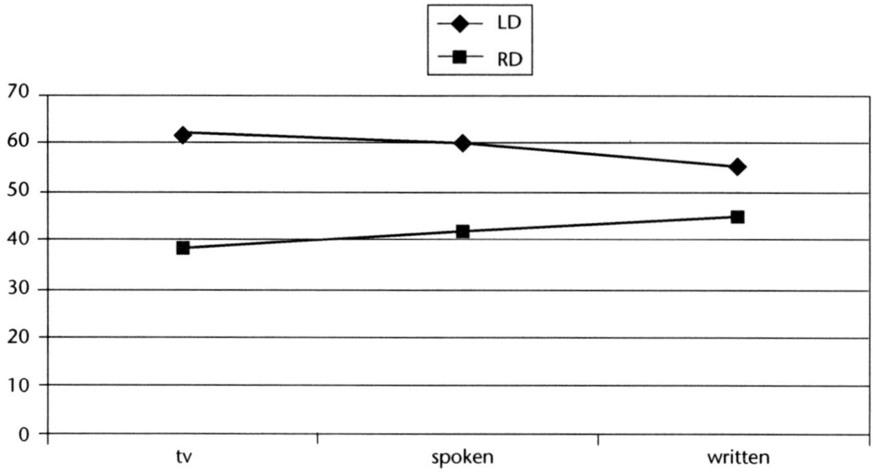

Fig. 9.5. Distribution of LD and RD in the three corpora.

two values are closer (55% and 45%, respectively), proceeding with speech, which has an intermediate position, and finally, TV, where the difference is the greatest (61% and 39%). WR is thus the corpus that contains the greater number of RD, and TV the one that has the greater number of LD.

Syntactic Parameters

FREQUENCIES OF FUNCTIONS. The three corpora show rather homogeneous tendencies in the syntactic categories of the dislocated constituents: the prevalence of DO is clear in the spoken, written, and broadcast language (even though in writing, values seem a little more distributed across all the other functions). IO slightly varies between 14% and 18%.

FREQUENCIES OF CATEGORIES. The salient evidence on dislocated grammatical categories concerns dislocations of CL (Fig. 9.6): TV is the only one of the three corpora in which they are more frequent than dislocations of PRO. The high frequency of dislocated clauses therefore seems a peculiar feature of Italian broadcast language; whereas the prevalence of dislocated N is stable in all the three corpora, with a particularly high frequency in writing, speech shows the highest frequency of dislocations of PRO. TV and WR data seem in general more homogeneous, whereas SP shows a more diversified curve.

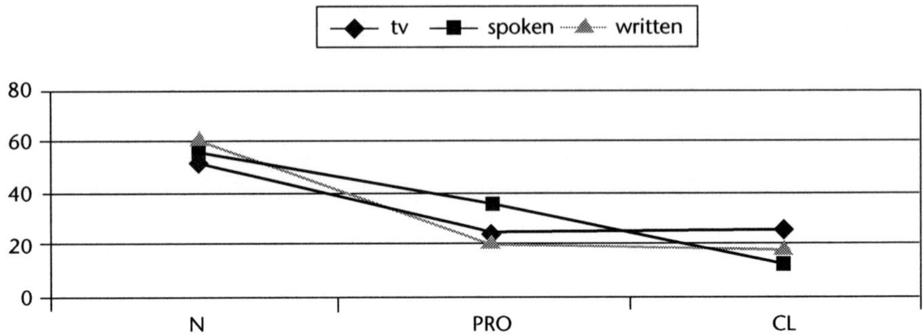

Fig. 9.6. Grammatical categories of dislocations in the three corpora.

N Semantic Features. The same trend as the TV corpus is confirmed in speech and writing: The great majority of dislocated N (about 80%) does not have a human referent (Fig. 9.7); concrete nouns are more frequent than abstract ones. Again TV, despite slight differences, occupies an intermediate position between SP and WR.

Types of PRO. Data extracted from TV and WR also has a very similar tendency as far as dislocated pronouns are concerned (Fig. 9.8): There are almost no first- and second-person dislocated pronouns; third-person pronouns increase, and neuter pronouns reach the highest frequency (they are more frequent in writing than in TV). The SP corpus shows on the contrary the highest frequency of third-person pronouns, much higher frequencies comparing to TV and WR of first- and second-person pronouns, and remarkably fewer occurrences of neuter pronouns.

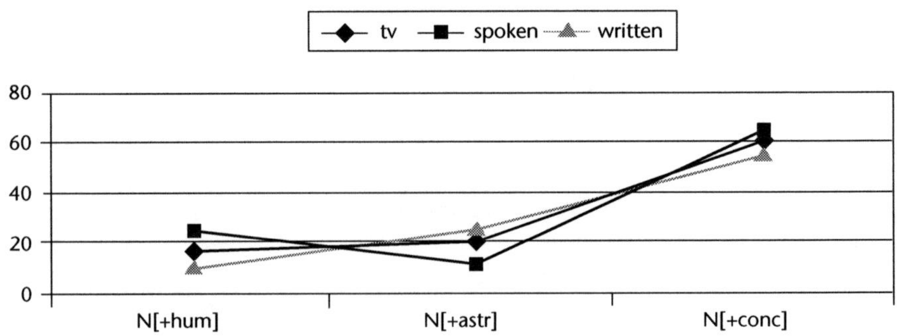

Fig. 9.7. Semantic features of N in the three corpora.

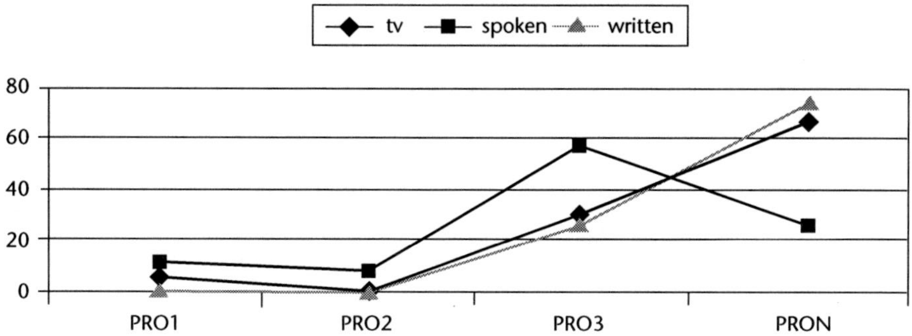

Fig. 9.8. Type of dislocated pronouns in the three corpora.

Subject Frequency. The language of television shows the highest frequencies with reference to the presence of subject in the sentences with dislocations; almost half the total number of dislocations presents an explicit subject; speech and writing have a frequency of 30%. Again, this strong presence of an explicit subject seems to be a peculiar feature of TV broadcast language.

Passives/Dislocations Ratio. Once again the TV data occupies an intermediate position between SP and WR (Fig. 9.9): TV has in fact a 1.3 passives/dislocations ratio, against WR 2.3 and SP that shows on the contrary a prevalence of dislocations (with a 0.1 ratio). The graph shows how the language of television is the only variety that does not have a complementary distribution of the two features, which have almost the same frequency. The data thus confirm the well-established finding that writing topicalizes sentence elements using many passive forms and few dislocations, and that exactly the opposite is true for speech (although dislocations in writing are more frequent than passives in speech); they also provide new evidence that broadcast language, differing from the other two varieties, equally uses both passives and dislocations.

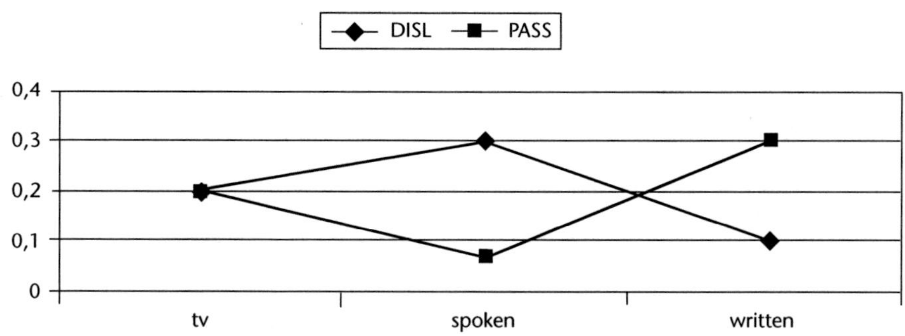

Fig. 9.9. Frequencies of passives and dislocations normed to 1,000.

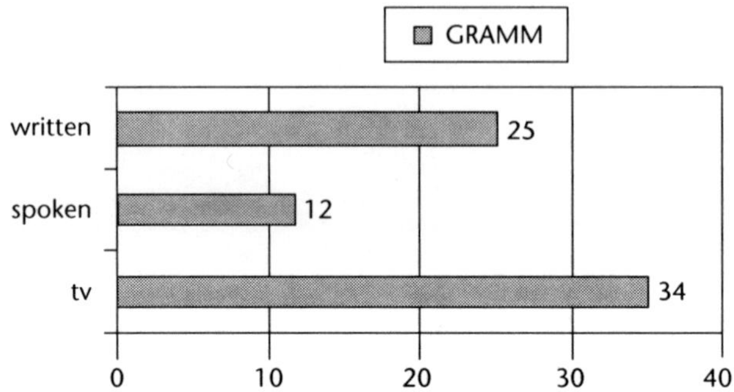

Fig. 9.10. Dislocations with grammaticalizations in the three corpora.

GRAMMATICALIZATION. The TV corpus contains the greatest number of grammaticalized dislocations (Fig. 9.10), with 9% more compared to WR and almost three times more compared to SP. Contrary to expectations, dislocations with grammaticalization are, based on these results, an unusual phenomenon in the spoken language (although they are basically fixed constructions), a very common phenomenon in television, and a phenomenon of intermediate frequency one in writing.

Pragmatic and Textual Parameters

As far as pragmatic functions are concerned, the data show a rather homogeneous trend among the three linguistic varieties (Table 9.6): the informational function prevails in all of them (mainly in the spoken language), but the expressive function is also well established (in this case writing has the highest values), whereas the control function is obviously present almost exclusively in the spoken language and at a low percentage.

Contrary to expectations, although with low differences, spoken language makes greater use of informational functions and written language makes greater use of expressive functions; again, the language of television occupies an intermediate position between the two traditional varieties.

Table 9.6. Pragmatic Functions Distribution in the Three Corpora (X 100)

	Informational	Expressive	Control
TV	55.5	43.5	0.9
SP	57.9	36.1	5.3
WR	52.7	47.3	0

Conclusions

The data and analyses briefly allow us to draw some conclusions on the use of dislocations in the three spoken, written, and broadcast varieties, starting with general frequencies of occurrence. Although the minor dimensions of the three corpora provide rather small numbers, the more deep-rooted presence of dislocations in speech than in writing appears to be clearly defined: Total occurrences are three times greater in SP than in WR. The hypothesis that dislocations are a typical spoken structure runs through all the literature on this phenomenon: with regard to English, Givón (2001) speaks of a "device restricted almost entirely to the informal spoken register" (266). Written Italian however noticeably turns to this syntactic device, in both the subvarieties included in the corpus (press and literature), undermining the assumption that it is exclusively typical of speech. The evidence more strictly connected to this study is the frequency of dislocations in the language of television, which, as mentioned earlier, is situated exactly half way between speech and writing; in television one turns to dislocations more often than writing but less often than speaking in the presence of the listener. This fact has a growing significance since it is strengthened by a range of other cases where TV has an intermediate position between SP and WR.

This study's assumption that television broadcast Italian cannot simply be identified with spoken Italian but needs to be distinguished as an independent variety is confirmed by data on a range of parameters in the use of dislocations that show striking differences between the two varieties (a one-way ANOVA test calculated for the following five parameters is statistically significant at $p<.0015$):

- Dislocated grammatical categories frequency. Apart from the prevalence of N, which is common to all the varieties, spoken Italian more frequently dislocates pronouns, whereas broadcast Italian more frequently dislocates clauses. They are sometimes relative clauses, but more often basically fixed expressions, depending on *sapere* 'know' or *vedere* 'see' verbs, as in Examples 7a and 7b. The language of television thus prefers dislocated sentences constituted by ready linguistic items, with a manifestly colloquial register as in Example 11;

- Dislocated pronouns types: speech dislocates most of all personal pronouns (mainly third-person pronouns), whereas the language of television dislocates mostly neuter pronouns, referred to an antecedent within the sentence. The spoken variety makes thus a prevalent use of pronouns' deictic function, referring to referents that are present in the communicative setting, and has therefore a more evident feature of interactivity; TV broadcast language uses more anaphoric pronouns, referring to something previously mentioned in the sentence, and has therefore a greater tendency to textuality. Two examples of this different use are Examples 14a and 14b:

(14a) ancora non l' ho aperta **quella tua**
'I still haven't opened yours' (referring to an envelop hold by one of the participants);

(14b) tutt' e due amiamo abbastanza a tal punto nostra figlia per non farla stare male e **di questo** LEI ne può stare certa
'both of us love our daughter so as not to hurt her and of this you can be certain' (referring to the fact, just mentioned, that 'both of us love our daughter').

- Subject frequency. The language of television, much more than spoken language, exploits the possibility of topicalizing two sentence elements: the one that is dislocated, and the subject, which is explicitly expressed in more than 44% cases (see Examples 8 and 10b);
- Passives/dislocations ratio. In contrast to speech and writing, TV broadcast language equally exploits two different devices to focus the sentence topic (or the center of interest): dislocations that are traditionally considered a speech specific feature; and passives, viewed as restricted to a written use;
- Grammaticalization. The language of television grammaticalizes the clitic pronoun within dislocations almost three times more than does canonical speech; the grammaticalized clitic loses its pronominal value to become part of the verb, often changing its meaning and developing basically fixed expressions as in Examples 4 or 10.

According to these data we can draw the conclusion that TV broadcast language, much more than canonical speech, tends to use ready linguistic segments of colloquial register (dislocations of CL and grammaticalizations), or double topic dynamic expressions, as well as more rigid forms like passives; and finally, within dislocated pronouns, it prefers the anaphoric to the deictic function, and therefore prefers textuality to interactivity. Together with these elements that distinguish it from typical spoken language, the language of television has important similarities to written language, for example in the dislocated pronouns typology (with a prevalence of neuter over personal pronouns), in the frequent use of grammaticalizations, and in the high number of dislocations with an expressive function. This last evidence also casts light on the extent and the typology of dislocations used in written language: they do not seem in fact an additional and scarcely used device, but a well-established structure that is used in writing mostly when a colloquial connotation close to speech is required. The strong presence of RD as well, which in the written corpus are more frequent than in the television and spoken ones, suggests a tendency of writing to become more like speech, making use of clearly colloquial devices.

This study is a first step towards a more accurate description of the language of television as an independent variety. The growing development of spoken new media genres such as podcasts and streaming internet video opens even further perspectives in a new definition of orality and in research on register variation in written, spoken, and broadcast means of production (see for example Hauesermann, this volume). Research questions concerning the linguistic specificity of spoken new media, or the difference among conversational structures in face-to-face interaction, television-mediated, and computer-mediated communication, are worthy of further investigation.

Notes

1. "*Mixed texts* is a label that can include three different meanings. *Mixed texts* can be defined those in which there is a combination of (a) forms typical of spoken and written language; (b) conversational devices; (c) fields of knowledge" (Dardano, 1994).

2. See also Luginbuehl (chap. 10, this volume), on the complex organization of television news.

3. Noun phrases are also typical of press language; Biber (2003) highlighted the frequency of this feature in newspaper prose and its internal complexity in comparison with noun phrases in other registers such as conversation or fiction.

4. Among the copious studies on dislocations, see Ashby (1988), Bally (1932), and Ziv (1994).

5. In Italian this clitic pronoun is not always mandatory in dislocations; in this study I have taken into account only clitic dislocations, for two different reasons: on the one hand because dislocations with clitics are by far the most frequent, as pointed out in the majority of research work, starting from Berruto (1985); on the other hand for methodological reasons connected to the use of corpora: the presence of clitics is in fact essential for the semi-automatic search of dislocations within the corpora, because clitics are the only searchable clues of dislocations.

6. For a detailed survey on previous studies on dislocations in Italian, see D'Achille (1990) and Rossi (1999); for the following remarks I refer particularly to Meriggi (1938), Gossen (1951), Duranti and Ochs (1979a, 1979b), Sornicola (1981, 2002), Berruto (1985, 1986), D'Achille (1990), Sala Gallini (1996), Rossi (1999, 2005), Messina (2002), and Zamora Muñoz (2002).

7. Passives have been computed considering the agent or agentless forms with the *essere* auxiliary 'to be'; forms with passive *si* and past participles without auxiliary has been excluded.

References

Accademia della Crusca. (1997). *Gli italiani trasmessi. La radio* [Broadcast Italians. The language of the radio]. Firenze: Accademia della Crusca.

Ashby, W. J. (1988). The syntax, pragmatics, and sociolinguistics of left- and right-dislocations in French. *Lingua, 75*, 203–229.

Bally, C. (1932). *Linguistique generale et linguistique francaise* [General linguistics and French linguistics]. Paris: Ernest Leroux.

Bazzanella, C. (1995). I segnali discorsivi [Discourse markers]. In L. Renzi, G. Salvi, & A. Cardinaletti (Eds.), *Grande grammatica italiana di consultazione* [Italian reference grammar] (pp. 225–257). Bologna: Il Mulino.

Bell, A. (1991). *The language of news media.* Oxford: Blackwell.

Berruto, G. (1985). "Dislocazioni a sinistra" e "grammatica" dell'italiano parlato [Left dislocations and spoken Italian grammar]. In A. Franchi De Bellis & L. Savoia (Eds.), *Sintassi e morfologia della lingua italiana d'uso. Teorie e applicazioni descrittive* [Syntax and morphology of Italian language usage. Theories and applications] (pp. 59–82). Roma: Bulzoni.

Berruto, G. (1986). Le dislocazioni a destra in italiano [Right dislocations in Italian]. In H. Stammerjohann (Ed.), *Tema-rema in italiano* [Theme-rheme in Italian] (pp. 55–70). Tübingen: Narr.

Biber, D. (1988). *Variation across speech and writing.* Cambridge: Cambridge University Press.

Biber, D. (2003). Compressed noun-phrase structures in newspaper discourse: The competing demands of popularization vs. economy. In J. Aitchison & D. M. Lewis (Eds.), *New media language* (pp. 169–181). London: Routledge.

Bozzone Costa, R. (1991). Tratti substandard nel parlato colloquiale [Substandard features in colloquial spoken language]. In C. Lavinio & A. Sobrero (Eds.), *La lingua degli studenti universitari* [The language of university students] (pp. 165–189). Firenze: La Nuova Italia.

Braun, P. (1998). Die Bilder und die Worter: Eine textlinguistische Annaherung an die Fussballsprache im Fernsehen [The images and the words: A textual approach to the language of football in television]. *Der Deutschunterricht 50*(2), 33–39.

D'Achille, P. (1990). *Sintassi del parlato e tradizione scritta della lingua italiana* [Syntax of spoken language and written tradition in Italian]. Roma: Bonacci.

Dardano, M. (1994). Testi misti [Mixed texts]. In T. De Mauro (Ed.), *Come parlano gli italiani* [How Italians speak] (pp. 145–181). Firenze: La Nuova Italia.

de Araujo, D. L. (2003). A língua falada na TV. Texto falado ou escrito? [Language spoken on TV. Spoken or written text?] *Linguagem & Ensino, 6*, 57–76.

De Mauro, T., Mancini, F., Vedovelli, M., & Voghera, M. (1993). *Lessico di frequenza dell'italiano parlato* [Lexicon of spoken Italian frequency]. Milano: ETAS libri.

Duranti, A., & Ochs, E. (1979a). "La pipa la fumi?" Uno studio sulla dislocazione a sinistra nelle conversazioni ["Do you smoke pipe?" A study on left dislocation in conversation]. *Studi di Grammatica Italiana, 7*, 269–301.

Duranti, A., & Ochs, E. (1979b). Left-dislocation in Italian conversation. In T. Givón (Ed.), *Syntax and semantics. Discourse and syntax* (pp. 377–416). New York: New York Academic Press.

Ennaji, M. (1995). A syntactico-semantic study of the language of news in Morocco. *International Journal of the Sociology of Language, 112*, 97–111.

Fairclough, N. (1995). *Media discourse.* London: Arnold.

Givón, T. (1976). Topic, pronoun, and grammatical agreement. In C. N. Li (Ed.), *Subject and topic* (pp. 149–188). London & New York: Academic Press.

Givón, T. (2001). *Syntax: An introduction.* Amsterdam & Philadelphia: John Benjamins.

Gossen, K. T. (1951). Quelques aspects de la mise en relief d'une idée en italien et en français [Emphasizing an idea in Italian and French]. *Zeitschrift für romanische Philologie, LXVII*, 147–166.

Goodman, S., & Manners, P. (1997). Making it "real": Words and pictures in television news. *Language and Communication, 17*, 53–66.

Herbert, J. (1997). The broadcast voice. *English Today, 13*, 18–23.

Herring, S. C., van Reenen, P., & Schøsler, L. (2000). *On textual parameters and older languages.* Amsterdam and Philadelphia: John Benjamins.

Ilie, C. (2001). Semi-institutional discourse: The case of talk shows. *Journal of Pragmatics, 33*, 209–254.

Marriott, S. (1997). The emergence of live television talk. *Text, 17*, 181–198.

Meriggi, P. (1938). La ripresa dell'oggetto in italiano [Direct object repetition in Italian]. *Volkstum und Kultur der Romanen, 11*, 1–30.

Messina, S. (2002). "Scrivere parlando" Alcune considerazioni sulle dislocazioni nel parlato seriale [Writing while speaking. Some remarks on dislocations in the spoken language of serials]. *Quaderni del Dipartimento di Scienze della Comunicazione—Università degli Studi di Salerno, 2*, 213–241.

Nacci. L. (2003). La lingua della televisione [The language of television]. In I. Bonomi, A. Masini, & S. Morgana (Eds.), *La lingua italiana e i mass media* [Italian language and mass-media] (pp. 67–92). Roma: Carocci.

Ong, W. SJ. (1982). *Orality and literacy: The technologizing of the word.* London/New York: Methuen.

Rossi, F. (1999). Non lo sai che ora è? Alcune considerazioni sull'intonazione e sul valore pragmatico degli enunciati con dislocazione a destra [You don't know what time it is? Some remarks on intonation and pragmatic value of utterances with right dislocation]. *Studi di Grammatica Italiana, 18*, 145–193.

Rossi, F. (2005). Tratti pragmatici e prosodici della dislocazione a destra nel parlato spontaneo [Pragmatic and prosodic features of right dislocation in spontaneous spoken language]. In E. Burr (Ed.), *Tradizione & Innovazione. Il parlato: teoria—corpora—linguistica dei corpora. Atti del VI Convegno SILFI* (pp. 307–321). Firenze: Franco Cesati.

Rossini Favretti, R. (1988). L'oralità del testo televisivo [Orality of television text]. *Studi italiani di linguistica teorica ed applicata, 17*, 203–218.

Sabatini, F. (1982). La comunicazione orale, scritta e trasmessa: la diversità del mezzo, della lingua e delle funzioni [Spoken, written and broadcast communication: Differences of medium, language and functions]. In A. M. Boccafurni & S. Serromani (Eds.), *Educazione linguistica nella scuola superiore* [Linguistic education in the high school] (pp. 103–127). Roma: Provincia di Roma—Istituto di Psicologia del CNR.

Sabatini, F. (1997). Prove per l'italiano «trasmesso» [Tests for broadcast Italian]. In *Gli italiani trasmessi. La radio* [Broadcast Italians. The radio] (pp. 11–30). Firenze: Accademia della Crusca.

Sala Gallini, M. (1996). Lo statuto del clitico nella dislocazione a destra: pronome vero o marca flessionale? [The status of clitics in right dislocation: True pronoun or inflectional marker?] *Archivio Glottologico Italiano, LXXXI*, 76–94.

Silverstone, R. (1991). Television, rhetoric, and the return of the unconscious in secondary oral culture. In B. Gronbeck, T. Farrell, & P. Soukup (Eds.), *Media, consciousness and culture* (pp. 147–159). Newbury Park, CA: Sage.

Skokut, N. (1993). An analysis of televised telephone conversations. *Issues and Developments in English and Applied Linguistics, 6*, 65–85.

Sobrero, A. (1997). Le telefonate in diretta: Struttura, scelte linguistiche e organizzazione conversazionale [Live telephone calls: structure, linguistic choices and conversational organization]. In *Gli italiani trasmessi. La radio* [Broadcast Italians. The radio] (pp. 505–524). Firenze: Accademia della Crusca.

Sornicola, R. (1981). *Sul parlato* [On spoken language]. Bologna: Il Mulino.

Sornicola, R. (2002). Crosslinguistic comparison and second language acquisition: An approach to topic and left-detachment constructions from the perspective of spoken language. In A. Giacalone Ramat (Ed.), *Typology and second language acquisition* (pp. 327–363). Berlin & New York: Mouton de Gruyter.

Spina, S. (2000). Che lingua si parla in televisione? Primi risultati di una ricerca [What kind of language is spoken on television? First results of a research study]. *Tuttitalia, 21*, 2–7.

Spina, S. (2005). Il Corpus di Italiano Televisivo (Cit): Struttura e annotazione [The corpus of television Italian: Structure and annotation]. In E. Burr (Ed.), *Tradizione & Innovazione. Il parlato: teoria—corpora—linguistica dei corpora. Atti del VI Convegno SILFI* (pp. 413–426). Firenze: Franco Cesati.

Swales, J. (1990). *Genre analysis*. Cambridge: Cambridge University Press.

Tannen, D. (1980). Spoken/written language and the oral/literate continuum. In *Proceedings of the Sixth Annual Meeting of the Berkeley Linguistics Society* (pp. 207–218). Berkeley, CA.

Wachtel, S. (1993). Zur Sprechkultur der Mattscheibe [The oral culture of television]. *Muttersprache, 103*, 342–352.

Zamora Muñoz, P. (2002). Dislocazioni a destra e a sinistra nell'italiano e nello spagnolo colloquiale parlato: Frequenza d'uso, funzioni e parametri linguistici [Right and left dislocations in spoken and colloquial Italian and Spanish: Frequency of use, functions and linguistic parameters]. *Studi Italiani di Linguistica Teorica e Applicata, 31*, 447–470.

Ziv, Y. (1994). Left and right dislocations: Discourse functions and anaphora. *Journal of Pragmatics, 22*, 629–645.

Chapter 10

Disclosing and Announcing, Interpreting and Entertaining

A Comparative Study of the History of TV News Presentation

MARTIN LUGINBÜHL

University of Zurich

Abstract

This chapter delivers a diachronic comparative study of the news presentation of an American and a European national TV news show (the American *CBS Evening News* and the Swiss *Tagesschau*). The shows, dating from 1949 to 2005, are analyzed within the linguistic framework of *text types*. Looking not only at the spoken text, but also at other semiotic resources of multimodal TV text like captions and graphics as well as the semiotics of studio design, important phases of the news presentation are presented.[1] The analysis illustrates on the one hand that there is a homogenization of the ways TV news is presented; on the other hand, I show that the forms have not become identical. Discussing the observed developments, I argue that the key site for understanding the form of news adequately is journalistic culture, understood as values and norms influencing the practices of the journalists. This culture again is influenced by a complex set of factors, commercialization being only one of them.

There probably is no doubt about the notion that the TV news anchor is the central person of the news show. Not only are they famous (especially in the United States) or at least well-known people (in the case of Switzerland), but they also have the most central role of all news people appearing on the screen: They open and close the show, read short news items, announce and wrap up correspondents' stories, introduce and link together new topics, and interview correspondents and experts.

In the literature at hand, the anchor's role is analyzed mostly in a synchronic way, thus giving the impression that this role is static or only changing in respect to some minor details (see Allan, 1998; Coupland, 2001; Landbeck, 1991; Morse, 1985; Oomen, 1985; Zelizer, 1990). The fact that national TV news largely covers the same events and that news is exchanged globally might easily lead to the assumption that there is a globalized TV news culture, with the same type of presentation all over the world.

In this chapter I present results of a comparative diachronic analysis of the *CBS Evening News* and its predecessor *Douglas Edwards with the News* on the one hand and the Swiss *Tagesschau* on the other. I describe important stages of the development of the anchor's and the newsreader's role by looking at the speech acts they perform.[2]

Additionally, I analyze the style of their presentation, integrating aspects of spoken and written language,[3] nonverbal communication, and pictures, as well as the semiotics of the studio. I am especially interested in the question of how the *presenters* (a term I use to refer to the anchors and newsreaders) address the audience; I also take a look at the question of the extent to which the presenters (or the over-the-shoulder graphics) give an interpretation of the reported event. In this regard I concentrate on passages where the anchor or the newsreader is greeting the audience, introducing or wrapping up a story, or linking stories.

The analysis here illustrates that the linguistic concept of text types (see Adamzik, 2004; Brinker, 2005; Fix, 2001) delivers a helpful framework for analyzing the forms of presentation in a detailed and inclusive way, especially by taking into account the multimodality of TV texts. I argue that there is a homogenization of the way in which TV news is presented. On the other hand, I also show that the forms are not identical and that there are cultural characteristics of the compared text types like lead-ins, greetings, and so on. Looking at the text types used to present news, different phases can be discerned that can be described by specific repertoires and styles of text types, thus characterizing the different formats of the shows and their way to (re-)present the news. Discussing the observed developments, I argue that the commercialization of the Swiss media system only delivers too simple and narrow an explanation for the changes that occurred. A key site for understanding the form of news adequately has to be the journalistic culture, understood as values and norms influencing journalistic practice. This culture again is influenced by a complex set of factors, commercialization being only one of them.

Corpus and Terminology

My analysis is based on 76 TV news shows dating from 1949 to 2005. In each decade, 1 week—or in case of a format change 2 weeks—was selected and the news shows of each station included in the corpus. The weeks were chosen according to an event that was reported over several days and extensively in both shows (e.g. the Soviet invasion of Czechoslovakia in 1968 or the Chernobyl nuclear accident in 1986).[4]

Some remarks regarding the terminology: I use the term *anchor* for a person who not only tells news but also introduces stories and connects them, organizes the news show, and holds interviews. The anchor leads the audience through the show.[5] A *newsreader*, by contrast, is the person who mainly reads news with hardly any personal involvement; he or she is a "mediator between the written word and the audience" (Oomen, 1985:596; see also Pufahl, 1988).

Regarding the genres or text types realized in the TV news shows I use the terms shown in Table 10.1 in this chapter.

Unlike the film item, in the case of a package the author of the package (usually the correspondent) is named. There are more text types in TV news, but I do not discuss them in this chapter.[6]

The concept of *multimodality* (Kress & van Leeuwen, 2001) can be used to relate text, image, and other semiotic resources within the expression of meaning. Kress and van Leeuwen pointed out that meaning can be expressed in "different semiotic modes" (1), whereby they understand *mode* as a semiotic resource that can be realized in different materializations (e.g., using written or spoken language, photography, or painting). Exponents of social semiotics prefer the term *semiotic resource* to terms like *code* or *sign* to stress that the semiotic resources used in texts only realize a meaning *potential* (Stöckl, 2004; van Leeuwen, 2005). But how a

Table 10.1. Text Types Analyzed

Headline, greeting, good-bye, lead-in, lead-out	Utterances of the presenter, e.g., to introduce ("lead-in") or to close a story ("lead-out")
Voice-over	Newsreader or anchor on screen reads the first part of a story, after which full-frame news footage can be seen
Anchor item/ Newsreader item	Story read by the anchor or the newsreader without news footage
Film item	Story read by the anchor or the newsreader with full-frame news footage
Package	Produced and told by a correspondent, and consisting of news footage, often with soundbites and sometimes with a correspondent on the screen (see White et al., 1984)

single individual actually will read a text is not predictable because it depends on a complex of different factors, static as well as situational. But as Hall (1980) argued, TV texts (as other media texts) usually stress a certain interpretation, a "preferred reading," usually fitting mainstream ideology. If, for example, a Swiss anchor announces all stories about the conflict between NATO and Yugoslavia in the show of March 24, 1999 in connection with the word "countdown," the entire situation is placed in a certain frame, stressing some features of this situation (the time remaining before the attack) and toning down others. Analyzing TV news texts, this preferred reading can be described—knowing that the audience also can develop "negotiated" or "oppositional" readings.

Analysis

CBS Evening News 1949: "Douglas Edwards saying good night, for Oldsmobile"[7]

In the *CBS News with Douglas Edwards* of April 7, 1949, Edwards was more of a newsreader than an anchor. Immediately following the credits—constituting a mixture of credits and an Oldsmobile commercial—Edwards greets the audience ("Good evening everybody,"; for transcript guidelines see footnote 5) announces the headlines ("Here is the news picture tonight,") and then reads them. Doing that, he is looking most of the time directly into the camera (i.e., to the audience), but occasionally glances to his left, where he seems to read the text. The headlines consist of complete sentences and always begin by naming the actor ("the administration announces," "the United Nations has voted," "the Athens government says" etc.). The headlines refer, not in every case, to long stories in the upcoming show, and not all long stories (such as a film item about unrest in Iceland or another about a military parade in Washington) are mentioned. The mentioned stories move from domestic to foreign policy and then to a crime topic. Having a good mix of stories in the headlines seems to be the deciding factor in a story's inclusion, rather than story length. After the headlines, he announces the first commercial break, addressing the audience directly: "Well, and just a moment, the rest of the story, and now a message from our sponsor, your Oldsmobile dealer. [Music of the commercial starts]." (Also note that "our" refers to the journalists, excluding the audience).

In the main part of the show, different stories are strung together. The beginning of a new story is often marked with a short gap and the discourse marker "well" ("Well it was a very big day in Washington"), by locating the reported event ("From China films of," "From Japan meantime a progress report," "at the United Nations today") or by using the formula "we have some" ("We have some conflicting reports from Greece . . .," "Well now we have some very dramatic pictures . . ."). Sometimes the transition between two stories is also marked by a glance down at the desk.

But the transitions between two stories always are quite rapid and sometimes it is difficult to discern that a new story has started ([End of story about European recovery bill:] "The final vote is still in {draft?}. The state department announces that the farm ministers of France, Britain, and our own Secretary of State have reached . . ." [story about Western Germany continues]). It only becomes clear by the end of the second sentence that a new story has begun; it could be that the state department announced something regarding the recovery bill. This is also owing to the fact that there is no glance down at the desk and no camera cut. Only rarely is the newsreader linking two stories together, be it by the time of the event ("Again today, . . ."), the topic ("other news concerning inter-American relations") or the "mood" of the story ("And now right in the same mood we have some new pictures just received of eh/ bonny prince Charlie, the future king of England, . . .").

The studio (Fig. 10.1) suggests a journalist's workplace. The newsreader is sitting at a desk with a typewriter and several piles of paper. There is also a microphone on the desk with the letters "CBS." On the back wall there is a barely recognizable map, above it large white letters reading "CBS-TV NEWS."

The studio pretends to show the working desk where the news is produced. You can see the place where the news comes from, and you know who is presenting it and where his desk is ("here in New York"). Person, place, and time are seemingly identified. The studio is to some degree loaded with what Coupland (2001) called "semiotic gravitas": The anchor is sitting behind a big desk, dressed formally, and speaking in a solemn and resolute tone. This helps "to create the impression of personal integrity and trustworthiness" (Allan, 1998:125), and it may help to "ratify the authenticity of the newsreader's own commitment to upholding the truth-value of the newscast" (125).

Fig. 10.1. Douglas Edwards in *Douglas Edwards with the News*, April 7, 1949 (reproduced with permission of CBS).

In summary, the main function of this presenter is that of a newsreader. There are at most very rudimentary lead-ins or linkings. The audience is addressed verbally at the beginning and the end (greeting and goodbye; "this is Douglas Edwards saying good NIGHT, for Oldsmobile"); in newsreader items the newsreader is looking to the audience most of the time. The only exception is the "little late news released just a few minutes ago in Washington," where the newsreader is looking down at his papers. This kind of presentation of course stresses (and partly stages) the actuality of the story presented.

Although the role of the presenter is mostly that of a newsreader, his personality is important: His name appears in the show's title and he mentions it in the sign-off; and he is most of the time looking directly at the camera, hence at the audience, thus staging a face-to-face (FTF) communication. This "intimacy" of communication is also stressed by the fact that the newsreader only can be seen in medium close and close shots, so the audience mostly sees the upper part of his body, but always his face. His contribution to the orientation of the audience within the show is restricted to the headlines and the announcements preceding the two commercial breaks. The semiotics of the studio setting stresses the integrity and trustworthiness of the show as well as the place where the news shows are produced.

Tagesschau 1958

There was no visible presenter in the Swiss *Tagesschau* of 1958. This is due to infrastructural circumstances: The show had to be broadcast in three languages (German, French, Italian), but as the show did not have its own studio, the texts had to be read at the same time during the transmission. In this format, it remains unclear who is reading the news and where it is produced. Although the voice of the newsreader may become familiar, it remains an anonymous, unidentified person disclosing the news.[8]

CBS Evening News 1968: "And that's the way it is"

In 1960 the ratings of the show fell behind the ratings of *The Huntley-Brinkley Report*, its competitor at NBC; thus a new format was developed. Then producer-director Don Hewitt said the show needed "a good, hard kick in the ass" (cited in Gates, 1978:91) and began to develop a new format. In April 1962, Walter Cronkite became presenter of the CBS news show (Gates, 1978), and in 1963 a new format, the *CBS Evening News with Walter Cronkite* was introduced with a "newsroom-studio" concept (Gates, 1978). In addition to reading anchor items, the presenter, Walter Cronkite, greets the audience at the beginning and the end; the formulations used are always the same: "Good Evening," "And that's the way it is, [date named, e.g., Friday August 23rd 1968], this is Walter Cronkite,

CBS News reporting from New York, good night." The anchor thus realized an important ritual to mark the beginning and the end of the show.[9] Sometimes he announces an upcoming story or another show, or interviews someone (although not in the studio).

The new important function of the presenter in this format is to lead in packages. There are other persons telling news (e.g., the correspondents), but they do not appear "omniscient" like the anchor, and they are not able to organize the show. Thus, the anchor's position in the show becomes central: The anchor reads or introduces all stories, is visually omnipresent, knows the rundown of the entire show, and seems to be familiar with every news story.

Leading in a story, the anchor summarizes the central facts (usually following the five W questions: *where, when, what, who, how*) and announces the correspondent who "has the story":

> Near Tay Ninh, US and South Vietnamese troops continue their hunt today—for a communist force which occupied THAT city for THIRTY HOURS—and then retreated into the bush. CBS news reporter George Syvertsen reports on that enemy attack. (Aug. 20, 1968)

> At the UN today the United States and six other countries introduced a security council resolution that would condemn the Soviet Union and demand its immediate withdrawal from Czechoslovakia. We have a report from CBS news correspondent Richard C. Hottelet. (Aug. 22, 1968)

> Demonstrators here[10] in Chicago—who claim the United Nation should be overseeing their runnings with Chicago police, now are charging police bruTALIty after a running clash last night. And new protests were staged today. Jack Laurence has the story. (Aug. 27, 1968)

It is typical for lead-ins to contain a frame interpreting the reported event. In the lead-in from August 20, 1968, the anchor uses the term *hunt* to describe the actions of the army. Using the metaphor of a *hunt* implicates other aspects belonging to this concept: where there is a hunt, there must be a hunter and a hunted, the hunt can be successful or not, and so on. This kind of framing (see Johnson-Cartee, 2005) can often be found in lead-ins, but less so in the packages themselves (at least at this time). So the anchor not only introduces the topic and gives the main information, but also interprets and often evaluates the event by framing the stories.

A slightly different way of leading in the stories is to announce the broader topic but then only hint at the event that will be reported in the following package:

> His party may be struggling over its Vietnam platform plank, but President Johnson has made it quite plain how he stands on the war. He spoke last night

to a convention of the veterans of foreign wars in Detroit. Not everybody appreciated his visit. As Robert Pierpoint reports. (Aug. 20, 1968)

And listen to those National Guard troops we mentioned earlier, the *Chicago Daily News* reports that SEVENthousandFIVEhundred regular army troops are training for possible Chicago duty next week. One of several reasons for official precautions next week—is evident in the following report from Ike Pappas. (Aug. 20, 1968)

Pope Paul today went to the people of Colombia, a trip that was both significant and in a way—disappointing. We have a satellite report from CBS News correspondent Burt Quint. (Aug. 23, 1968)

In the first lead-in it may be clear to an informed audience of that time how President Johnson stands on the war—but it remains unclear who did not appreciate his visit. Also in the second example it only becomes clear that some information will follow about one reason for the precautions by the police—which reason this is, is not said. The third lead-in leaves the question open of why the Pope's trip was disappointing. This form of lead-in is less redundant and obviously aims at building up tension by not giving all central information. This kind of orientation towards the attention of the audience can also be seen in the request to the audience at the beginning of the second lead-in ("And listen to," similar in another lead-in of the same day: "And listen to this . . .").

The studio was in the newsroom. At the beginning of every show the anchor could be seen from the right side sitting at a desk. To his left there were several telephones and news ticker machines in front of a big map of the world (see Fig. 10.2). The sound of the news tickers and/or typewriters could be heard when the anchor was on screen. During the lead-ins and the anchor items, the anchor could be seen from the front, looking into the camera. There were two different camera

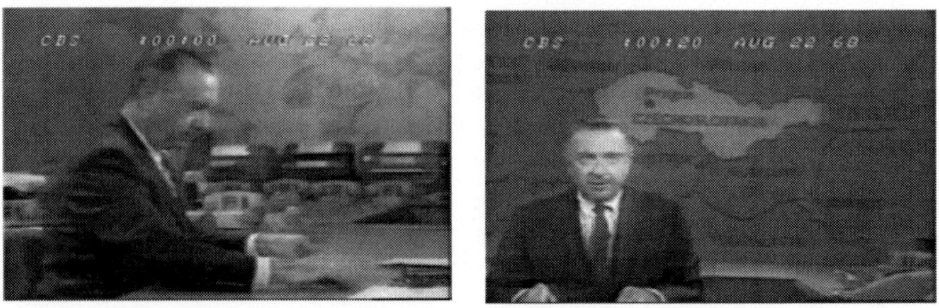

Fig. 10.2. Walter Cronkite in the show of August 22, 1968 (reproduced with permission of CBS).

positions. In the first, used for lead-ins and anchor items without an illustrating picture in the background, some outlines of newsroom devices and a world map on the wall were recognizable behind the anchor. In the other, there sometimes was part of a world map visible, sometimes an illustrating picture (mostly a portrayal) or a map corresponding to the main actor or place of the event, thus emphasizing this main actor or the location of the event.

Again, the anchor is located in the place where the news show is produced. And as before, the anchor is dressed formally, but his desk has become even bigger and there are several devices that symbolize his connection to news sources. The fact that the news ticker machines and typewriters can be heard is important: the news picture presented can change every moment, the flow of news never runs dry; the news show is an intermediate product of an everlasting process of investigation and news gathering.

In this format, the presenter is an anchor, leading in stories by summarizing them, sometimes only hinting at their content. Often he also interprets the reported events by using framing metaphors (like "hunt" for war operations). The audience is addressed directly not only in the lead-ins; the growing orientation toward the audience can also be seen in the fact that the anchor sometimes builds up tension. The presentation of the news has become more complex given the fact that graphics can be added to the lead-ins; these mostly stress the main actor or the location of the event. The studio—aside from its seriousness—stresses the actuality and the production of the news.

Tagesschau 1968: "Here is the Tagesschau" ("Hier ist die Tagesschau")[11]

The presenter in the Swiss *Tagesschau* of 1968—introduced in 1966, when the show got its own studio—mainly has the function of a newsreader. After opening the show and announcing the headlines ("Here is the Tagesschau. First the most important news in headlines"; "Hier ist die Tagesschau. Zunächst die wichtigsten Meldungen in Schlagzeilen." Aug. 20, 1968), he reads the headlines over still pictures. The headlines consist of one complete sentence each (e.g., "Dramatic loss of life due to major material damage is the result today by the fighting in South Vietnam"; "Sehr schwere Menschenverluste durch grosse Sachschäden brachten heute die Kämpfe in Südvietnam." Aug. 20, 1968), announcing the main stories from the sections international news, national news, and the last section with news about criminal, cultural, or technical topics. The word order is marked, and the keyword of the story can usually be found in the focus position of the sentence (e.g., "Still critical is the condition of former U.S. president Eisenhower."; "Weiterhin kritisch ist der Zustand des früheren U.S.-Präsidenten Eisenhower." Aug. 20, 1968) instead of the unmarked word order "The condition of former president Eisenhower is still critical"). The existence and the form of these headlines are almost the only obvious signs of an intention to make the show compelling for the audience.

On the average there are only two lead-ins in one show. Most of the news items are strung together without lead-in. Lead-ins can be found together with film items consisting of one long sound bite; in the lead-in the person speaking is named and the context explained (the German original is followed by a translation):

> In wenigen Stunden wird in Chicago der mit Spannung erwartete Konvent der Demokratischen Partei eröffnet, der den demokratischen Präsidentschaftskandidaten bestimmen wird. Der amerikanische Vizepräsident Hubert Humphrey gab sich optimistisch, als er gestern in Chicago eintraf. (Aug. 26, 1968)

> In Chicago in just a few hours the long awaited convention of the Democratic Party, which will name the democratic presidential candidate, will be opened. The American vice president Hubert Humphrey was optimistic when he arrived in Chicago yesterday.

Other lead-ins contain information not given in the following story—so they are more than just a summary of the related news item:

> Mit einem Tumult ging die erste Sitzung des gestern abend in Chikago eröffneten Nationalkonvents der amerikanischen Demokratischen Partei vorzeitig zu Ende. Der Kongress hatte einen Streit unter den Delegationen dreier Bundesstaaten zu schlichten, von denen immer zwei Gruppen Anspruch darauf erhoben, die rechtmässigen Vertreter ihres Staates zu sein. Heute abend—etwa um Mitternacht Schweizer Zeit—tritt der Konvent zur zweiten Sitzung zusammen. (Aug. 27, 1968)

> The first meeting of the national convention of the American Democratic Party, which opened yesterday evening in Chicago, ended with a tumult. The congress had a dispute to settle between the delegations of three states, of which two groups in each case claimed to be the legitimate representative of their state. This evening—about midnight Swiss time—the convention will convene for the second meeting.

This lead-in of the following film item also could be an independent anchor item, followed by a story about another topic. There is no announcement of this film item. Nevertheless it also introduces the topic of the item. But the audience never knows in advance whether a film item following such an item is about a different topic. Only in rare cases is the content of a film item summarized (and in rather complex syntactic structures):

> Die zahlreichen und meist von vielen Tausend Menschen besuchten gestrigen Demonstrationen in der Schweiz verliefen überall sehr eindrucksvoll und ruhig. Zu schweren Zusammenstössen kam es allein vor der sowjetischen Botschaft in Bern. (Aug. 23, 1968)

> There were numerous demonstrations yesterday in Switzerland, some of them attended by several thousand persons. All of them have passed off very impressively and quietly. Severe riots occurred only in front of the Soviet embassy in Berne.

This announcement follows a film item about the mentioned demonstration and the riots in Berne.

The *Tagesschau* newsreader is present throughout the entire show—nevertheless the audience orientation is much more distanced that in the *CBS Evening News* of the same week. The *Tagesschau* newsreader is hardly looking into the camera, but usually at the manuscript in front of him. His inflection is very monotonous, not showing any involvement in the text read and thus suggesting impartiality; there are fewer lead-ins, no overall framings of news stories, and no wordplays. A maximum of detachment is staged; in one entire week the audience is only addressed four times (besides the greetings), mostly to lead-in the weather forecast ("And now you will see . . ."; "Sehen Sie nun . . ."). There are evaluations (which conform to the bias of the sphere of consensus [see Hallin, 1994] like the "impressive and quiet" ["eindrucksvoll und ruhig"] demonstrations), but they are more seldom and more discreet than the ones realized in the CBS show.

The fact that the newsreader reads from the manuscript is not only due to technical equipment; Douglas Edwards also did not have an autocue—yet he did not read from a visible manuscript. Also, in the late 1970s, before the autocue was introduced to the Swiss *Tagesschau*, the newsreader tried to look as much as possible into the camera. The autocue then just perfected the eye contact to the audience. As the newsreader's text and the way it is pronounced is not archived (see footnote 1), I can only conclude from an archived show of July 11, 1968 that the text is articulated clearly, but also prosodically flat. This fits the detached way of presenting the news.

The *Tagesschau* studio design is simple (Fig. 10.3). The newsreader can always and only be seen in a medium close shot in front of a world map. He is seated,

Fig. 10.3. Léon Huber in the *Tagesschau* of October 11, 1968 (reproduced with permission of SF).

but the desk or table at which he is seated remains invisible. While in the CBS newsroom studio the production of the news is indicated, the studio's location named and the anchor identified, the Swiss newsreader seems to be reporting from a virtual place and his name is not mentioned. More important seems to be the written word, which seems to declare an absolute truth. I will call this kind of studio design the "studio set."

The presenter of the 1968 *Tagesschau* format is a newsreader, often leading in stories only when they consist of a long soundbite or, if important information is not given, in the film item. The style of the presentation is detached, lacking any kind of evaluation (except for some discreet evaluations according to the sphere of consensus, e.g., that the communists' "invasion" is bad[12]). Also the studio setting fits this staging: Neither the production process nor the location of the studio is recognizable.

CBS 1986: Storytelling

In 1981, Dan Rather became anchor of the *CBS Evening News*. As the ratings dropped in the following months, the new president of CBS News, Van Gordon Sauter, started to change the content of the show, advocating a soft news approach ("We emphasized storytelling, both verbally and visually"; cited in Hamilton, 2004:165). Rentability became important; in 1985 10% of the CBS news staff was cut (Barkin, 2003; Hallin, 1994; Hamilton, 2004). On the other hand, the fact that a lot of money was spent in technical equipment to enhance the graphic possibilities in the show (Ludes, 2001) speaks of a "graphication" (Zettl, 1989, cited in Johnson-Cartee, 2005) of TV news.

As in 1968, in the shows analyzed from 1986 the anchor greets the audience, introduces every package, and reads the text of film items and occasionally of anchor items. So, although these functions and duties of the anchor remained more or less the same, the style changed. Specifically, most of the lead-ins have become longer, not only summarizing the central facts of the following story, but also giving additional information, which is not repeated or elaborated in the package. Here is a typical example from April 30, 1986:

> Anchor [straight on medium close shot, graphic over his right shoulder showing an atom symbol with hammer and sickle in the middle and the caption "Nuclear Accident"]: Even as the Soviets stuck to their official story of only two dead,—no more fire,—no more radiation, Moscow did acknowledge for the first time today that there were what is called 197 people injured and hospitalized. In addition.to the regular satellite photographs CBS news tonight obtained—ENHANCED satellite photos—showing what appear to be—TWO nuclear reactor hot spots—where the Soviets insist there are NONE. Defense Department correspondent David Martin reports.

The following package comments on the satellite photographs and the knowledge of U.S. officials about killed and injured people. This example not only illustrates the fact that the lead-ins have become more than simple summaries, it also illustrates how the texts have become designed to be more dynamic. Although the sentences in the lead-ins of 1968 were shorter and less complex (i.e., with fewer subordinate clauses, clearly marked sentence ends by intonation, etc.), the lead-ins of 1986 usually are read in a fast-paced manner, often without short gaps or gaps at unusual positions in the sentences ("CBS news tonight obtained—ENHANCED satellite photos—showing what . . ."). Also the intonation does not always follow the sentence structure ("In addition.to the regular . . ."). This gives the news an agitated, even hounded atmosphere, stressing the actuality of the news, as well as its changeability. Already after the introducing "Good evening" there is no falling intonation and no gap, but a speedy transition to the first news item.

The style of the lead-ins from 1986 is also different from that in 1968 when it comes to the multimodal character of the text. More lead-ins—four-fifths—include an over-the-shoulder graphic of the anchor. Although photographs and maps could be seen in 1968, computer-generated pictures, some of them animated (zoom-in and zoom-out), became very popular by 1986. A very prominent graphic during the coverage of the nuclear accident was a kind of "logo" for this coverage, showing an atom symbol with hammer and sickle in the middle and the caption "Nuclear Accident" (see Fig. 10.4).

This type of graphic helps to organize a "cluster text." Püschel (1992) defined *cluster texts* as the sum of texts in one show that relate to the coverage of the same event. Cluster texts already existed in the *CBS Evening News* of the 1960s, but there was no recognizable organization; the single stories were just strung together. In the 1986 shows, the beginning of the coverage is explicitly marked ("Our coverage of the nuclear cloud begins with Wyatt Andrews in Moscow," April 28, 1986; "Our correspondent reports begin with Wyatt Andrews in Moscow," April 29, 1986). Organizing the cluster, again the multimodality of the TV news text becomes important (April 30, 1986, see Table 10.2).

Again, the content is not summarized, but the topics to be reported become clear—and tension is built up by posing questions. Also note—and this is different from the 1960s—the dramatic and emotional formulation "almost incredible nuclear disaster." Although the text formulates questions, the captions state the upcoming topics directly, in one case suggesting an answer to the question ("What happens to the people"—"Nuclear Sickness"), again dramatizing the topic.

Although some lead-ins, as mentioned earlier, include more information than just a summary, there are more lead-ins now only giving a hint at the content of an upcoming story. Although this kind of lead-in almost without exception could be found in the 1960s with "soft news" topics, it now is a common way to lead in all kinds of stories. Often this is realized by posing a question ("Now, is there a potential Chernobyl-type disaster in America's own future? Correspondent Bob Schieffer has been investigating THAT." April 29, 1986).

Most of the time the anchor can be seen in a close shot (or in a medium close shot in the case of an over the shoulder graphic) in front of a grey world

Table 10.2. Transcript of April 30, 1986

Time	Text	Footage
0:00	Anchor on[a]: In a moment we'll continue our extended worldwide look tonight at this still unfolding, almost incredible nuclear disaster.	Close shot: Anchor on the left side, over his right shoulder 'logo' of the accident with atom symbol, hammer and sickle in the middle and caption "Nuclear Accident"
0:07	Anchor off: What does it take to douse the fire of a nuclear reactor	Full-frame graphic: blue-black background, in the middle the 'accident logo.' The logo moves to the upper left corner, to the right of the 'logo' the caption "Nuclear Accident" appears. Under this caption the following
0:11	gone awry? What happens to the people	captions appear one after the other:
	and the land closest to the source of the	'accident logo', caption "Nuclear Fire" on a red bar
0:13	spewing radiation in the Soviet Ukraine? AND	'accident logo', caption "Nuclear Sickeness" on a red bar
0:18	what happens when and if the fallout reaches the skies over the United STATES?	'accident logo,' caption "Nuclear Fallout" on a red bar.
0:24		

[a] I mark the speaking person with "on" when he or she can be seen on screen, I speak of an "off"-voice when the speaking person cannot be seen.

Fig. 10.4. Dan Rather in the show of April 30, 1986 (reproduced with permission of CBS).

map. Only at the beginning of the show when the opening credits are inserted to the left of the anchor can the anchor's desk be seen. This is also the case when a commentator is in the studio and the anchor gives him the word. Here you also can see a back view of the anchor. This format is more anchor-centered than in 1968 or in 2005: The anchor reads or announces every news item, and a close relationship to the audience is staged. This central position is also expressed in the fact that the anchor can be seen in an anchor studio set with no traces of the news production process.

The style of the news presentation of 1986 is faster and more dynamic, and its multimodal character is more complex than in 1968. Also the content of the lead-ins is more variable (summary, additional info, hints, etc.) and often has a tendency to dramatize the reported event. The anchor also organizes the entire show. There is no newsroom studio, but rather, the anchor is shown in close shot dominating the entire show.

Tagesschau 1986: "More Immediate and More Lively" ("direkter, unmittelbarer, lebendiger")

The format of the Swiss *Tagesschau* changed drastically by January 1980. Instead of a newsreader, an anchor was presenting the news. He was seated in an armchair, leading in and out stories as well as linking them and commenting on them. A newsreader sometimes was presenting a so-called "studio report" ("Studiobericht"), a series of stories about a certain topic. The official aim of this format was more analysis, more background, and more national news (SF, January 1, 1980).

This format was not successful; it was controversially debated in public and was replaced by another format after just 1 year (see Luginbühl, 2005). Nevertheless, the show with just one newsreader proclaiming the news in a detached manner disappeared forever. In the new format, introduced in 1981, the anchor no longer commented on the stories and was limited to leading them in.

In 1985, the *Tagesschau* was redesigned again. The aim of the new show was to be "more direct, more immediate, and more lively" ("direkter, unmittelbarer, lebendiger"), as the then program chief Ulrich Kündig said (SF, January 1, 1985). In the middle of an editorial office there were now two desks with an anchor, a newsreader, and a sport journalist. The anchor realized about eight lead-ins each show with an average length of 24 seconds (1968: 19 seconds). The lead-ins announced the longer stories (packages and longer film items); the shorter stories were realized without lead-in. Often the lead-ins did not just summarize the story but gave a short introduction to the topic. And this happened often in a quite didactic style of explanation:

> Obgleich die Radioaktivität in der Schweiz nicht weiter angestiegen ist, wurde heute die Nationale Alarmzentrale in Betrieb genommen. Dieser Schritt ist so

zu verstehen: in der Ostschweiz haben wir zur Zeit acht mal höhere Werte an Radioaktivität gegenüber normalen Zeiten, während in den übrigen Landesteilen die entsprechenden Werte auf zwei bis fünf mal höher liegen als üblich. Diese Werte wären an sich nicht alarmierend, aber anderseits wurde die Untersuchung von Gras und Milch angeordnet, und dort hat man kleine Strahlungswerte festgestellt. Die Daten dieser Untersuchungen müssen nun systematische erfasst werden, und genau diesem Zweck dient die heute in Kraft gesetzte Nationale Alarmzentrale. Charles Clerc berichtet. (May 2, 1986)

Although the radioactivity in Switzerland did not rise, the National Alarm Center has been put into operation today. This step has to be understood as follows: in the eastern part of Switzerland we have levels of radioactivity eight times higher than usual, while in the other parts of Switzerland the respective levels are two to five times higher than usual. These levels would not be alarming in and of themselves, except for the fact that the analysis of grass and milk has been ordered, and small levels of radiation have been detected there. The data of these investigations have to be collected now systematically, and this is exactly the aim of the National Alarm Center that opened today. Charles Clerc is reporting.

The anchor here not only announces the topic of the next story, but also explains why the institution in question is necessary and how the reported event can be understood. Thus, he delivers an explanation as well as assistance to understand the event. This kind of didactic explanation can also be found in lead-outs (May 2, 1986):

Gewiss hat die Flüchtlingsbewegung aus der Dritten Welt zugenommen, aber eine Zahl sollten wir uns dennoch vergegenwärtigen: 90 Prozent der Flüchtlinge halten sich nach wie vor in Ländern der Dritten Welt selbst auf, während 10 Prozent in Westeuropa oder in den USA um Asyl nachgesucht haben. (Gy wendet sich zu Huber)

Of course the refugee movement from the Third World did grow immensely in the last few years, but one NUMBER we have to keep in mind: 90 percent of the refugees still stay in countries of the Third World itself, while 10 percent have applied for asylum in Western Europe or in the USA. (Gy turns to Huber)

Explanations like this often come together with the accentuation of the investigative activities of the journalistic staff themselves ("Christoph Häberlin inquired about the safety measures in Swiss atomic power plants in Leibstadt {location of a Swiss atomic power plant}"; "Ueber die Sicherheitsmassnahmen bei Schweizer Kernkraftwerken hat sich Christoph Haeberlin in Leibstadt erkundigt," April 29, 1986; "Hansjörg Erny surveyed in Berne:" ("Hansjörg Erny hörte sich in Bern um") April 30, 1986; "We asked how long the current radiation will be going on and which consequences it has. Madeleine Hirsiger interviewed the radiation biologist Werner Burkhard of the Federal Institute of Reactor Research." "Wir sind der Frage nachgegangen, wie lange sich die gegenwärtige Strahlung auswirken kann und welche Folgen sie hat. Madeleine Hirsiger befragte den Strahlenbiologen Werner Burkhard vom Eidgenössischen Institut für Reaktorforschung," May 1, 1986) and the use of "we" including the audience ("And the Russians still don't bring themselves to an open information policy, to that of the foreign countries, to which we are also entitled"; "Und die Russen ringen sich immer noch nicht dazu durch, eine offene Informationspolitik zu führen, auf die auch das Ausland, auf die auch wir ein Anrecht haben." May 2, 1986; "Since the 50s and the 60s . . . not only have the measuring instruments been improved considerably, we also have become aware of health dangers caused by environmental impact"; "Seit den 50er und 60er Jahren . . . sind nicht nur die Messinstrumente erheblich verbessert worden; wir sind auch hellhörig geworden für Gefährdungen der Gesundheit durch Umwelteinflüsse." May 1, 1986). New is also the use of the "we" referring to the editorial staff (as in "Everything we know about the odds later in this show—as usual—in the sport section"; "Alles was wir über die Gewinnquoten wissen, später in dieser Sendung—wie gewohnt—im Sportteil." April 28, 1986). Besides this kind of lead-in there still are a lot of short lead-ins summarizing the reported event in the form of the inverted pyramid. The lead-ins of regular film items often look like anchor items; that is, without any indication that more on this topic will follow, the lead-ins of packages mostly name the respective correspondent.

Although the didactic explanations are typical for some old stories of the 1950s and lead-ins of the 1980s, another feature that appeared in the 1980s remains today: a more casual language use. This can be seen in the growing variety of formulations used to greet the audience at the end of the show ("Have a nice evening and goodbye"; "Schönen Abend und auf Wiedersehen," April 26, 1986; "Have a pleasant evening. Goodbye"; "Angenehmen Abend. Auf Wiedersehen," April 30, 1986; "I wish you a good evening and bid you goodbye"; "Ich sage Ihnen guten Abend und auf Wiedersehen," May 2, 1986; "Have a pleasant evening. Goodbye"; "Ich wünsche Ihnen einen vergnügten Abend. Auf Wiedersehen," Saturday, May 3, 1986; "Have a good start of the week. Goodbye"; "Ich wünsche Ihnen einen guten Wochenbeginn. Auf Wiedersehen," Sunday, May 4, 1986). In addition to the greetings and goodbyes, the audience still is hardly ever directly addressed (as in the exceptional case of May 2, 1986: "We'll come to the weather forecast in a few moments—but you'll first see . . ."; "Zum Wetterbericht kommen wir in einigen Augenblicken—doch sehen Sie zunächst noch . . .").

Additionally, the anchor sometimes summarizes the main events by bringing them together in a comical metaphor ("Sunday was election day, Sunday was lotto day. Payday for politicians and players—with open questions until today"; "Der Sonntag war Wahltag, der Sonntag war Lottotag. Zahltag für Politiker und

Spieler—mit offenen Fragen bis heute," April 28, 1986). Regularly the lead-ins and linkings of soft news stories make use of wordplays or are held in a casual tone (after a story about some physical training of then U.S. president Reagan: "He who remains fit will resist all storms. The weather here was inhospitable."; "Wer fit bleibt, trotzt allen Stürmen. Das Wetter bei uns war unwirtlich," April 28, 1986; leading in a story about stud tires "Because the weather is like it is, . . ."; "Weil das Wetter ist, wie es ist, . . ." April 29, 1986). The trend to a more casual way of presenting the news also becomes visible in the way the presenters in the studio address each other. Not only does the anchor address the other presenters with their names ("Also in the news section, Leon Huber {= newsreader in the studio, M.L.}, next up: about the consequences of the reactor accident in Chernobyl"; "Auch im Nachrichtenteil, Leon Huber, geht es zunächst um die Folgen des Reaktorunfalls in Tschernobyl"; "Upcoming in the sport news, Thomas von Grüningen {= sport journalist in the studio}, a tragic story"; "Beim Sport, Thomas von Grüningen, steht eine tragische Meldung an der Spitze.") or by using nonverbal signs (see the note in the manuscript cited above: "Gy {= abbr. for the anchor's name Gysling} turns to Huber"), the manuscript also gives evidence of planned improvisation when it reads after an anchor item about an avalanche: "(improvisational question to the weatherman)"; "(improv. Frage an Wetterfrosch)" (all quotes from May 2, 1986).

There are more different text types in the new format, more topics addressed, and a rather complex structure of the news format; thus the anchor also spends more time in organizing the entire show. This becomes especially evident in metacommunicative utterances regarding the main sections of the show (international news, national news, sports, weather): "We now come to the world news in the Tagesschau"; "Wir kommen zu den Auslandnachrichten in der Tagesschau" (April 30, 1986) or "A summary of other national news follows"; "Es folgt eine Zusammenfassung weiterer Nachrichten aus dem Inland" (April 26, 1986).

Fig. 10.5. Robert Ruoff, Léon Huber, and Matthias Hüppi in the show of November 24, 1986 (reproduced with permission of SF).

The studio was in the newsroom: The anchor, the newsreader, and the sport journalist were sitting at one curved desk, and the newsroom with additional desks could be seen in the background (see Fig. 10.5). There were no graphics next to the anchor. Sometimes people were seemingly working at these desks (e.g., typing on a typewriter). This typing was staged, as the sound would have disturbed the show (current editor-in-chief of the *Tagesschau*, Heiner Hug, personal communicator). But obviously it was important to stage the work of journalists during the show; this pointed to the production of the news, which were thus conceptualized as an intermediate product of an everlasting process of investigation.

To summarize, since 1980 there is an anchor in the *Tagesschau*, leading in packages and film items, organizing the show, and so on. The audience is sometimes addressed by the anchor's use of an inclusive "we." The often didactic tone is striking, as well as the tendency to give some stories an investigative touch, which is typical for anchors in the 1980s. There also is a shift toward a more casual style of presentation, which can be found in later formats, too.

CBS 2005: "More Interesting and Compelling"

In 1991, the *CBS Evening News* dropped to third place in the ratings of national TV news (Hamilton, 2004) and stayed there until 2006, when it regained second place. The media market had changed drastically since 1986, mostly because of more competition in the market (see Ludes, Schütte, & Staab, 2002; Schudson, 2003).

In March 2005, Dan Rather stepped down as anchor. During the following months, CBS was working on the format of the *CBS Evening News* with interim anchor Bob Schieffer. The analyzed shows of December 2005 are consistent in themselves, but probably the format had not yet found its definite shape. Katie Couric, former co-anchor of NBC News' *Today* show, took the anchor chair of the show in September 2006.

Marcy McGinnis, senior vice president for news gathering at CBS, said about the redesign of the *CBS Evening News* in July 2005: "We're experimenting this summer with new, interesting ideas for how to tell stories in a more interesting and compelling way" (Gold, 2005).

In the December 2005 format, the anchor's function to announce news is more important than that of giving the information. This can be seen in the decline of these text types that are realized by the anchor and aim at delivering information, like anchor items or voice-overs. Anchor items are nearly nonexistent (0.36% of the show's duration); the voice-overs' share fell below 10% (9.52%; 1986: 14.1%). On the other hand, text types primarily aimed at leading in stories, organizing the show (like headlines, previews, advance notices, etc.) gained about 50% compared with 1986 (from 12.07% to 18.07%). The function of the anchor to present the news and organize the show became more important than delivering information.

Moreover, interviews led by the anchor became much more important, usually involving questioning a correspondent on the scene.

Although the greeting at the beginning of the show is the same ("Good evening. I'm Bob Schieffer"), the sign-off varies sometimes and is occasionally rather casual in style, as in the following example that follows a package about new dog breeds (December 5, 2005):

> Bob Schieffer, straight on medium close shot, laughing: Which is our cue to say TOODLE-OO To YOU. I'm Bob Schieffer, CBS News in New York. See you tomorrow.

Headlines—still rare in 1986—are now a well-established section at the beginning of every show. Whereas Dan Rather used to read the headlines himself, things were different in the new format (transcript from the show of December 5, 2005; see Table 10.3).

The anchor identifies himself and announces the first story, then the rest of the headlines. These headlines—and this is an important change—are told by the correspondents themselves while they are on the screen (together with footage of the event reported in a split screen). The anchor still is the central person, but as his duties concentrate on announcing and organizing the news, the correspondents become more important in telling the stories. By interviewing the correspondents, the anchor again is not telling the news himself, but asking for it. Looking at the style of the lead-ins, the addressing of the audience is striking (transcript of December 5, 2005):

> Bob Schieffer on: You have to wonder now just how long this is going on. Saddam Hussein's trial was convened for the third time today, and once again Saddam screamed, complained, threatened the judge, and said he was not afraid to die. But again, the proceedings went NOWHERE and the trial was adjourned until TOMORROW. For the latest on today's developments, we go to Kimberley Dozier in Baghdad.

The anchor is addressing the audience directly ("you have to wonder") and then is accompanying it to another place ("we go to") and giving the narrator role to the correspondent. Whereas in 1986 the anchor often gave additional information in the lead-in, the anchor of 2005 limits himself to the most important facts of the following story (see also the above example; transcript of December 6, 2005):

> Bob Schieffer on: In a war where there have not been very MANY good days, this was surely one of the worst. 43 Iraqis dead, 70 others wounded in one of the deadliest suicide bomb attacks of the war,—this one, incredibly, inside Baghdad's main police academy.—And another American has been kidnapped by terrorists in Baghdad. We go first to national security correspondent—David Martin.

Table 10.3. Transcript of December 5, 2005

Time	Text	Footage
0:00	Moderator on: I'm Bob Schieffer	Medium close shot: Anchor in the studio, screen in the background with caption "CBS Evening News December 5 2005."
0:02	Saddam Hussein says he is not afraid to die as his trial is reconvened for the third time. – We'll start with that tonight, as we cover these stories. –	Medium close shot: Anchor in the studio, on the screen in the background four different shots of Saddam Hussein talking in the courtroom. Slow zoom in on the anchor.
0:13	Gloria Borger on: A Texas judge rules in the case against Tom DeLay. But what	Split screen with two frames in front of a blue background, caption at the bottom: "Delay Case." Right frame: Group of people walking into the direction of the camera, and then turning to the right. On the right side a graphic of a scale. Left frame: close shot of the correspondent on the scene.
0:15	will it mean to his job in Washington? I'm Gloria Borger.—	Right frame: Close shot, two men go to a car, which is surrounded by a crowd, camera pan to the left, both men enter the car. On the right side a graphic of a scale. Left frame: close shot of the correspondent on the scene.
0:20	{. . . more headlines from correspondents on camera follow}	

Again, the mentioned facts are repeated in the following package. It is typical that the event is not just summarized in the lead-in, but also dramatized and evaluated ("one of the worst," "incredibly"). Next to this type of lead-in, there are lead-ins that summarize some facts but only hint at some others.

A new development is the mix of formerly different text types, like voice-over and lead-in. In some cases of the 2005 analyzed week, the anchor is not on screen during the entire lead-in, but only at the beginning. In the second part—similar to a voice-over—he reads the lead-in to news footage. Given this combination of anchor on screen and news footage, the lead-in becomes more complex—and probably more visually appealing for the audience. The following transcript is from the show of December 8, 2005 (see Table 10.4).

The studio is again a "newsroom studio," the anchor sitting at a huge elevated desk with the journalists' desks in the background with people working at them. At the wall behind the journalist's desks are three big screens, showing a turning globe or captions and graphics during lead-ins (see Fig. 10.6). To the left and right of these screens are two panels with the CBS logo. Toward the left of the anchor's desk there is a grey world map and a screen for interviews. At the beginning of the show and in the second part the anchor is standing at a standing desk in front of a screen.

All lead-ins—without exception—are accompanied by a caption and/or a graphic. This can be seen on the screens behind the anchor or—in the case of a close shot—as an over-the-shoulder graphic. Often the graphics are animated (e.g., falling snowflakes, waving flags, moving vehicles) and the fonts of the captions are carefully designed and integrated in the graphic (in "Pakistan Earthquake" "earth" is separated from "quake" and is somewhat tilted, written between to lines of a seismograph) or stay the same—as a kind of logo for an event ("THE BATTLE FOR IRAQ," "battle" and "Iraq" in all capital yellow letters).

The anchor's duties concentrate on announcing the news, not on delivering them—his main job is not reading entire stories but introducing them. Thus, text items like lead-ins, previews, and advance notices become more important. The

Table 10.4. Transcript of December 8, 2005

Time	Text	Footage
0:00	Bob Schieffer on: It was 25 years ago today that former Beatle John Lennon was shot and killed. And tonight	Medium close shot: Anchor in the studio, in the background working desks with screens and one person working. At the studio wall behind the anchor three screens, on the left side caption "John Lennon," on the right side portrait photo of John Lennon.
0:07	Bob Schieffer off: {sounds of the original situation} there is a huge turnout here in New York at the Strawberry Fields memorial in Central Park, just across the street from where Lennon lived and died. Here's our report from Jim Axelrod.—{4 seconds}	Medium close shot: Crowd in a park at night, some holding photos of Lennon. Fire engines in the background. Bar with the CBS logo appears with the caption "New York City Live EST." Slow zoom out.
0:16		

Fig. 10.6. Bob Schieffer in the show of December 6, 2005 (reproduced with permission of CBS).

type of lead-in only summarizing the main facts of a story is almost completely gone; most lead-ins only hint at the event reported, evaluate, or even dramatize it. The audience is often addressed directly. The visual presentation of the news has become more complex (e.g., with animated graphics or mixed-text types). The studio is again a newsroom studio, stressing not only the production of news, but also the electronic availability of it.

Tagesschau 2005: "A New Way of Staging"

The format of the Swiss *Tagesschau* has been changed twice since 1986. The 1990–1992 format had an anchor presenting the news in front of a grey wall. From 1992 to 2005, there was an anchor in front of a bluish wall presenting the news. In both formats there was only one newsperson in the studio, sometimes together with a guest for an interview (Luginbühl, 2005).

In December 2005, the Swiss *Tagesschau* was redesigned. The anchors remained the same, but the studio and also some aspects of the stories changed. Editor-in-chief Ueli Haldimann said that the usual style should be varied more often, and art director Alex Hefti spoke of "a new way of staging" ("eine neue Art der Inszenierung") the news that is necessary (Impact, 2005:5).

As in 1986, the anchor announces all longer stories. Although the amount of lead-ins is the same, the average length declined slightly. In most cases, the central facts are summarized. This is the case in lead-ins of film items, where often no explicit transition to the film item can be found. There is no information about the background of these events, no didactic style, no striking rhetoric figure. The lead-ins of packages usually have an explicit transition to the package, naming the correspondent on the scene. Also, these lead-ins sometimes only hint at the following content (December 6, 2005):

Moderator on: Die konservative Partei Grossbritanniens hat einen neuen Parteichef gewählt, und damit auch einen neuen Oppositionsführer. Er heisst David Cameron und das ist—doch einigermassen überraschend. Aus London der Bericht von Florian Inhauser.

Anchor on: The conservative party of Great Britain has chosen its new party president and with him a new opposition leader. His name is David Cameron and this is quite surprising. From London the report from Florian Inhauser.

Similar to the *CBS Evening News*, lead-in and voice-over are sometimes blended (see Table 10.5).

Table 10.5. Transcript of December 7, 2005

Time	Text	Footage
0:00	Anchor on: Today exactly 25 years ago he was shot in the open street. – John Lennon, *Heute vor genau 25 Jahren ist er auf offener Straße erschossen worden. – John Lennon,*	Medium close shot: Anchor in front of purple blue wall with unclearly visible screens of the control room and with a world map. On the left side a big TV screen with a portrait screenshot of John Lennon.
	{song "Imagine" is played} who had composed with Imagine the hymn of the peace activists. In Imagine, John Lennon *der mit "Imagine" die Hymne der Friedensaktivisten komponiert hatte. In Imagine sang John Lennon* sang of a world full of peace.	Anchor turns head toward screen, begins to play; John Lennon is singing "Imagine."
0:11	*von einer Welt voller Frieden.* John Lennon on {sings}: Living life in peace, uuuuh Correspondent over {John Lennon singing in background}: A utopia, which John Lennon also uses as his political agenda. [. . .] *Eine Utopie, die John Lennon auch zu seinem politischen Programm macht. [. . .]*	(fading) Close shot: John Lennon is singing.
0:22		

In cases such as these, a fluent transition from lead-in to package is realized; there is no clear transition point anymore. This lead-in does not follow the style of the inverted pyramid. There is an epiphrasis at the beginning, building up tension somewhat, then naming one of John Lennon's most known songs and its content—all of which allows a smooth transition to the following song passage that can then be heard. In such lead-ins, the different modes (spoken text–music–picture) are carefully coordinated to a lead-in that is innovative in its form and probably leads to a new text type, which has the potential of bringing closer together the different parts of cluster texts. But most of the lead-ins summarize the content of the following story.

In the new studio setting the anchor is standing (no longer sitting) behind a very long wooden desk (see Fig. 10.7). The floor is red, behind the anchor there is a white wall with a big window showing the screens from the control room. On the left end in this window, usually on the left side of the anchor, there is a semitransparent world map. On the left side of the window there is a big screen for news footage and interviews.

The anchors become physically more present. This is due to tracking shots and the use of three different cameras, but also because the entire upper part of the anchor's body can be seen, in the case of interviews sometimes from behind. Also the greetings and sign-offs are intensified ("Good Evening my ladies and gentlemen, welcome to the *Tagesschau*"; "Guten Abend, meine Damen und Herren, willkommen zur *Tagesschau*"; "This has been the main edition of the *Tagesschau*, THANK you for joining in, have a nice evening and see you soon," "Das war die Hauptausgabe der *Tagesschau*, DANKE, dass Sie dabei waren, einen schönen Abend und bis bald"). In summary, the anchor appears as a more dynamic person.

Whereas this kind of staging of the anchor usually is interpreted as typical for sensationalized news shows (see Grabe, Zhou, & Barnett, 2001), other features

Fig. 10.7. Franz Fischlin in the show of December 8, 2005 (reproduced with permission of SF).

of this new format are decidedly simple. The graphics only show portrait photos, no computer graphics; the captions next to the anchor are very simple and plain as well.

In this format, the anchor is physically more present than ever before. The lead-ins have partly become more complex (mixed-text types), but in most cases they follow the traditional pattern of summarizing the main facts. This coexistence of innovative and traditional forms is replicated in the studio, which is a mixture of a newsroom studio (with the screen of the control room in the back) and a studio set (no other newspersons or newsgathering devices visible). Even the two dominant colors—blue and red—refer to the more "serious" type of studio (blue) and the more sensationalized style of news (red). The design of the studio also mixes a modern design of the studio set with very plain graphics.

Discussion

The overview in Table 10.6 summarizes the findings of my analysis and illustrates some main features of the news presentation.

The chart demonstrates that the compared shows' presentations were never and are not identical; on the other hand there is a certain trend to homogenization. Some trends can be observed in both shows—although at different points in time (like the changing approach to the audience from (partly) detachment to closeness, the trend towards more news announcing than newsreading or the growing use of complex multimodal signs, etc.). Other characteristics remain specific to one of the compared shows (like the didactic style in the *Tagesschau* of the 1980s or the presence of an identified narrator in every story that is told in the *CBS Evening News*).

Looking at the long-term development of TV news it is no longer possible to claim that "since the mid-1950s to the present it has remained essentially stable" (Connell, 1992:144). Additionally, the preceding analysis of text types produced for news presentation suggests that there is not just a simple trend over time, but that there are changes of tendencies and local differences.

In the case of the *CBS Evening News* the style of the show in the 1940s was characterized by detached announcing and reading of the news, only slightly oriented to the audience. This changed with the new format introduced in the late 1960s: With an anchor as host and graphics during the lead-ins there is a more intense orientation toward the audience; emotional and physical closeness (e.g., via more close shots) becomes important. Also, there is more embedded interpretation in the lead-ins and a visible newsroom, indicating the production of the news. In the 1980s, the style has been designed to be more entertaining for the audience, with animated graphics and a very dynamic and fast speech style. There is even more interpretation embedded in the lead-ins. In the 2005 format, the

Table 10.6. Features of Compared Shows' Presentations

Broadcast Year	Topic	*CBS Evening News*	*Tagesschau*
1948–58	Role	Newsreader, neutral reporter	n/a
	Approach toward audience	Staged face-to-face (FTF) communication, nearly no direct addressing, detachment	
	Speech acts/style	Greetings, headlines, news reading, rudimentary lead-ins	n/a
	Multimodality	Studio background, static	n/a
	Studio design	Journalist's workplace	n/a
1968	Role	Anchor, interpreting host	Newsreader, neutral reporter
	Approach toward audience	Staged FTF communication; sometimes tension-building; some direct addressing	Detached disclosing information; very rare direct addressing
	Speech acts/style	Greetings, news reading, leading-in by summarizing, sometimes hinting and/or framing	Greetings, headlines, news reading, only seldom lead-ins, mostly in special cases
	Multimodality	Graphics: portraits and maps	Studio background
	Studio design	Newsroom studio	Studio set

continued on next page

Table 10.6. Features of Compared Shows' Presentations *(continued)*

Broadcast Year	Topic	*CBS Evening News*	*Tagesschau*
1986	Role	Anchor; interpreting and entertaining host	Anchor; didactic and partly casual host
	Approach toward audience	Staged FTF communication; often tension-building; some direct addressing	Staged FTF communication; shaping of critical awareness; very rare direct addressing
	Speech acts/style	Greetings, news reading (mostly to news footage), leading-in by summarizing, giving extra information and hinting and/or framing; dynamic style	Greetings, news reading, leading-in by summarizing, giving extra information; partly casual, didactic style
	Multimodality	Graphics: animated computer graphics, "logos"	Newsroom background
	Studio design	Studio set	Newsroom studio
2005	Role	Anchor; interpreting and entertaining host	Anchor; neutral, dynamic host
	Approach toward audience	More intensely staged FTF communication; often tension-building; more direct addressing	More intensely staged FTF communication; sometimes tension-building; very rare direct addressing

Speech acts/style	Greetings (partly varied), leading-in with evaluating summaries or hints; organizing by announcing headlines, giving previews, etc.; news reading is declining	Greetings (often varied), headlines, leading-in with classical summaries, only sometimes with hints; news reading
Multimodality	Graphics: animated computer graphics, "logos"; mixture of text types	Graphics: very plain; portraits and maps, sometimes news footage; mixture of text types
Studio design	Newsroom studio	Studio set, traces of newsroom

anchor's role is focused on the organization of the show by announcing headlines and stories. The anchor is mostly evaluating the reported events and building up tension; the stories themselves are told by the correspondents on the scene, thus emphasizing the importance of (at least pretended) eyewitnesses. In summary, after a first, short phase of detached announcing, the news presentation shows a continuous trend toward more closeness to the audience and more embedded interpretation of the reported events. The decidedly technical design of the 1980s (with a studio set) seems to be a trend—at least in the *CBS Evening News*—that has been alleviated since the 1990s, but nevertheless the presentation becomes more multimodal over time.

There has been no visible newsreader in the Swiss *Tagesschau* until 1966. The newsreader of the 1960s disclosed the information like an official statement in very detached way, giving the impression of disclosing "the" objective truth. In the Swiss *Tagesschau*, the paradigm change from the newsreader to the anchor occurred in 1980, together with an accentuation of the closeness to the audience. During the following decade, especially in a format produced from 1985 to 1990, the anchor not only addressed the audience casually, he also had a didactic attitude, aiming at the creation of a critical awareness on the side of the audience. This (sometimes pushing) didactic attitude can no longer be found in the formats since the 1990s. Today, the anchor is a neutral, dynamic host, but the style of presentation is quite plain when it comes to the graphics (unlike the *CBS Evening News* or the Swiss format produced in the 1990s).

The convergences and divergences that can be observed analyzing the presentation of the news can also be found analyzing the news stories of these two shows (see Luginbühl, 2006). In the CBS packages, for example, the correspondents can usually be seen on the scene, looking and speaking directly to the audience. Since the 1990s, the correspondents can often be seen in a live stand up, stressing closeness to the event and actuality. In the Swiss *Tagesschau* there are no packages until the late 1970s, but its percentage is rising rapidly in the 1980s. In the newest format, however, the correspondents can mostly no longer be seen in their packages.

The different and changing news styles not only constitute different news aesthetics, they establish different representations of the world and imply different norms and values. "The power of the media," Michael Schudson (1995) wrote, "lies not only (and not even primarily) in its power to declare things to be true, but in its power to provide the forms in which the declarations appear" (54).

To explain the homogenization and the differences found, various aspects could be discussed. Of course, my linguistic analysis cannot deliver a conclusive explanation, but I think it can contribute some import arguments.

One possible explanation for the changes in TV news presentation is *technological development*. But this explanation remains unsatisfying: Not all new technologies available are used, and the way they are used is not explained by the development of a new technology. For example, the *Tagesschau* of 2005 does not realize animated over-the-shoulder graphics or different font types in their captions while the *CBS Evening News* of 2005 does; neither the *CBS Evening News* nor the *Tagesschau* uses rolling caption bars at the bottom of the screen during lead-ins, whereas this is used in the completely virtual studio set that can be seen in other TV news shows. Nevertheless, technological development is of course important due to the fact that it makes possible auto cueing, over-the-shoulder graphics, or split screens. But technical developments cannot completely explain the form of news presentation.

Other aspects of the media system can explain the mentioned differences more plausibly. The U.S. *media system* is commercialized, whereas the Swiss *Tagesschau* is part of a public TV network. The *CBS Evening News*, one could explain, uses audience-appealing features (like staged face-to-face communication, tension-building, direct addressing, and complex graphics) to keep the audience watching and foster economic success. The Swiss *Tagesschau* on the other hand, as a public broadcasting company without direct competitors, does not use such forms in the 1960s—but it does so (although less intensively than the *CBS Evening News*) since the 1980s. This could be explained through the changing media market in Europe since the 1980s, which often is described as "Americanization" of the media and their texts—understood as commercialization and a trend to more audience appealing forms of reporting ("infotainment") (see Genz, Schönbach, & Semetko, 2001; Thussu, 2003).

However, there are a few problems with the explanation of commercialization. A trend to more "attractive," more audience-appealing forms of reporting exists in the Swiss news already in a format of the show introduced in 1980, and also in the formats thereafter. Since 1980, an anchor presents the news, looking directly at the audience, and in the format from 1985 to 1990 (analyzed earlier), it was an explicitly formulated goal to be "more direct, more immediate and more lively." But the "dual system" with public and private TV stations was not established in Switzerland until 1998, while in neighboring Germany private stations did not become important until the late 1980s. So the changes in the Swiss *Tagesschau* toward closeness and a more casual style started *before* commercialization and deregulation of the media.[13] And commercialization does not necessarily lead to more infotainment in all TV news, as the development of the news shows in the German TV stations showed (see Bruns & Marcinkowski, 1997). The deregulation of the media market surely had and has an influence on the text design, but this process alone cannot explain the changes observed (see also Schudson, 2003), since the deregulation reinforced already existing trends in reporting.

Another resource for explanations is the complex set of *political and social developments* during the 20th century. These changes cannot be fully discussed here; only a few keywords can be mentioned (see Hallin & Mancini, 2004): changes in journalism education and professionalization; a process of secularization, understood as "a separation of citizens from attachments to religious and ideological 'faiths' " (Hallin & Mancini, 2004:263) like the church, parties, or trade unions and leading to a more fragmented and individualized society with media targeting individuals; and a shift toward "postmaterialist" values of participation and free expression. Of course, the changes caused by the globalization and commercialization of the media play an important role here, too. It is important to note that many of these changes first took place in the United States, so the *CBS Evening News* started at another point than did the *Tagesschau*. This perhaps could explain the earlier staging of closeness to the audience in the *CBS Evening News*.

Furthermore, explanations through different *general communication practices* (different "intellectual styles," see Clyne, 1993; Galtung, 1981), or different forms of politeness (see Hickey & Stewart, 2005; Watts, 2003) could explain certain trends like a more intense "reader orientation" or a more intense staging of closeness in the CBS texts (see Luginbühl, 2006).

As an overall concept of TV news style and its change, I suggest a focus on what has been called *journalist culture* (Djerff-Pierre, 2000). "Culture" can be understood as an ensemble of practices (e.g., the language and footage use by journalists; see Hall, 1997), which are influenced by values, beliefs, and assumptions of a group (see Luginbühl, 2006). Djerff-Pierre (2000) wrote:

> These ideals and values include overarching conceptions of how journalism should relate to the audience . . . and society at large . . . on which journalists are supposed to be reporting events and processes. (252)

A look now at the way the editors themselves formulate their guiding values (at least the official ones) gives an impression of the norms and values the journalist culture incorporates. I illustrate this by looking at utterances of chief editors of the *Tagesschau* and of press releases of the same show. In 1970—when the Swiss news presenter had the role of a distant reporter and disclosed the information in a detached way—Dario Robbiani, the then editor-in-chief, wrote that the *Tagesschau* would present the news in a "truthful, accurate, balanced and neutral way" (Robbiani, 1970). In the 1980s—the presenter meanwhile turned into a didactic and casual host—the journalists aimed at "more analysis" and "background stories" (press release January 1, 1980) and the show should become more "direct, immediate and live" (press release February 7, 1985). Today, with a dynamic and close anchor, the studio has been designed to be "dynamic" and allowing "more flexible forms of presentation" (Impact, 2005:6).

The differences observed in my analysis can be conceptualized and understood as differences in journalist cultures, deriving from different values and role models, which leave traces in the journalistic artifacts, the texts (Djerf-Pierre, 2000). Thus, the different roles of the presenter (e.g., neutral reporter, interpreting or didactic host) can be explained by changing and different audience perceptions.[14] Of course, this culture is shaped by a complex set of different factors: media system and media market, technological development, general communication practices. Thus, the form of news always comprises traces of other broader and surrounding cultures—be it communication culture, general media culture, and so on. However, it is important to see that the form of news cannot be derived *directly* from these factors: The forms of news and with them the journalistic cultures are not (only) the logical or mandatory consequence of certain circumstances—and thus the circumstances alone cannot explain (or predict) all the changes observed. The concept of journalistic culture allows us to see the form of news presentation as what it is: the result of choices, the symbolic formulation by the journalistic staff. Even if we can name external factors that influence the form of news, we have to take into account that these factors are able to influence the form of news in very different ways. *How* these influences leave their traces in the form of news is a question of the journalist culture—and we also have to take into account the possibility that the journalists *own* (and also the nonofficial or even unconscious) norms and values leave their traces. Thus, the journalist culture plays a decisive role regarding the form of news—and its meaning. This becomes apparent when we look at the history of news presentation.

Notes

All figures reprinted with kindly permission of CBS and SF.

1. For other studies of interfaces of "visual language," multimodality, and processes of semiosis, see chapters 1, 6, 12, and 14, this volume.

2. The terminological difference is explained later.

3. For a discussion of spoken and written aspects of television discourse, see also Spina (chap. 9, this volume).

4. A serious problem regarding the data is the fact that in the case of the Swiss *Tagesschau*, until the late 1980s only the news footage and the newsreader's and anchor's manuscript have been archived; there is no footage in the archive of passages leading in a story (the "lead-ins") or the news stories read by a person on screen (the "newsreader items" or "anchor items"). The newsroom setting can be seen, however, in a few archived photos and in some copies of the show from 1979, 1981, 1982 and 1984 archived at the German Department at the University of Zurich.

5. The term *anchor* (or *anchorperson*, *anchorman*, *anchorwoman*) was borrowed at the presidential conventions in 1952 by the CBS News producer Don Hewitt from the name of the strongest runner of a relay team, the anchorman, who runs the final leg of the race. The name was used to describe Walter Cronkite's role—suggesting that he is the strongest runner, but also that he is centering the discourse of the news (Diamond, 1991; Morse, 1985).

6. For the terminology, see Pufahl (1988) and the TV jargon glossaries on the Web sites http://journalism.ukings.ns.ca/journalism_3925.html and http://scrippsjschool.org/producermore.php?story_id=339.

7. My comments, music, sound, and barely comprehensible words are written in {braces}. Accentuated words are written in CAPITALS. Short gaps are indicated with dashes ("-" 0.5 second, "- -" 1 second), longer gaps with three dashes and the duration in {braces}. A period (.) indicates falling, a comma (,) rising intonation.

8. An anonymous male voice can also often be heard in commercials, where it stands for neutrality (Wyss, 1998).

9. On the importance of this separation, see Graddol (1994).

10. During important events, the U.S. anchors sometimes leave the studio and present the show from another place, like here from Chicago, where the Democratic convention took place in 1968. Note that with this deixis the anchor is referring to the place where he is presenting from—an information never given in the Swiss *Tagesschau.*

11. All quotes from the *Tagesschau* in my translation from the German original.

12. In a film item of August 22, 1968, for example, the reader is talking of a "malicious invasion."

13. See also Hauesermann (chap. 8, this volume), about issues of public and private ownership and regulation of mass media with regard to radio.

14. With another perspective we also could talk of audience design. The notion of audience design (Bell, 1984) delivers a framework to understand the design of the shows analyzed. Although I do not analyze microstyle here like Bell does (I concentrate on aspects of macrostyle), the style of the news could be explained by the fact that media language is initiative style shift (instead of a responsive style shift), it "creates the relationship between communicator and audience" (Bell, 1984:192). The communicators "use style as an expressive instrument, a declaration of identity, saying to the audience 'you and I are ingroup' " (Bell, 1984:192).

References

Adamzik, K. (2004). *Textlinguistik. Eine einführende Darstellung* [Text linguistics: An introduction] (Vol. 40). Tübingen: Niemeyer.

Allan, S. (1998). News from nowhere: Televisual news discourse and the construction of hegemony. In A. Bell & P. Garrett (Eds.), *Approaches to media discourse* (pp. 105–141). Oxford: Blackwell.

Barkin, S. M. (2003). *American television news. The media marketplace and the public interest.* Armonk, NY: M. E. Sharpe.

Bell, A. (1984). Language style as audience design. *Language in Society, 13*(2), 145–204.

Brinker, K. (2005). *Linguistische Textanalyse: Eine Einführung in Grundbegriffe und Methoden. 6. überarb. und erw. Auflage* [Linguistic text analysis: Introduction, fundamental terms and methods, 6th edition, revised and extended] (Vol. 29). Berlin: Erich Schmidt.

Bruns, T., & Marcinkowski, F. (1997). *Politische Information im Fernsehen. Eine Längsschnittstudie zur Veränderung der Politikvermittlung in Nachrichten und politischen Informationssendungen* [Political information on TV: A longitudinal study on changes in the depiction of politics in news and political information broadcasts] (Vol. 22). Opladen: Leske + Budrich.

Clyne, M. (1993). Pragmatik, Textstruktur und kulturelle Werte. Eine interkulturelle Perspektive [Pragmatics, text structure and cultural values: An intercultural perspective].

In H. Schröder (Ed.), *Fachtextpragmatik* [Pragmatics of LSP texts] (Vol. 19, pp. 3–18). Tübingen: Narr.

Connell, J. (1992). Television news and the social contract. In S. Hall, D. Hobson, A. Lowe, & P. Willis (Eds.), *Culture, media, language: Working papers in cultural studies, 1972–1979* (pp. 139–156). London: Routledge.

Coupland, N. (2001). Stylization, authenticity and TV news review. *Discourse studies: An interdisciplinary journal for the study of text & talk [= DiSt], 3*(4), 413–442.

Diamond, E. (1991). *The media show. The changing face of the news, 1985–1990.* Cambridge, MA: MIT Press.

Djerf-Pierre, M. (2000). Squaring the circle: Public service and commercial news on Swedish television 1946–99. *Journalism Studies*, 1/2, 2000, 239–260.

Fix, U. (2001). Grundzüge der Textlinguistik [Basics of text linguistics]. In W. Fleischer, G. Helbig, & G. Lerchner (Eds.), *Kleine Enzyklopädie Deutsche Sprache* [Small encyclopedia of the German language] (pp. 470–511). Frankfurt am Main: Lang.

Galtung, J. (1981). Structure, culture, and intellectual style: An essay comparing Saxonic, Teutonic, Gallic and Nipponic approaches. *Social Science Information, 20*(6), 817–856.

Gates, G. P. (1978). *Air time: The inside story of CBS News.* New York: Harper & Row.

Genz, A., Schönbach, K., & Semetko, H. A. (2001). "Amerikanisierung"? Politik in den Fernsehnachrichten während der Bundestagswahlkämpfe 1990–1998 ["Americanization?" Politics in TV news during the election campaign for the Federal Diet 1990–1998]. In H.-D. Klingemann & M. Kaase (Eds.), *Wahlen und Wähler. Analysen aus Anlass der Bundestagswahl 1998* [Election and voters: Analysis of the Federal Diet Election 1998] (pp. 401–413). Wiesbaden: Westdeutscher Verlag.

Gold, M. (2005, July 6). CBS News explores storytelling. *Los Angeles Times*, p. E1.

Grabe, M. E., Zhou, S., & Barnett, B. (2001). Explicating sensationalism in television news: Content and the bells of whistles of form. *Journal of Broadcasting & Electronic Media, 45*(4), 635–655.

Graddol, D. (1994). The visual accomplishment of factuality. In D. Graddol & O. Boyd-Barrett (Eds.), *Media texts. Authors and readers* (pp. 136–160). Clevendon: Mulitlingual Matters.

Hall, S. (1980). Encoding/decoding. In S. Hall, D. Hobson, A. Lowe, & P. Willis (Eds.), *Culture, media, language* (pp. 128–138). London: Hutchinson.

Hall, S. (1997). Introduction. In S. Hall (Ed.), *Representation: Cultural representation and signifying practices* (pp. 1–11). London & Thousand Oaks, CA: Sage.

Hallin, D. C. (1994). *We keep America on top of the world: Television journalism and the public sphere.* London & New York: Routledge.

Hallin, D. C., & Mancini, P. (2004). *Comparing media systems. Three models of media and politics.* Cambridge, UK: Cambridge University Press.

Hamilton, J. (2004). *All the news that's fit to sell: How the market transforms information into news.* Princeton, NJ: Princeton University Press.

Hickey, L., & Stewart, M. (Eds.). (2005). *Politeness in Europe* (Vol. 127). Clevedon: Multilingual Matters.

Impact. (2005 September). TV-News. Der tägliche Blick in die Welt erhält ein frisches Gesicht [TV news: The daily glance at the world gets a fresh face]. *Impact. Medien, Märkte, Macher*, pp. 4–5.

Johnson-Cartee, K. S. (2005). *News narratives and news framing. Constructing political reality.* Lanham, MD: Rowman & Littlefield.

Kress, G., & van Leeuwen, T. (2001). *Multimodal discourse: The modes and media of contemporary communication.* London: Arnold.

Landbeck, H. (1991). *Medienkultur im nationalen Vergleich: Inszenierungsstrategien von Fernsehnachrichten am Beispiel der Bundesrepublik Deutschland und Frankreichs* (Vol. 33) [Media culture in national comparison: Strategies of staging in TV news of Germany and France]. Tübingen: Niemeyer.

Leeuwen, T. van (2005). *Introducing social semiotics.* London: Routledge.

Ludes, P. (2001). *Multimedia und Multi-Moderne: Schlüsselbilder. Fernsehnachrichten und Word Wide Web—Medienzivilisierung in der europäischen Währungsunion* [Multimedia and multi-modernity: Key pictures. TV news and world wide web—media civilization in the European Monetary Union]. Wiesbaden: Westdeutscher Verlag.

Ludes, P., Schütte, G., & Staab, J. F. (2002). Entwicklung, Funktion, Präsentationsformen und Texttypen der Fernsehnachrichten [Development, function, forms of presentation and text types in TV news]. In J-F. Leonhard, H.-W. Schwarze, L. Dietrich, & E. Strassner (Eds.), *Medienwissenschaft. Ein Handbuch zur Entwicklung der Medien und Kommunikationsformen* [Media studies: A manual of the development of media and forms of communication] (Vol. 15.3, pp. 2308–2321). Berlin & New York: de Gruyter.

Luginbühl, M. (2005). Changing faces—Changing cultures. The Swiss TV news show Tagesschau from the 50s to today. *TMG (Tijdschrift voo Mediageschiedenis), 2,* 159–169.

Luginbühl, M. (2006). Cultural differences in TV news—and in pragmatic practices or: How pragmatic culture shapes "globalized" TV news. *LAUD-paper. Series A: General and Theoretical Papers, A664,* 1–28.

Morse, M. (1985). The television news personality and credibility: Reflections on the news in transition. In T. Modleski (Ed.), *Studies in entertainment: Critical approaches to mass culture* (pp. 55–79). Bloomington: Indiana University Press.

Oomen, U. (1985). Oral and literate strategies in American and German television news. *Studia linguistica, diachronica et synchronica,* 593–610.

Pufahl, I. (1988). *Informing the public: A comparison of television news discourse in the United States and the Federal Republic of Germany.* Unpublished doctoral dissertation, Georgetown University, Washington, DC.

Püschel, U. (1992). Von der Pyramide zum Cluster. Textsorten und Textsortenmischung in Fernsehnachrichten [From pyramid to cluster: Text types and text type mixtures in TV news]. In E. W. B. Hess-Lüttich (Ed.), *Medienkultur—Kulturkonflikt* [Media culture—cultural conflict] (pp. 233–258). Opladen: Westdeutscher Verlag.

Robbiani, D. (1970). Tagesschau—Téléjournal—Telegiornale. In SRG (Ed.). *40. Jahresbericht 1970* [40th Annual Report] (p. 37). Berne: SRG.

Schudson, M. (1995). *The power of news.* Cambridge: Harvard University Press.

Schudson, M. (2003). *The sociology of news.* New York: Norton.

SF (1939–2007). SF Press releases of the public TV station of Switzerland. Retrieved July 27, 2007, from http://www.medienportal.sf.tv/sf/mainframe.aspx?public=yes&cat=77422&cattype=chronik_current_message&sto=1.

Stöckl, H. (2004). *Die Sprache im Bild—Das Bild in der Sprache. Zur Verknüpfung von Sprache und Bild im massenmedialen Text* [The language in the picture—the picture in the language. The linking of language and picture in mass media texts] (Vol. 3). Berlin & New York: deGruyter.

Thussu, D. K. (2003). Live TV and bloodless deaths: War, infotainment and 24/7 news. In D. K. Thussu & D. Freedman (Eds.), *War and the media: Reporting conflict 24/7* (pp. 117–132). London & Thousand Oaks, CA: Sage.

Watts, R. (2003). *Politeness.* Cambridge, UK: Cambridge University Press.

Wyss, E. L. (1998). *Werbespot als Fernsehtext. Mimikry, Adaptation und kulturelle Variation* [TV commercials as TV text: Mimicry, adaptation and cultural variation] (Vol. A 49). Tübingen: Niemeyer.

Zelizer, B. (1990). Where is the author in American TV news? On the construction and presentation of proximity, authorship, and journalistic authority. *Semiotica, 80*, 37–48.

III

Evolving New Media Genre Norms

Chapter 11

Technologies and Techniques

The (N)etiquette of Telephone and E-Mail Communication

SHARON MILLAR

University of Southern Denmark

Abstract

The introduction of new technologies inevitably raises various questions of know-how, both technical and communicative: How does one use the new medium? What for? When? With whom? This chapter considers the codification of norms in relation to the telephone (including telephone answering machines and voice-mail) and e-mail communication, using a range of materials relating to telephone etiquette and netiquette. Of particular interest is the extent to which new technologies involve the creation of new norms or the adaption or adoption of existing ones. Three aspects of telephone communication are first considered, namely contexts of use, the voice, and expressions to be used, before etiquette in relation to telephone answering devices and voice-mail is explored. Netiquette in relation to e-mail is examined from the perspective of e-mail style and electronic norms of correctness. It is argued that both telephone etiquette and netiquette aim to regulate the informality that was/is encouraged by, but not necessarily caused by, the technology of the telephone and e-mail, respectively. In so doing, they share many characteristics and enduring normative concerns (e.g., worries about courtesy and clarity), despite differences in format and rhetorical style.

The introduction of new technology has always been accompanied by claims and counterclaims regarding its social, linguistic, cultural, and cognitive effects. So, for example, the telephone was seen as a tool in the service of social and linguistic homogeneity, helping to realize the "united" in the United States of America and to remove dialectal variation in speech (see Casson, 1910; Dilts, 1941; Fischer, 1992); it was seen as a means of spreading English as an international language (Young, 1991); or as a means of transforming how people communicated, socialized, and conducted business, giving a new speed to everyday life (Casson, 1910). Similar claims have been made about the Internet and its creation of new communicative and social spaces populated by virtual communities using new varieties of language and text types in their interactions (Crystal, 2001; Smith & Kollock, 1999).[1] Yet, as noted by Standage (1999), such claims and reactions may be the "direct consequences of human nature, rather than technology" (199). In short, the effects of the medium cannot be assumed, but must be both teased apart from other influences and seen in the context of those influences (Herring, 2003; Marvin, 1988).

Bearing this in mind, this chapter examines the effects of new technologies on normative notions in relation to language use and communication (i.e., the focus is on communicative norms and not technical know-how). Of interest is to what extent new norms are deliberately created for new media as opposed to existing ones being adapted or simply adopted (see also Orlikowski & Yates, 1994; Orlikowski, Yates, Okamura, & Fujimoto, 1995; Yates & Orlikowski, 2002). The perspective taken here is that of prescribed or codified norms as found in materials relating to etiquette and netiquette. Such norms do not necessarily reflect or affect actual usage, but they give an insight into how the "experts," be they in technical or social matters, try to regulate the use of new means of communication and harmonise them with existing pragmatic practices. The technologies under scrutiny are the telephone (and associated developments in relation to recording telephone messages) and e-mail. Choosing technologies from different eras—in particular those that use differing channels of communication (spoken and written)—allows for comparison of normative attitudes and approaches across time and medium.

Theoretical and Methodological Considerations

The notion of a norm is widely used in linguistic literature, but its exact meaning is not always made explicit. As understandings of the concept differ, it is useful to outline what is meant here by a norm. Given my concerns with prescribing norms for new media, emphasis here is on norms that are explicit, that is, those that can operate at the level of conscious awareness, those that are associated with authority of some sort, and those that aim to either capture perceived regularity (Bartsch, 1987) or create it (Cameron, 1995). Many theoretical accounts of norms include sanctions as an integral part of a norm (e.g., Bartsch, 1987; Taylor, 1990); that is, nonadherence to a norm must be met with criticism and correction. As Bartsch argued, part of the definition of a norm is the enforcement of the norm itself because without normative force, one is left with "social habits and customs or usages" (169). Using a criterion of enforcement to distinguish between norms, customs, conventions, and usages has empirical difficulties, however, because enforcement measures (criticism, correction, sanction) are not always readily observable. In the context of this chapter, one could ask whether the data—that is, the contents of etiquette and netiquette manuals—actually deal with norms at all. Or can we only define such materials as norms, as opposed to customs and conventions, once we have evidence of others enforcing the behaviours recommended? Following Bartsch (1987), it is assumed here that manuals of etiquette and netiquette are dealing with norms, in that they are instruments of normalisation and enforcement. Bartsch, for example, regarded books of style and etiquette as the codification of "norms of use of communicative means," which aim to "achieve a normalizing effect within the language use of a population" (171). Similarly, Pattaro and Sartor (2002) view netiquette as regulating cyberspace and consisting of norms in the sense of "shared beliefs that certain patterns of behaviour ought to be followed by each individual (in the interest of a community or to achieve the common aims of its members). . . . Such shared normative beliefs are usually combined with informal sanctions" (12). Crystal (2001) argued that the novelty of the Internet and the language used over the Internet has caused people to seek guidance from the netiquette literature, which he sees as a blend of descriptivism and prescriptivism resulting in an "interesting kind of semi-prescriptivism" (63). This implication that the character of normative literature itself may have been changed by new technology is explored later in the chapter.

The data consist primarily of general etiquette manuals, materials on telephone etiquette and skills, and offline and online manuals on netiquette. These materials were chosen on the basis of their availability in libraries and on the Internet, and thus they were not randomly selected in the strict sense of the term. Reference is also made to secondary sources. All of these materials relate to English and many originate in the United States. The impact of the United States in matters relating to telephone etiquette and netiquette should be noted. Of the 62 general etiquette books surveyed, 27 were from the United States, 32 from Britain, 2 from Australia and 1 from Canada. Of the U.S. books (and the 1 Canadian book), 81.5% mentioned the telephone compared with 37.5% of the British (and none of the Australian) books, a difference that cannot be explained by the date of publication of the books. Rather, explanations lie in the different reception to the telephone in North American and British societies, combined with the link between etiquette and business as forged in the United States. As Schlesinger (1947) noted, "stereotyped politeness, having been ejected from the drawing room and the dance floor, found an unforeseen asylum in the marts of trade" (61). Regarding netiquette, many European and non-European languages have borrowed the actual coinage from English, as can be gauged from advanced searches in Google where language can be chosen from a given list; for instance, a search of "netiquette" gives results for languages such as Arabic, Danish, Italian, Hebrew, Hungarian, Korean, and Turkish. Moreover, the concept itself is often explicitly linked to netiquette literature from the United States, particularly Shea (1994) and Rinaldi (1998), be this in translated form or in the original English. Rinaldi, for example, has been translated into 10 languages, including Japanese and Bahasa Indonesian.[2]

It is difficult to make any claims about the impact of etiquette and netiquette literature, but its popularity and availability can be estimated. It is clear that books on etiquette were extremely popular in the United States in the first half of the 20th century; for instance, Emily Post's first book on etiquette from 1922 sold more than 1 million copies and was at the top of the nonfiction bestseller list (Arresty, 1970). As for netiquette, a Google search of "netiquette manuals" in 2007 gives 1,310,000 hits, with fewer hits (1,210,000) for "online netiquette manuals." Such numbers, despite the inherent problem of repetition with search engines, give some idea of the availability of online and offline netiquette literature.

The Telephone

In the early days of the telephone, prescribing norms about its use lay in the hands of the telephone companies (in the United States) or Post Office (in Great Britain). Materials emanating from these sources were directed to users, or "subscribers" as they were then referred to, and to employees (i.e., the telephone operators). It should be remembered that, initially, communication by telephone involved not just two parties, but three (i.e., including the operator). Indeed, in cases of shared telephone lines, there could be quite a number of people on the "party line" and listening in on others' conversations was an acceptable form of social diversion. From the early 20th century, general etiquette books in the United States began to make mention of telephone communication; in Britain, mention of the telephone in etiquette manuals before the 1950s appears to be infrequent and brief. Specific books devoted entirely to telephone etiquette and skills generally emerge much later. It is such books, along with abundant Web sites, especially from the United States, which are current and often relate to self-development, career development, impression management and enhanced business, or institutional or corporate image. The notion of telephone communication as a skill to be acquired for self-enhancement or for commercial purposes, however, should not be seen as recent; the concept dates back to the early 20th century and is probably an extension of even older ideas about self-presentation and rhetorical skills.

For purposes here, I focus on three aspects of telephone communication that have attracted normative comment: contexts of use, the voice, and expressions to be used. I then consider etiquette in relation to telephone answering machines and voice-mail.

Contexts of Use

Although the telephone was first envisioned by Alexander Graham Bell as a form of broadcasting tool, it was generally perceived and marketed initially as an instrument of business. Its role in the business context, although not unproblematic, was broadly accepted. Acknowledged too was that business uses of the telephone required specific training. So, for instance, Scammell (1924) noted in the preface that "realizing that productive use of the telephone involves the understanding and use of a definite technique, hundreds of business concerns are now giving their employees special telephone training." Etiquette manuals often included sections on business and the telephone, making distinctions between such usage and that of the social domain. Today, the distinction between social and business use is still maintained, the latter being marketed as an acquired skill despite widespread familiarity with the telephone in the home. Trim (1997) complained that

> Many people have never learned the basics of telephone courtesy and effectiveness. They've been using the phone since childhood, yet they have never modified or polished phone habits for appropriate business or organizational use. Although it's okay at home, casual telephone usage is totally inappropriate for business. (9)

Historically, pragmatic issues were considerably more salient in relation to the social rather than business uses of the telephone. The main concern was how to regulate the increasing informality of social practices, in evidence before the use of the telephone, but increased by the open and spontaneous nature of that technology. Webster and Hopkins (1930) approvingly noted that

> Social telephone calls . . . to express congratulations or some good fortune or simply to inquire about health and general welfare, have become common as the telephone has come into general use. The custom is a pleasant one, much more human than the old way of leaving cards. (51)

Interestingly, other forms of technology were also related to greater intimacy in the social domain; so, for example, typewritten letters were seen as appropriate for informal, but not formal, social purposes (*Etiquette for Ladies*, 1923; Quigley & Clarke, 1926).

The speech act of paramount importance from the perspective of social etiquette was that of inviting. Etiquette manuals often suggested the types of activities suited to telephoned invitations, for example, bridge parties, informal teas, theater, opera (Davison, 1928; Troubridge, 1931) and even more formal dinners (Rives, 1926; *Vogue's Book of Etiquette*, 1935). However, the kudos of the written word was still emphasized in clear social terms; A. Richardson (1927), for example, claimed that "In spite of telephoning and other informalities of the day, well-bred persons still write certain notes and letters punctiliously" (91). Even today, certain types of more formal invitation, such as wedding invitations, are rarely telephoned and, although printed, the font chosen often resembles handwriting and the names of guests are frequently written by hand.

Issuing and accepting invitations by telephone were initially more acceptable in the United States than Britain as the etiquette of good breeding in Edwardian England forbade such behavior (Perry, 1977). Telephoned invitations raised concerns about creating awkward situations for people wishing to decline the invitation (Allen & Briggs, 1937; Quigley & Clarke, 1926). Webster and Hopkins (1930), whose etiquette manual is based on questionnaires sent out to young people from all over the United States, noted the solution preferred by one of their female informants: "Telephone invitations are the hardest because they catch you unprepared. I have typed and placed beside my telephone a firm but courteous refusal, which I read off, adapting it, if necessary to the occasion" (108). Many etiquette books addressed the problem by prescribing the form of telephoned invitations, acceptances, and refusals. Today, advances in technology offer the option of text messaging for those who wish to avoid the awkwardness of giving or receiving verbal refusals; this option seems to be particularly exploited by teenagers, for instance when asking someone out on a date or breaking up with someone.

It would be a simplification to assume that the telephone straightforwardly replaced written genres. Written invitations for certain informal functions were already being replaced by (nonmediated) verbal, impromptu ones in the late 19th century (e.g., Armstrong 1889; Saville 1882). Burleigh (1925) assured her readers that "Invitations and acceptances . . . instead of being invariably a matter of formal notes or cards, are often as not made verbally in happy-go-lucky style amongst more intimate friends without any slight being intended or suspected" (23). The telephone would have fitted into this trend of moving from the written to the spoken channel in certain contexts. Neither did the telephone involve default informality; rather, it was given what Wilson (1947) referred to as "pseudo formality" through the use of servants. Van de Water (1936) noted that "More and more are invitations by telephone taking the place of informal written invitations. The big house telephones by its butler, the little house by its parlor maid, or perhaps one hostess calls another directly" (37). This pseudo formality is reflected in many of the early etiquette books, which deal with how servants should use the telephone.

Voice

The characteristics of voice most usually mentioned in the etiquette literature include "tone," "rate," "pitch," "enunciation," and "pronunciation," that is a mixture of prosodic and articulatory features. So, for example, Allen and Briggs (1937) recommend "pleasant tones, clear enunciation and correct pronunciation" (145). These are all familiar elocutionary terms and precepts, but they were frequently given a technological twist: They were important because of the nature of the technology itself. For instance, in the training of telephone operators, sounds were not to be omitted and words were not to be run together as such speech "does not carry well over the phone" (Dilts, 1941:115). Similarly, operators were taught proper breathing habits and voice projection because "if she were to speak too loudly, her good clear diction would be blurred by the volume of voice going through the phone" (Dilts, 1941:116). According to Scammell (1924), "the sound of the voice as conveyed by the telephone is subdued, certain letters are almost entirely muffled. This distortion must be taken into account as well as the probability of noisy conditions at the other end of the line" (25). In the 1920s British subscribers were given similar advice:

> the distinctive sounds of consonants become blurred in the transmission of speech by telephone and words containing the same vowels are apt to sound alike. Greater care is therefore necessary in speaking by telephone than is required in ordinary speech if mistakes are to be avoided. (Young, 1991:108)

Later recommendations about speaking clearly and distinctly tend not to refer to the technical aspects of telephone communication, but to the lack of visual cues for the listener: "He can't read your lips" to quote *Esquire Etiquette* (1954:26). One obvious normative motivation for such advice was clarity or intelligibility, an acknowledged overarching norm of communication, so it is not surprising that many of the recommendations followed standard elocutionary principles in this regard, including voice projection, rate of speech, and avoidance of connected speech phenomena, such as elision, assimilation, and coalescence (Millar, 1993). However, the concept of clarity is not asocial and is often entangled with notions of politeness and social appropriateness. Scammell (1924), for instance, condemned utterances such as "wotcha gona tellum" or "yuh gotta eatcha dinner 'n ketcha train" as "coarse and slovenly" (28). Criticisms of telephone operators were often couched in similar terms. Electrical World (1885) noted that "a gentleman of fine ear, who uses the telephone frequently" had suggested to them that operators should be given elocution lessons to rid them of "nasality, shrillness and snappiness of utterance" (cited in Marvin, 1988:29). According to Dilts (1941), women with "flat, tinny or nasal voices" were not hired as operators for fear of customers thinking them to be "irritating and disagreeable" (115).

The normative mixture of clarity and social courtesy is also evident in the advice about not shouting or yelling when speaking on the telephone (e.g.. Miall, 1950; Post, 1950; Scammell, 1924; W. Stevens, 1934). People seemed to have had difficulties judging appropriate levels of volume; as one commentator from 1891 said, "the instrument is not to be compared to a deaf man" (Marvin, 1988:90). But regulating loudness of voice would not have just been a question of technique; loud and shrill voices have attracted negative comment for centuries and the ideal recommended by Victorian etiquette for the well mannered was a low, soft voice: "There is a certain distinct but subdued tone of voice which is peculiar to only well-bred persons" advised *Routledge's Manual of Etiquette* (1875:7). Technology was not to be permitted to interfere with that cherished notion: On the telephone, civilized people were not to shout. In addition to its status connotations, the voice was seen as a conveyor of personality and this was made all the more important by a technology that allowed one to speak without ever being seen. A familiar maxim was to put a smile in one's voice (Bosticco, 1967; Daly, 1946): "A voice with a smile gets business" counseled A. Richardson (1927:18).

What is noteworthy in the older, general etiquette literature is that regional or social accent is not explicitly mentioned when dealing with telephone communication. What does get occasional mention are "offensive expressions or violations of grammar" such as "ain't," "he seen," "it done him no good nohow," "yup," "nope," "all-righty" (Scammell, 1924:34). There is also evidence to suggest that telephone operators, at least in Britain, were chosen with care and speaking the "King's English" was an asset (Young, 1991). Salient phonological concerns were prosodics and articulation rather than finer points of accent *per se*. It is from such concerns that the idea of a "telephone voice" might have arisen in the popular mind and been produced as a distinct style by some. A great deal of the advice in the etiquette manuals distinguishes telephone speech from ordinary speech on the basis of quantity or degree (i.e., you have to enunciate more clearly, use more varied inflection, be more polite, etc.). Little surprise then if some people were tempted to produce rather exaggerated speech as their telephone voices.

More modern approaches to telephone skills do not seem to have substantially changed with regard to the voice, which is dealt with in similar ways, using a mixture of technological, business, and social arguments. Today, telephone skills are generally marketed for the business professional, particularly in relation to customer service and career management. The "disembodiment" property of the technology necessitates a nonvisual projection of a telephone persona through positive attitude, appropriate voice, and competent handling of the customer (Friedman, 2000).[3] Old advice about smiling remains. Hewitt (2001) in an online piece advised:

> It has been said that you should smile into the phone when dealing with customers. Don't overdo it. A slight grin is all you really need. If you walk away from this article remembering nothing more than the Smile Rule, you've accomplished a great deal. However, another point to remember is that your choice of words, as well as your inflection should "smile."

Ingram (2005), discussing the hire of call center professionals, referred to "voice tone presence" and advised employers to "keep en ear on" elements such as "rate and speed," "diction," "volume," and "overall tone" as well as "confidence projection" and "honesty and sincerity."

Reigh (2006), a voice image specialist, claimed that

> The sound of the voice is a physical experience for the listener and accounts for up to 97% of the impression we make when on the telephone. On top of that, 30% of our voice energy is lost over the telephone equipment making us sound more impersonal. (1)

Her tips include the traditional smiling advice as well as admonitions to use moderate volume and "good" grammar. Similarly, Hattaway (1999) listed tips for telephoning that are motivated by the technology itself and reflect old ideas. He advised his readers to "speak clearly and more slowly than usual" because the lack of visual cues necessitate more comprehension time on the part of the listener, and to "speak with more energy" because the poor "hi-fidelity" of telephones "rob your voice of expression" (12). Interestingly, in one of the tips relating to use of jargon, mention is made of accent, in this case foreign-accented English. "With all due respect to people for whom English is a second language, staff can't afford to have unintelligible foreign accents" advised Hattaway (12) and recounted a tale of a Chilean call center operative in Melbourne shouting "You lie, you lie" down the phone, much to the consternation of the customer who did not realize the intended meaning was "July." Accents do not necessarily marry well with globalization.

However, as with the older, general literature, explicit references to accent in the modern (i.e., specifically telephone skills) literature are few. Wage (1973) provided one example, favoring absolutely "correct English usage," but seeing no need to "banish all traces of regional accents which have a charm of their own" (69). Gough and Grier (1970) did not refer to accent as such, but highlighted some well-known variables in British English, such as alveolar realizations of /ŋ/ ("singing" does not become "singin' ") and vocalization of /l/ ("altogether" does not become "awtogether"). A course on telephone skills currently offered by "Business training works.com," a U.S. onsite training company with clients such as Microsoft and the U.S. Coast Guard, includes a module on accents and dialects, arguing that

> A thick accent or regional dialect can make a telephone conversation more difficult. This module looks at ways to better understand heavy accents and dialects, along with skills for reducing the speech markers of your organization's staff. (www.businesstrainingworks.com, telephone customer service training, accents and dialects)

These negative and positive attitudes to accent are not of course unique to telephone communication and have simply been transferred to that medium.

Conversational Routines

The invention of the telephone raised a myriad of issues in relation to both conversational structure (e.g., what to say when you answered the phone, or what to say to the telephone operator) and conversational courtesy. There were concerns that the lack of a physical presence led to abuses of courtesy on the part of telephone operators and users alike (e.g., swearing, shouting). *Vogue's Book of Etiquette* (1935) warned that "A telephone is too easy a shield for a short temper. . . . The way in which a message is given or taken, and the manner in which the conversation is conducted, are all certain signs of character and breeding" (9). In the early days, the telephone companies sought to control the language used by training their operators in the use of codified expressions or routines. For instance, Katherine Schmidt, a supervisor in the Bell system, devised a method of diction and drew up a list of standard phrases (e.g., "I'm sorry that li-on is busy"), which formed the basis for operator training (Maddox, 1977:271). The use of such expressions was strictly enforced and any deviation met with reprimand (Aronson, 1977). The Post Office in Britain followed the U.S. example and in 1908 introduced standard operating expressions, such as "trunk number please," "sorry to have kept you waiting," "your six minutes is up. Should I book again?" (Hall Ellis, 1990:41).

The codification measures of the telephone companies were undoubtedly effective in establishing and regulating not only an occupational register, but a repertoire of "conversational routines" for their employees (Coulmas, 1981). Laver (1981:290) claimed that "routine behavior is polite behavior" and certainly, one of the major motivations for creating telephone routines was courtesy. The general public too were to be educated in the use of routine expressions for both social and business uses of the telephone. Most of these routines centered on those speech acts most commonly associated with telephone use (i.e., requesting, inviting, and greeting). Etiquette manuals promoted indirect formulations for frequently occurring telephone requests, such as asking people to speak louder, or asking to speak to a certain person, or asking who is calling. Allen and Briggs (1937) advised that "When you can't hear distinctly, you should say 'I beg your pardon' or 'Will you please repeat' " (145). For E. Barker (1956), it is better to ask the question "may I please speak to Mr. Brown," rather than to use the imperative "Give me Mr. Brown" even when this is accompanied by the word "please" (87). The modern literature continues this trend of teaching formulaic use. According to Trim (1997), one should not say "Who is this?," "What's your name?," "Speak up please," but should say "May I ask who's calling?," "I'm sorry I didn't get your name," "I'm sorry I can't hear you. Could you speak a little louder." Phonepro.com, an online company providing telephone training as part of customer service call center training in the United States, United Kingdom, and Canada to clients such as McDonalds, Coca-Cola, and the Ford Motor Company, offers "phone etiquette skills webinars" which include courses on "wordsmything" where one is introduced to "motivating phrases," "affirming phrases" and "visual words" (www.phonepro.com, training courses, Wordsmything II: Gems for conversation). Indeed the company promotes itself as offering "formulas and techniques to assist companies in establishing effective standards for telephone, written and face-to-face interaction" (S. Richardson, 2005:1).

In addition to model requests, early etiquette manuals also gave model telephoned invitations. Post (1922:128) suggested the following formulation for a telephoned dinner invitation using a servant:

> Is this Lennox 0000? Will you please ask Mr. and Mrs. Smith if they will dine with Mr. Grantham Jones next Tuesday the tenth at eight o' clock? Mr. Jones's telephone number is Plaza one two ring two.

And if no servant:

> Is that you, Mrs. Smith? (or Sarah?). This is Mrs. Jones (if she is elderly) or Alice Jones (if she is fairly young) or Alice (if an intimate friend). Will you and your husband (or John) dine with us at eight o' clock? (Emily Post, 1930:179).

These pragmatic models included not only the invitation, but identification behaviour (checking identity, identifying oneself) and terms of address.

Perhaps the best example of training the public in telephone routines is that of how to answer the telephone. Initially, habits varied among operators and public alike. According to Young (1991:16), the first public exchange in Connecticut used "Ahoy, ahoy" and Bell Canada operators used "well" or "are you there" until 1895. The formulation "number please" was eventually suggested and this became the standardized opener. A sense of how the public dealt with the telephone can be gauged from a report in the Norwich Argus 1885, which described the experiences of two journalists while staying at a hotel:

> The hotel clerk has possession of the instrument. "Who speaks?" came distinctly from the wires into the office. "2577" was the reply—it was the hotel number. "Put me into communication with 1200." "What is it?" was the instantaneous rejoinder. "Can we have two stalls for the Princess's, two stalls, Covent Garden." (taken from Occomore, 1995:42)

From this anecdotal report, it seems that telephone openers could be a form of "wh" question (Who speaks? What is it?) and did not at this time necessarily involve "hello." The telephone companies disapproved of the greeting "hello" (as indeed did Bell) as this hinted strongly of the noncommercial and was seen as "vulgar" (Fischer, 1992). Etiquette books for the most part, however, favored this greeting in the social context. Rives (1926:272) argued:

> The American salutation "Hello" will not readily be abandoned by the public. There are even signs that it will replace the British "Are you there?" The most usual form of address is "Hello. Is this Barclay 4500?" (272)

Allen and Briggs (1937) and Stevens (1934) discouraged openers such as "who is it?" or "what do you want?" or "who are you?" in favour of a simple "hello." Also discouraged was the use of one's name as an opener in the home context (*Esquire Etiquette*, 1954; Post, 1950). Roosevelt (1962) argued against the telephone companies in this regard:

> In spite of the Telephone Company's insistence that the proper way to answer a telephone is to speak your name, I find that it is a little pompous when you are answering the telephone in your own home. "Hello" or even a questioning "Yes?" spoken in a pleasant and friendly way seem to me to be preferable. (516)

In contrast, the use of "hello" was proscribed in the business context (*Esquire Etiquette*, 1954; Norwood, 1937; A. Richardson, 1927; Van de Water, 1936); Carney (1948:154) referred to the "useless 'hello' habit." The proper greeting in business was always the name of the company, but opinion differed as to what else should be said. Some advised the use of a greeting, such as "good morning," "good afternoon," and so on, after the name of the company (E. Barker, 1956; Post, 1950; Roosevelt, 1962); others prescribed the use of one's own name (Norwood, 1937; Post, 1937; Pringle, 1949). Modern advice favors the format of company name and one's own name (Trim, 1997; Walker, 1998). Interestingly, as the importance of identifying oneself in a telephone opener has become established, the frequent use of the caller's name throughout the conversation is being recommended. Tulane University technology services, for instance, directs its staff to "pronounce the caller's name distinctly and repeat it frequently. The sweetest music to a person is the sound of his/her own name" (Telephone etiquette, identify yourself section). This trend of personalization in the business context, however, is probably best seen as a positive politeness strategy rather than an automatized routine.

The controversies surrounding telephone conversational openers seemed to be motivated primarily by concerns about courtesy. "Hello" was worrisome because it lacked deference: "Some people seem to feel that it is undignified and are constantly tying to find some way to open a telephone conversation that would be less colloquial" suggested Wilson (1947:164). Direct "wh" questions were too abrupt, and giving names was too formal for the home—and not necessarily safe, because one did not know who was on the other end of the line. That "hello" won the day in social contexts probably reflected the contemporary loosening of social relations. It is also likely that the use of "hello" in Britain was influenced by telephone practices in the United States. What is curious is that no British manual makes mention of a telephone opener that includes the telephone number. Yet this was a common routine in social contexts in Britain, where the person called answered with their telephone number, usually accompanied by a greeting (e.g. "Hello. Oxford 12345"). It could be that giving a phone number was viewed as an established, technical ritual that identified the number reached, rather than an aspect of social interaction that required consideration in relation to politeness issues.

It is clear that both the telephone authorities and the etiquette literature aimed (and still aim) to regulate telephone communication through the use of routines. Their method of providing pragmatic blueprints is in itself unremarkable as this has a long history in conduct and etiquette literature as well as foreign language-learning materials (Millar, 2002). With the exception of greetings, it is difficult to know the impact of model formulations on actual usage by the public, especially in social contexts. It may be that in providing the patterns, etiquette literature had the effect of containing the range of alternatives or promoting certain formulaic uses over others.

Technological Developments: Telephone Answering Machines and Voice-Mail

An enduring problem with telephone communication is the need for the person being rung to be actually present and available to receive the message (servants and secretaries not withstanding). Indeed this disadvantage works to the advantage of e-mail, which does not require synchronous interaction. This synchronicity problem for the telephone has long been recognized and was countered by numerous developments in technology, starting with the telegraphone in 1898, which could record telephone messages. It was not, however, until the 1960s and 1970s that commercially viable telephone answering machines were available in the United States and it was the 1980s that marked the beginnings of their widespread popularity, especially in the residential market (Morton, 2000). Such technology was not ideally suited for the corporate sector, given that, for example, each telephone required an answering machine. The preferred solution here was voice-mail, patented in 1982 in the United States, and which was a centralized management system of incoming calls. Voice-mail has two major functions: telephone answering and voice messaging (where one calls a voice-mail system and not the individual's phone to leave a message). Further developments in technology have increasingly integrated voice-mail with e-mail, giving rise to unified messaging systems (for a brief overview of the history and development of voice-mail see http://en.wikipedia.org/wiki/Voicemail).

From the perspective of etiquette, telephone answering machines and voice-mail have to confront similar issues, namely recording an outgoing message that the caller will hear, and leaving a message on a machine or system that the receiver will hear. Online advice, thus, tends to deal with answering machines and voice-mail together. Obviously, recommendations on such matters are linked to the business context for voice-mail, but even techniques in relation to telephone answering machines are given a "business" slant in that they are frequently related to careers advice and job-hunting. The sources of most online guidelines are companies in the training and customer services domain, employment agencies and educational institutions. The advice given tends to be very similar. The characteristics stressed for both the outgoing message (greeting) and the incoming message (response) are brevity, clarity, politeness, and professionalism, all to be reflected in appropriate tones. According to Apple One, the largest privately owned employment agency in North America, the outgoing message should be "short and simple," recorded in a "businesslike but pleasant" voice and when leaving a message one should give your name and number and "speak clearly, slowly and be brief" (Core Newsletter, 2003). Similarly, Customer Focus, a customer services and call center training company, which boasts IBM and Dell as clients, recommended brevity and suggests a "professional and warm voice mail message" (voice-mail etiquette, point 4). Familiar advice about smiling features in etiquette guidelines from Basebuilders.com, a company providing software solutions for project management in the fields of engineering and architecture, which tells us to "use a natural, friendly, and pleasant voice—don't speak too loudly. Smile while recording your message—it reflects in your tone of voice. Enunciate clearly; speak slowly" (Basebuilders.com, 2004: What is the professional etiquette for your voice-mail message section).

With regard to the actual content of the outgoing message, model formulations are often provided, a well-established practice in general telephone etiquette as discussed earlier. For instance, the University of Southern Illinois suggests a range of formats for their employees, including the following for an ongoing voice-mail greeting:

> Hello you have reached the (DEPARTMENT) of Southern Illinois University. I'm sorry I cannot answer your call at the present time please leave a message at the tone. I will pass your message on to the appropriate person or call you back. Thank you for calling. Bye. (voice-mail etiquette, some suggested personal voice-mail greetings section)

York Technical College in Rock Hill, South Carolina proposes that students use a similar format on their personal answering machines:

> Hello this is (phone number). I am sorry I am not available to take your call right now. Please leave your name, telephone number, a brief message, and the best time to reach you. I will get back to you as soon as possible. (Telephone etiquette, answering machines/voice-mail section)

The aim of such advice is to provide blueprints for greetings, identification behavior, and request formulation. As is the case with telephone etiquette in business contexts, the extent to which telephone answering machines and voice-mail require greetings (hello, hi, etc.) seems to vary and the status of closings (farewells such as goodbye, bye, etc.) is not clear. For instance, all the model formats given by the University of Southern Illinois (above) include greetings (good morning, hello, hi, etc.) but only a few include farewells (bye, cheers now). The omission of greetings and farewells reflects the practices of the telephone companies themselves in the standardized, recorded messages they provide in their answering service for customers. Telephone companies and manufacturers of telephone answering devices have tended to include technical instructions in relation to the "tone" or the "beep." Morton (2000) noted that equipment manufacturers in the United States were responsible for the "beep" at the end of the outgoing message and instructed users to ask callers not to talk until after the "beep" or "tone."

As noted earlier, current technological developments are increasingly integrating various types of media, such as the telephone and e-mail in unified messaging systems or cell phones and the worldwide web. Another development is that of internet telephony (voice-over IP[VoIP]/IP telephony) as provided by, for example, Skype, a company that promotes its brand name as a verb in the language (to skype, skype me). Given its relation to the Internet, it is perhaps not surprising that Skype has its own etiquette for its users. Skype.com etiquette (2007) informs us that "Skype is a community, and members of a community must respect each other as human beings." For such humanistic reasons, certain behaviors are proscribed, for example, exploitation of a minor or discussion of illegal activities (e.g., how to get drugs or how to make bombs). Linguistic proscriptions relate to "vulgar, profane, offensive or obscene language" and "racially, religiously or sexually offensive language." In many ways, there are no surprises here; etiquette throughout history has been grounded in contemporary politeness norms and has attempted to regulate content (e.g. suitable topics for conversation) and form (e.g., no swearing, use of politeness formulae). What should be noted, in relation to Skype, is that the sublink to etiquette is placed under "terms and policies," making an explicit connection between etiquette and legality. Given that etiquette has always had links to ethics and morality, it is not overly surprising that matters of (n)etiquette should have legal implications and, as oral/aural dimensions are added to the Internet (such as Skypecasts that allow for moderated group conversations), questions of (n)etiquette may become increasingly issues of legal concern and controversy.

Netiquette: E-Mails

To some extent, etiquette has had a new lease of life with the coming of the Internet: Net etiquette has been launched into cyberspace as the morphological blend netiquette.[4] The concept relates to a wide range of Internet use ranging from e-mails to multi-user dungeons (MUDs). For instance, the RFC 1855 memo on netiquette guidelines from 1995 deals with three types of communication: one-to-one communication, including e-mail and talk; one-to-many communication, including mailing lists and usenet; information services, including MUDs and multi object oriented (MOOs). So netiquette itself has historical dimensions with roots in what are now older Internet facilities; indeed the usenet predates the Internet itself. In this chapter, the focus is on e-mails only, the predominant form of Internet usage (Castells, 2001), although the advice referred to may also apply to other uses. According to Shea (1994), "some rules of Netiquette are based on common courtesy, but have been adapted for a computer culture. Others are based on the technological limits of cyberspace. Still others are matters of convention" (23). Focusing on the linguistic and pragmatic aspects of netiquette, I investigate what these adaptions and conventions are. Two aspects of e-mail communication are considered: e-mail style and electronic norms of correctness. Such issues generally never fail to make an appearance in the many popular books and primers on e-mail management, communication, and use (for recent examples see Chan, 2005; Kerr & Doody, 2007; Shipley & Schwalbe, 2007; Steele, 2006). Indeed, modern publications have been likened to their traditional forebears; for instance, reviews of Shipley and Schwalbe (2007) on the Amazon.com Web site compare them to etiquette and style gurus Miss Manners, Emily Post, and Strunk and White.

E-Mail Style

It is generally recommended that e-mails should be brief: "Keep paragraphs and messages short and to the point" (Rinaldi, 1998); "Short sentences are easier to write, read and understand" (Flynn & Flynn, 2000:30); "Keep everything short. Keep your lines short, keep your paragraphs short, and keep the message short" (Sherwood, 2001). Brevity is generally linked to the nature of the technology, particularly its speed, limited disk space, size of document window, and cost (Sherwood, 2001), but some manuals on how to write e-mails associate brevity with notions of clarity of structure and ease of comprehension, which apply to writing generally (e.g., Flynn & Flynn, 2000; Whelan, 2000). In addition to brevity, informality is promoted as the ideal e-mail style: "The e-mail medium encourages an informal, conversational style of writing because it is an instantaneous and one-on-one medium of communication" (Angell & Heslop, 1994:56); "Electronic mail is known for its informality. This is partly because of the hacker culture whence email springs and partly because email is so easy to send" (Shea, 1994:47). Some manuals stipulate the type of linguistic constructions that denote a conversational manner, such as active voice, contractions, and ending sentences with a preposition (Chan, 2005; Flynn & Flynn, 2000).

Questions of brevity and informality raise a number of related issues including the use of greetings and closings, abbreviations (such as "thx" [thanks], "pls" [please], "NE1" [anyone]), acronyms (such as "AFAIK" [as far as I know], "IMO" [in my opinion], "BTW" [by the way]), and emoticons (such as ☹; ☺). With regard to greetings and closings, there would not appear to be any fixed norms as such (as indeed is the case with telephone answering devices and voice-mail). Whelan (2000) noted that some people see greetings as informal and some as formal, an observation that is borne out by the literature. Sherwood (2001), for instance, sees lack of greeting as informal, at least in the United States, whereas

Angell and Heslop (1994) consider that a greeting adds a "friendly touch" and recommended the use of a simple closing, such as "regards" (21). As with terms of address in business telephone interaction, there seems to be a move toward promoting positive politeness in business e-mails, but concerns about respect and deference remain. Certainly, in practice, corporate e-mail signatures may include a standardized closing formula; my own university, for example, uses "best regards." As for individuals' actual use of greetings and closings, a preliminary analysis of corporate e-mails collected from two global Danish companies, primarily in relation to internal communication, indicates that the use of greetings and closings varies considerably and with all possible permutations, be the language English or Danish (i.e., both greeting and closing, only greeting, only closing, and neither greeting nor closing).[5] Forms are generally limited to "hi," "hello," or "dear," and "best regards" or "(with) kind regards." On the whole, actual usage seems to point to somewhat fluid norms, as is also evidenced by the netiquette literature, although some form of greeting and/or closing does seem to be favored, probably as a positive politeness strategy. It is worth noting that these users are operating in a global context, often in a lingua franca (English), and hence other permutations can be observed on occasion, such as the greeting in one language and the closing in another, where the level of formality, at least by native speaker norms, need not match (e.g., beginning the e-mail with "hej" [an informal, colloquial Danish greeting similar to English "hi"] and ending with "you [*sic*] faithfully" [a rather formal closing expression in English, correctly rendered as "yours faithfully"]).

The use of abbreviations, acronyms, and emoticons is met with varied responses in the netiquette literature. Manuals addressed to the business user tend to be wary because of the potential for misunderstanding and fears of unprofessionalism: "Avoid emoticons. Don't use symbols or silly abbreviations. Use English" (Barker, 2000:132). "Although smileys may have their place in an informal setting, you should not use them in business messages." "Although abbreviations may reduce the amount of keying . . . you should be careful about using them in business messages" (Whelan, 2000:156–157). Netiquette guru Rinaldi (1998) warned of overuse of acronyms as they can confuse and annoy the reader.

Yet, despite their uncertain normative status in netiquette, these phenomena have been subject to normalization in that they have been informally codified in online and printed dictionaries. From a normative perspective, it is illuminating to consider responses to such material. For instance, press and web reviews of the internet dictionary "Netlingo" (available at www. netlingo.com, in the news section) are favorable about the dictionary, but refer to "Internet gibberish" (*Reader's Digest*), "confusing Internet jargon" (*The Times*), and "the often cryptic and comedic vocabulary of the Internet" (*PC Magazine*). A review by Apple, on the same Web site, refers to emoticons as "the punctuation graffiti littering chat rooms and e-mail" and describes the "shorthand language" of chat rooms and e-mail as "a particularly noxious method of communicating heretofore mostly limited to the armed services." It would seem that jargon, acronyms, and emoticons have yet to attain complete acceptability and respectability. Nonetheless, in the corporate e-mail database referred to above, abbreviations, such as "pls" (please), "br" (best regards), "wkr" (with kind regards) do occur, as do the occasional acronym (e.g., c u) and emoticon, both smiley icons and ascii smiley characters.

As with the telephone, it is important to contextualize historically the notions of brevity and informality in relation to e-mails. These trends in writing were in existence long before the appearance of e-mail communication. Etiquette manuals from the early 20th century noted a change in the style of letter writing. For instance, *Etiquette for Ladies* (1923) complained that "Modern correspondence is characterised by a brusqueness and brevity which do not compare favourably with the leisurely grace, neatly-turned sentences and ceremonious style of our slower-going ancestors" (163). Even earlier, there were complaints in the literature about the use of abbreviations in writing: F. Stevens (1884:195) forbade the use of abbreviations, such as "sh. be glad to see you" in written invitations; the *New Book of Etiquette* (1907:36) censured the "fashion of abbreviating words" and the "mutilation of phrases," referring to usages such as an "invite" and "thanks." Abbreviations were thought to denote laziness (Thornborough, 1924).

Such complaints indicate that a conversational style of writing was in use. Good examples of this kind of writing are provided by postcards, which to some extent were the e-mail equivalent of the early 20th century (see Diekmannshenke, chap. 6, this volume). At that time, a frequent and rapid daily postal service, at least in metropolitan areas in Britain, meant that, for example, invitations for lunch could be safely sent on a postcard that same morning and a relatively speedy form of asynchronous interaction was possible. Postcard correspondence from 1900 and 1922, respectively, is shown below:

> Dr B You see I have kept my promise. Aren't these cards lovely, they are the best we can buy. Paris is too glorious, there is no place like it, but I wish I had one of you to see the shops they are indescribable—the exhibition is too wonderful, we have just returned dead beat. yr. loving aunt. (Phillips, 2000:35)

> Ma chère Cissily. How are you? Yes I'll see you usual place Friday 6.20pm. Have booked Hippodrome. Yours as B4. (Phillips, 2000:124)

The postcard from 1900 includes the abbreviations *dr* "dear" and *yr* "your" and punctuation is rhetorical rather than grammatical. The postcard from 1922 has instances of ellipsis (e.g., omitted first-person pronoun before "have"), an abbreviation (e.g., B4 for "before") and an anaphoric "yes" and is markedly interactive in style. The postcard was a realistic, and often preferred, alternative to the telephone, but as postal services became slower and the telephone networks expanded and became cheaper, its potential as a fast, interactive medium could not be sustained (Phillips, 2000). What is interesting is that certain of the linguistic and pragmatic trends thought to characterize e-mail communication were already in operation in postcard writing at the beginning of the 20th century, if in embryonic form. It was also at this time that the memo was developing as a genre in corporate settings in the United States, fueled by the organizational needs of ever-expanding corporations and technological developments, such as the typewriter and carbon paper; the memo was characterized by a standardized format (to, from, subject) as well as brevity, traits transferred to e-mail communication (see Yates, 1989a, 1989b).

Norms of Correctness

As with traditional etiquette, netiquette directs attention toward linguistic correctness and standards in writing, be this in relation to spelling, punctuation, or grammar. As always, the salience of such matters has a base in both usage and ideology. The following, for example, is an extract from an e-mail I received from a student in 2004, which ignores some of the conventions of punctuation (e.g., apostrophes and capitalization):

> Thank you very much for your email, i applied for the Diploma as i thought i didnt qualify for the MA because of my terrible result! however i did feel i might be considered for the MA.

One's normative reaction to this e-mail partly depends on how one views the electronic medium. Should e-mails follow the linguistic and pragmatic conventions of print media? The literature is not agreed on this point. Some manuals regard e-mail as simply a form of writing: "E-mail is writing. Treat it as you would any other kind of writing: plan; write; edit" (A. Barker, 2000:131). "Your ability to write a clear, concise, non-electronic message the good old-fashioned way will determine the effectiveness of your e-mail documents" (Flynn & Flynn, 2000:1). Others see it as a distinct genre and "not merely the bastard child of the print medium" (Angell & Heslop, 1994:3). Nonetheless, many of their recommendations follow those of general writing manuals regarding structure, punctuation, choice of words, and grammatical forms. The correctness norms promoted for the business and professional user follow those of the print medium (see Crystal, 2001).

All of the materials I sampled recommended that care be taken with linguistic form; in short, they endorse the editing process. Sherwood (2001) argued that language competence is a status cue online: "If you have lots of misspellings, your subjects do not agree with your verbs, or you use the wrong word, people may assume that you are uneducated" (4). Similarly, Shea (1994) argued that "spelling and grammar do count" because they are part of "looking good online" (41). Flynn and Flynn (2000) warned that "An e-mail document full of grammar, punctuation and spelling errors will tax the reader's patience and lessen your credibility" (1). Chan (2005) urged her readers to plan and proofread and linked questions of punctuation and grammar to presenting a professional image. Etiquette guidelines for e-mail written by and copyrighted to David Harris (1990-2006) and used by the Pegasus e-mail system tell users to

> Use correct grammar and spelling. Electronic mail is all about communication—poorly-worded and misspelled messages are hard to read and potentially confusing. Just because electronic mail is fast does not mean that it should be slipshod, yet the worst language-mashing I have ever seen has been done in e-mail messages. If your words are important enough to write, then surely they are also important enough to write properly? (Etiquette guidelines for electronic mail, message length, content and format section)

Yet, there is controversy about applying such editing standards to e-mail as this runs counter to the speed and spontaneity of the medium (Angell & Heslop, 1994; Shea, 1994). The same is true of instant messaging, chat and text messaging. Crystal (2001) analyzed in some detail a specific guide to language use on the Internet generally, namely Hale and Scanlon's (1999) edition of *Wired Style*, which advocates linguistic play and general irreverence for the rules of the standard language. Such advice, with its emphasis on the personal rather than the social, would seem to signal a change in direction from that of traditional etiquette and might indicate that netiquette involves some form of ideological shift in relation to norms and correctness. I argue, however, that irreverence for the rules is not as yet typical for netiquette. Hale and Scanlon (1999) reflected hacker or geek culture and ideals, but even in this realm the communicative norm of clarity is paramount. The online hacker "Jargon File," (v. 4.4.7 December 2003), which is essentially a catalogue of hacker slang and practices, notes that the hacker love of overgeneralizing grammatical rules

> is not "poor grammar," as hackers are generally quite well aware of what they are doing when they distort the language. It is grammatical creativity, a form of playfulness. It is done not to impress but to amuse, and never at the expense of clarity. (Chapter 4, overgenerlization, final paragraph)

It would seem that any changes in ideology, possibly originating from the hacker/geek realm, are destined to be constrained by the overarching norm of clarity. An "anything goes" scenario for netiquette is as yet unlikely. Note, for instance, that Hale and Scanlon (1999) modified a number of their guidelines from a previous 1997 edition of the manual, including "screw the rules," which was transformed to the milder "be irreverent." In similar vein, Storrer and Waldenberger (1998), in their analysis of netiquette, identified what they referred to as Gricean norms (dealing with matters relating to the Gricean maxims of quantity, quality, relevance and manner) and etiquette, in German "knigge," norms (dealing with matters of politeness, morality and ethics). It would seem that such superordinate norms still hold considerable sway in the regulation of cyberspace, hence the continued concern with not only linguistic but also pragmatic editing (i.e., check your grammar, check your tone, and think before you press the "send" button; see Shipley & Schwalbe, 2007). New technological developments uphold rather than challenge these norms.

Conclusion

It is to be expected that normative literature should lean toward the conservative and the status quo. Etiquette and netiquette are not designed to be instruments of revolution, but regulation. That netiquette and modern telephone etiquette in many ways still resemble traditional etiquette is testament to the endurance of certain overarching linguistic, communicative, and cultural values (e.g., linguistic correctness, clarity, and politeness). The rhetoric of netiquette and modern etiquette may be more upbeat and apologetic (stressing that the recommendations are not dogma or mandates), but imperatives and the modal "should" remain, and the metalinguistic vocabulary is familiar ("sloppy," "proper," "slipshod"). I argue that in relation to new technologies, etiquette and netiquette share the aim of regulating the informality that these technologies have not necessarily caused, but have certainly encouraged and further promoted. Use of the telephone required the creation of suitably polite conversational routines for workers in the industry and for members of the public when telephoning in both private and business contexts. Other normative concerns, such as the voice, although given a technological motivation, were dealt with by adopting existing cultural norms, themselves not neutral with respect to social class, ethnic background and gender, which were long propagated by elocution and public speaking manuals. Modern telephone etiquette (including telephone answering devices and voice-mail) continue this tradition. Prescribed styles of e-mail communication have been influenced strongly by print media in relation to formal matters of spelling, punctuation, and grammar, and the tendency in the normative literature has been to view e-mail as a rather static, unitary text type, which is often determined by the nature of the technology itself and generally characterised as brief and informal, either with or without greetings and with or without editing, depending on viewpoint. This monostylistic conceptualisation underlies theoretical discussions in recent years regarding the development of e-mail styles or a stylistic range for e-mail to incorporate greater formality (Baron, 2000; Crystal, 2001). Empirical research, however, suggests that some of the normative assumptions about e-mail may not always be well-founded. Herring (1998), in a study of e-mail style over an 11-year period from 1975 to 1986, found that the trend was toward increased informality, but not increased errors or decreased syntactic complexity. However, variability, particularly in relation to formality and informality was the norm, underlining the importance of contextual factors, such as participants and purpose of the communication, rather than the technology itself. In terms of actual language use, there is no single, predetermined e-mail style.

From a normative perspective, there has been an interesting development in that some netiquette materials, albeit a minority, are no longer embracing the traditional literate norms of the formal language with regard to spelling, punctuation, and grammar. Is this a preliminary sign that perceptions of literacy are in flux? Baron (2003:92) argued that "writing in general has become more speech-like" and that we are currently in a period where writing is not subject to the traditional rules or monitoring. Indeed she doubts whether a "more normative (and contemplative) writing will return to fashion" any time soon because people generally seem to relish the more free, unedited style (Baron, 2003:93). Whether we will see the appearance of fundamentally different norms of literacy based on more speech-like characteristics, including unedited informality, is an open question. Any such developments, however, will not necessarily be due alone to technologies such as e-mail, but to social and cultural trends, which since the 19th century have been permitting greater linguistic informality across differing genres and media. Technological developments, however, may affect the nature of e-mail as a medium, which can now be more widely used in functions once traditionally denied it (e.g., legal contracts) through the availability of legally binding, digital, and electronic signatures.[6] E-mail as a medium may, thus, achieve equal prestige to the print medium in these more formal domains. Add to this advances and improvements relating to IP telephony, which may eventually allow this technology to become the primary informal medium of interaction on the Internet, then the future of the e-mail medium may be more tied to formal contexts and functions or simply those requiring the transfer of documents. One thing is certain, however; whatever the future holds in relation to existing and new technologies, someone will be telling us how and how not to use them.

Acknowledgments

My thanks to Ronald E. Rice, Charley Rowe, and Eva L. Wyss for providing useful comments on an earlier draft of this chapter.

Notes

1. As well as being a medium in itself, the Internet is a metamedium, including other types of media (film, radio, television, etc.).

2. The extent to which netiquette norms are global is an interesting question: Are there global norms for a global medium? That netiquette literature from the United States is being translated for Asian cultures might suggest some degree of globalization, but the possibility that such translations may include cultural adaptations should not be overlooked.

3. It is now possible to download voice changers that can change the voice completely (e.g., male to female, adult to child) or enhance your own voice (e.g., to sound sexier, sophisticated), all in the pursuit of a virtual personality. This adds a novel technological twist to the notion of "telephone voice," but is based on the old idea that the voice expresses personality.

4. The term *etiquette* is still of course widely used, either alone or in hyphenated versions such as e-etiquette or techno-etiquette.

5. As part of a project on Global Communication in Danish companies, financed by the Danish Research Council for the Humanities (2003–2007), corporate e-mails have been collected from two global Danish companies, which have suboffices around the world.

6. In U.S. law an electronic signature is defined as "an electronic sound, symbol, or process, attached to or logically associated with a record and executed or adopted by a person with the intent to sign the record" (www.wikpedia.com). A digital signature is an encrypted (cryptographic) electronic signature to ensure the authenticity of both the message and the sender. Electronic and digital signatures should not be confused with e-mail signatures that can be automatically generated by e-mail systems, giving name, address, and so on, often with corporate logos in the business arena, or that can be downloaded from various Web sites in the form of witty or sagacious quotations, so-called siglets or sigs.

References

Allen, B., & Briggs, M. (1937). *Behave yourself. Etiquette for American youth.* Chicago: Lippincott.

Angell, D., & Heslop, B. (1994). *The elements of e-mail style.* New York: Addison-Wesley.

Apple One. Core Newsletter (2003.). Electronic etiquette—manners for modern day communication. Retrieved March 14, 2006, from http://www.appleone.com/Core/0303/etiquette.aspx.

Armstrong, L. (1889). *Etiquette and entertaining.* London: F.V. White.

Aronson, S. (1977). Bell's electrical toy: What's the use? The sociology of early telephone usage. In I. de Sola Pool (Ed.), *The social impact of the telephone* (pp. 18–33). Cambridge, MA: MIT Press.

Arresty, E. (1970). *The best behavior. The course of good manners from antiquity to the present as seen through courtesy and etiquette books.* New York: Simon & Schuster.

Barker, A. (2000). *The Sunday Times. Improve your communication skills.* London: Kogan Page.

Barker, E. (1956). *Short cut to etiquette.* Kingswood: The World's Work.

Baron, N. (2000). *From alphabet to email.* London: Routledge.

Baron, N. (2003). Why email looks like speech. In J. Aitchison & D. M. Lewis (Eds.), *New media language* (pp. 85–94). London: Routledge.

Bartsch, R. (1987). *Norms of language.* London: Longman.

Basebuilders.com. (2004). Quick guide to voicemail. Retrieved March 14, 2006, from http://www.basebuilders.com/downloads/voicemailetiquette.pdf.

Bosticco, M. (1967). *Etiquette for the businessman at home and abroad.* London: Business Publications.

Burleigh, C. (1925). *Etiquette up to date.* London: T. Werner Laurie.

Business training works. (n.d). Getting beyond "hello." Professional telephone courtesy and customer service. Retrieved March 14, 2006, from http://www.businesstrainingworks.com/Onsite%20Training%20Web/Course%20Outlines/Telephone%20Customer%20Service%20Onsite%20Training%20Program%20Outline.htm.

Cameron, D. (1995). *Verbal hygiene.* London: Routledge.

Carney, M. (1948). *Etiquette in business.* New York: McGraw Hill.

Casson, H. (1910). *The history of the telephone.* Chicago: A.C. McClurg.

Castells, M. (2001). *The internet galaxy.* Oxford: Oxford University Press.

Chan, J. F. (2005). *E-mail: A write it well guide. How to write and manage e-mail in the workplace.* Oakland, CA: Write it Well.

Coulmas, F. (Ed.). (1981). *Conversational routine.* The Hague: Mouton.

Crystal, D. (2001). *Language and the internet.* Cambridge: Cambridge University Press.

Customer Focus. Voice mail etiquette. Retrieved July 30, 2007, from http://www.customerfocusinc.com/images/stories/pdfs/Voice%20Mail%20Etiquette.pdf.

Daly, S. J. (1946). *Personality plus.* New York: Dodd, Mead.

Davison, I. (1928). *Etiquette for women.* London: C. Arthur Pearson.

Dilts, M. (1941). *The telephone in a changing world.* New York: Longman, Green.

Esquire etiquette. (1954). London: Angus & Robertson.

Etiquette for ladies. (1923). London: Ward, Lock.

Fischer, C. (1992). *America calling. A social history of the telephone to 1940.* Berkeley: University of California Press.

Flynn, N., & Flynn, T. (2000). *The Sunday Times. Writing effective e-mail.* London: Kogan Press.

Friedman, N. (2000). *Telephone skills from A–Z: The telephone "doctor" phone book.* Boston: Crisp Learning.

Gough, V., & Grier, B. R. (1970). *Better telephoning. A plan to improve your telephone techniques.* Oxford: Pergamon.

Hale, C., & Wired. (Eds.). (1997). *Wired style. Principles of English usage in the digital age* (revised ed.). Berkeley, CA: Publishers Group West.

Hale, C., & Scanlon, J. (Eds.). (1999). *Wired style. Principles of English usage in the digital age.* New York: Broadway Books.

Harris, D. (2006). Pegasus mail. Etiquette guidelines for electronic mail. Retrieved March 14, 2006, from http://www.pmail.com/etqtte.htm.

Hall Ellis, M. J. (1990). *"Weston 100." A century of telephone service in Weston-super-Mare.* Bristol: British Telecom Severnside.

Hattaway, P. (1999). Telephone manners improve your image. *MG Business Mercantile Gazette,* 122.5333, 12.

Herring, S. (1998). Le style du courrier électronique: Variabilité et changement. *Terminogramme, 84–85,* 9–16.

Herring, S. (2003). Media and language change. Introduction. *Journal of Historical Pragmatics, 4*(1), 1–17.

Hewitt, B. (2001). Good business begins with good phone skills. Retrieved March 14, 2006, from http://www.bpic.co.uk/articles/phone_skills.htm.

Ingram, M. B. (2005). *Hiring call center professionals . . . when the process starts.* Retrieved July 30, 2007, from http://www.phonepro.org/hiring%20call%20center%20workers.html.

Jargon File Version 4.4.7. (2003, December). Retrieved March 14, 2006, from http://catb.org/~esr/jargon/.

Kerr, C., & Doody, J. (2007). *The bliss or "diss" connection? Email etiquette for the business professional.* Santa Ana: Execuprov Pr.

Laver, J. (1981). Linguistic routines and politeness in greeting and parting. In F. Coulmas (Ed.), *Conversational routine* (pp. 289–304). The Hague: Mouton.

Maddox, B. (1977). Women and the switchboard. In I. de Sola Pool (Ed.), *The social impact of the telephone* (pp. 262–280). Cambridge, MA: MIT Press.

Marvin, C. (1988). *When old technologies were new.* Oxford: Oxford University Press.

Miall, A. (1950). *Modern etiquette.* London: C. Arthur Pearson.

Millar, S. (1993). In pursuit of clarity: An analysis of speech education manuals. *Language and Communication, 13*(4), 287–303.

Millar, S. (2002). Eloquence and elegance: Ideals of communicative competence in spoken English. In R.Watts & P. Trudgill (Eds.), *Alternative histories of English* (pp. 173–190). London: Routledge

Morton, D. (2000). *Off the record: The technology and culture of sound recording in America.* Piscataway, NJ: Rutgers University Press.

Netlingo. (n.d.). The internet dictionary. Retrieved March 14, 2006, from http://www.netlingo.com.

New book of etiquette. (1907). London: Cassell.

Norwood, H. (1937). *Common sense etiquette dictionary.* Emaus, PA: Rodale.

Occomore, D. (1995). *Number please. A history of the early London telephone exchanges from 1880 to 1912.* Romford: Ian Henry.

Orlikowski, W. J., & Yates, J. (1994). Genre repertoire: The structuring of communicative practices in organizations. *Administrative Science Quarterly, 39*(4), 541–574.

Orlikowski, W. J., Yates, J., Okamura, K., & Fujimoto, M. (1995). Shaping electronic communication: The metastructuring of technology in the context of use. *Organization Science, 6*(4), 423–444.

Pattaro, E., & Sartor, G. (2002). *Norms, laws and the internet.* Retrieved March 14, 2006, from http://www.ieid.org/congreso/ponencias/sartorpattaro.pdf.

Perry, C. (1977). The British experience 1876–1912: The impact of the telephone during the years of delay. In I. de Sola Pool (Ed.), *The social impact of the telephone* (pp. 69–96). Cambridge, MA: MIT Press.

Phillips, T. (2000). *The postcard century.* London: Thames & Hudson.

Phonepro.com. (n.d.). Training courses. Retrieved March 2006, from http://www.phonepro.com/training.html.

Post, E. (1922). *Etiquette in society, in business, in politics and at home.* New York: Funk & Wagnalls.

Post, E. (1937). *Etiquette. The blue book of social usage.* New York: Funk & Wagnalls.

Post, E. (1950). *Etiquette. The blue book of social usage* (new ed.). New York: Funk & Wagnalls.

Pringle, G. (1949). *Etiquette in Canada* (2nd ed.) Toronto: McClelland & Stewart.

Quigley, M., & Clarke, M. (1926). *Etiquette jr.* London: Heinemann.

Reigh, M. (2006). *Telephone tip—delivering great customer service with your voice.* Retrieved July 30, 2007, from http://www.myvoicespeakssuccess.com/?cat=4.

RFC. (1995). RFC 1855: Netiquette guidelines. Retrieved March 14, 2006, from http://www.stanton.dtcc.edu/stanton/cs/rfc1855.html.

Richardson, A. S. (1927). *Etiquette at a glance.* New York: Appleton & Co.

Richardson, S. (2005). *Impressive first impressions.* Retrieved March 14, 2006, from http://www.phonepro.org/ImpressiveFirstimpressions.html.

Rinaldi, A. (1998). *The net: User guidelines and netiquette.* Retrieved March 14, 2006, from http://www.cs.biu.ac.il/home/leagal/netguide/index.html.

Rives, H. W. (1926). *The complete book of etiquette.* Philadelphia: J.C. Winston.

Roosevelt, A. E. (1962). *Eleanor Roosevelt's book of common sense etiquette.* New York: MacMillan.

Routledge's manual of etiquette. (1875). London: Routledge.

Saville, W. A. (1882). *Social etiquette for ladies and gentlemen.* New York: J. W. Lovell.

Scammell, J. (1924). *Use of the telephone in business.* New York: Ronald Press.

Schlesinger, A. (1947). *Learning how to behave. A historical study of American etiquette books.* New York: Macmillan.

Shea, V. (1994). *Netiquette.* San Francisco: Albion Books.

Sherwood, K. D. (2001). *A beginner's guide to effective email.* Retrieved March 14, 2006, from www.webfoot.com/advice/email.top.html.

Shipley, D., & Schwalbe, W. (2007). *Send: The essential guide to email for office and home.* New York: Knopf.

Skype.com. (2007). *Skype etiquette.* Retrieved July 30, 2007, from http://www.skype.com/company/legal/terms/etiquette.html

Smith, M., & Kollock, P. (Eds.). (1999). *Communities in cyberspace.* London: Routledge.

Southern Illinois University at Carbondale. (n.d.). Voice mail etiquette. Retrieved March 14, 2006, from http://infotech.siu.edu/telecom/vm_etq.html.

Standage, T. (1999). *The Victorian internet.* London: Phoenix.

Steele, J. (2006). *Email: The manual: Everything you should know about email etiquette, policies and legal liability before you hit send.* Oak Park, IL: Marion Street Press.

Stevens, F. (1884). *The usages of best society. A complete manual of social etiquette.* New York: A.C. Burt.

Stevens, W. O. (1934). *The correct thing. A guide book of etiquette for young men.* New York: Seers.

Storrer, A., & Waldenberger, S. (1998). Zwischen Grice und knigge. In H. Strohner, L. Sichelschmidt, & M. Hielscher (Eds.), *Medium sprache* (pp. 63–77). Frankfurt am Main: Peter Lang. Retrieved August 10, 2006, from www.ids-mannheim.de/grammis/orbis/net/netfram2.html.

Taylor, T. J. (1990). Normativity and linguistic form. In H. Davis & T. Taylor (Eds.), *Redefining linguistics* (pp. 118–148). London: Routledge.

Thornborough, L. (1924). *The etiquette of letter writing.* New York: Baran Hopkins.

Trim, P. (1997). *Winning telephone tips.* Franklin Lakes, NJ: Career Press.

Troubridge, Lady L. (1931). *The book of etiquette.* London: Assoc. Bookbuyers.

Tulane University, Technology Services. (n.d.) Telephone etiquette. Retrieved July 30, 2006, from http://ts.tulane.edu/services/telecommunications/phone_etiquette.cfm.

Van de Water, V. (1936). *Present day etiquette.* New York: A.C. Burt.

Vogue's book of etiquette. (1935). Garden City, NY: Doubleday, Doran.

Wage, J. L. (1973). *How to use the telephone in selling.* Epping: Gower Press.

Walker, L. (1998). *A self-development programme. Telephone techniques.* London: Marshall.

Webster, D., & Hopkins, M. (1930). *Mrs. Grundy is dead. A code of etiquette for young people.* New York: Century.

Whelan, J. (2000). *E-mail @ work.* Business.ft.com books. Edinburgh: Pearson Educational.

Wilson, M. (1947). *The new etiquette. The modern code of social behavior.* Philadelphia: Lippincott & Co.

Yates, J. (1989a). *Control through communication: The rise of system in American management.* Baltimore: Johns Hopkins University Press.

Yates, J. (1989b). The emergence of the memo as a managerial genre. *Management Communication Quarterly, 2,* 485–510.

Yates, J., & Orlikowski, W. J. (2002). Genre systems: Structuring interaction through communicative norms. *Journal of Business Communication, 39*(1), 13–35.

York Technical College, South Carolina. (n.d.). Telephone etiquette. Retrieved March 14, 2006, from http://www.yorktech.com/department/jobplacement/Phoneetiquette.html.

Young, P. (1991). *Person to person. The international impact of the telephone.* Cambridge: Granta Editions.

Chapter 12

Diary 2.0?

A Genre Moves from Page to Screen

LAURIE McNEILL

University of British Columbia

Abstract

This chapter analyzes the Internet diary, examining its relation to both print culture practices (the "offline" diary) and digital texts (in particular, "blogs"). Focusing on understandings of genre and pragmatic elements of text (presuppositions and long spans), I explore how and if the move from page to screen has changed the diary, and with what implications. The study argues that, after a brief dalliance with newness, Internet diarists have re-embraced the traditional diary, reproducing print diary conventions in the online writing world.

Since the late 1990s, the diary has become one of the most popular online genres to grace computer screens. Hundreds of thousands of people—some estimates suggest millions (Herring, Scheidt, Bonus, & Wright, 2005)—are writing and (instantly) publishing serial chronicles of daily life. Running the gamut from the stereotypical schoolgirl's catalogue of crushes to the daily musings of Hollywood starlets, Internet diaries (also known as "blogs") give access to lives in progress to a potential audience of thousands. Likely more people than ever are now writing, reading, and talking about diaries, and many Internet diarists have used their "amateur" narratives to achieve "professional" writing acclaim. Actor and blogger Wil Wheaton, for instance, has published two books based on his blog, and Julie Powell turned her diary blog into the best-selling cookbook-memoir *Julie and Julia: 365 Days, 524 Recipes, 1 Tiny Apartment Kitchen*. Online diaries have attracted media attention (and derision), and now no product roll-out or political career can be complete without the requisite Weblog (blog) to accompany it. The Internet diary has ignited firestorms of controversy among users, particularly those who maintain that blogs and diaries are separate entities; although in practice these two forms have intermarried, many would argue that their bloodlines are and should remain distinct. The speed and relative ease with which diaries, a centuries-old print form, have adapted to the World Wide Web, make this genre an excellent case study for examining generic transitions in the age of the Internet. I chart the history of this particular genre's digitization and explore some reasons for its successful cross-over. As well, I consider how online publication has affected language and generic features, allowing users to simultaneously uphold and subvert traditional diary elements.

What's in a Name: Blogs and Diaries

This analysis begins with a question: Can we—and should we—read the online diary as another form of the genre, its most recent evolution, or, as my title suggests, the latest updated version? Such characterizations suggest improvement, significant developments in form that render previous versions obsolete, just as the seemingly endless stream of new software makes our old systems digital dinosaurs. New theories of genre invite this kind of thinking, as they shift the conceptual framework for understanding "genre" away from reading through rigid classification to appreciating formal dynamism, flexibility, and change over time. Kathleen Jamieson (1973) drew explicit attention to the evolutionary metaphor by encouraging critics "to approach the study of genres with a Darwinian rather than a Platonic perspective" (168), and the notion of evolution has become foundational to revised theorizations of genre. I concur quite heartily with these conceptualizations, and the diary genre could be a poster child demonstrating the validity of these claims. Over its centuries of existence, it has shown remarkable resiliency, adapting and adopting many other modes into its own as it has changed to suit the changing situations to which it responds, incorporating the (unsent) letter, for example, when the diarist is away from home, especially through (forced) immigration, exile, or war (e.g., most famously, the *Diary of Anne Frank*), or the fieldnotes of a participant-observer in an ethnographic study (e.g., *The Ordeal of Elizabeth Vaughan*). I am, however, slightly wary of the narrative of progress that has become (inaccurately) attached to terms such as *evolution*, particularly when we look at genres at the intersection of print and digital culture. Does going online actually change the genre itself, at the genetic level, or simply alter its features and functions, giving the genre the textual equivalent of blue or brown eyes or curly or straight hair? Exploring this issue can help us tease out the relationship between print and online texts, as well as between types of Internet texts (blogs and diaries), and, in particular, determine how to position digital diaries in the long history of the diary genre.

In this analysis, I focus on Internet diaries, however, since researchers have shown (e.g., Herring et al., 2005; Herring, Scheidt, Kouper, & Wright, 2006; Su, Wang, & Mark, 2005; Trammell, Tarkowski, & Sapp, 2006) that many so-called "blogs" ("web pages whose content are periodic, reverse chronologically ordered posts" [Su et al., 2005:1]) operate as "diaries." The politics of naming remain fraught, with those firmly in the blog-is-not-a-diary camp remaining particularly vehement about generic distinctiveness.[1] I suggest this generic wagon-circling testifies to the aura of newness that surrounded the Internet, especially in the nascence of the World Wide Web, when the possibilities for social change and community seemed ripe. It was important, then, to develop unique forms that responded to this new medium, situation, and terrain, and the blog, as seen later, was part of that development. Keeping a "diary" online may have seemed—and may continue to seem—counter to this ideology of newness, a de-evolution. For, even in this new environment, print culture associations persist, particularly around gender and genre. Studies by Herring and Paolillo have shown that more men than women produce filter-style blogs, that is, annotated lists of links to particular Web content. Although these forms of blogs are the minority in the blogosphere, they receive a disproportionate amount of media attention, and thus perpetuate the notion that the important, the "A-list," the cutting-edge online writing takes place in filter blogs instead of in diary blogs, the silent majority (Herring et al., 2005; Herring & Paolillo, 2006). This disconnect between theory and practice, or hype versus reality, suggests the depth of cultural investments in some positions: that women are more introspective than men, more interested in the interior, domestic, and private life, for instance, whereas men engage with the public sphere, and their life narratives have a particular relevance or import because of that milieu. Similar associations once were attached to the "diary" versus the "journal" before those distinctions became critically unfashionable, and it is significant that the Internet reignites rather than dispenses with such old debates. Despite Web pioneers' attempts to create a communicative situation that would dispense with the need for old responses, as well as their often problematic cultural associations, actual users have brought over their practices and the prejudices, meaning that they read and write with sensibilities forged both in the past and the present.

Generic History and (Re-)Invention: Diaries On- and Offline

This combination of "newness" and "tradition" means that the diary genre comes to the Internet with a great deal of cultural baggage, loaded down with popular perceptions that are not particularly flattering. Associated in the 20th century with the intensely personal and confessional, the sorts of secrets that should be hidden away under lock and key, the diary has been cast as an amateur mode meant only for the diarist's eyes. The ordinary person, in ordinary circumstances (i.e., not in the midst of war or working for a celebrity), would not be likely to write, and certainly would be unable to publish, a diary meant for a wide audience; indeed, doing so would be read as narcissistic or hubristic at worst, and at least as generically invalid. However, for the English-language diary, this ban on readers is a relatively recent development that began mid-19th century, tying into the rise of Romanticism and its valorizing of the introspective emotional life of the individual. At the same time, new developments in the sciences, particularly in the emerging field of psychoanalysis, drew attention to the psyche and the potential secrets locked within the human mind, and suggested that hidden aspects of the self could be unleashed through talk therapy or writing (Gannett, 1992)—an idea that appears, unchanged, in contemporary self-help "journaling" guides. The publication in 1825 of Samuel Pepys' decoded diary added to public interest in reading diaries at the same time that it reinforced the idea that one shouldn't *be able* to read them; this paradox gave the act of reading others' diaries a certain frisson of the forbidden.

Before this time, however, the diary had many communal functions that raised no eyebrows: Travellers and immigrants, away from home and reliable mail services, kept diaries and sent them back home to update loved ones; individuals wrote spiritual diaries recounting—and accounting for—their devotion, and shared these documents with their religious advisors; family members contributed narratives to shared diaries that became clan chronicles to pass on to future generations. Overlapping with the letter, the commonplace book, and the travel narrative, among other forms, the diary has always *in practice* contained a public element. Only in the 20th century did the diary become synonymous with secrecy and privacy; the now-traditional book that comes with lock and key (underscoring both the clandestine nature of the contents and their value) is a product of the early 1900s (Culley, 1985). That 20th-century readers resisted this reality, insisting that the "authentic" diary—that is, the unaltered, unmediated, and truly private text that operated as a window in the writer's "true self"—was personal and not meant for circulation, suggests changing social and cultural values, rather than any particular fidelity to a "pure" form of the genre. Current blog research that argues against reading blogs as diaries based on the public element of texts (e.g., Miller & Sheperd, 2004; Nardi, Schiano, & Gumbrecht, 2004) reflects this relatively recent understanding of the diary as "written and stored in secret [and] secured by lock and key" (Nardi et al., 2004) without recognizing the genre's much longer and more complex history.

With at least four centuries of production in English, the diary has demonstrated tremendous staying power, perhaps due to in part to its seemingly endless flexibility. It seamlessly absorbs, sponge-like, the functions and audiences of other forms (e.g., the list, the dialogue, the almanac, the report) in order to serve the needs of the individual diarist, the rhetorical purposes for which the individual has turned to this particular form, while the writing retains enough "diary-ness" to remain recognizable as a diary (i.e., as a first-person, cumulative account, typically told chronologically and episodically, with entries often, although not always, separated by date, and concerned primarily with the eyewitness experiences of the diarist). If we think in Darwinian terms, the diary demonstrates the genetic wiliness that ensures the survival of the fittest. What the diary is "fittest" for, however, has changed over time and practically with each use, as diarists employ this everyday form in every possible circumstance, recording the most mundane details—the daily life of a midwife, for instance, as Laurel Thatcher Ulrich chronicled—and the most extraordinary circumstances—the invasion of China by Japanese forces, for example, as recorded in the instant by Peggy Abkhazi. With its ability to capture the small details of life, it privileges the viewpoint of the private citizen as well as the famous: because anyone can keep a diary, anyone can become part of history. However, it is only very recently that the public has felt any interest at all in the history most diary narratives represent: the ordinary, the quotidian, the domestic. Moreover, access to such records has remained spotty, with the traditional mechanisms of the research and publishing world—archivists, agents, editors, and publishers—determining whose diaries are worth our reading.

The explosion of diaries onto the World Wide Web has, of course, changed and challenged even these systems. Contemporary readers might expect the diary to be quite remote from the needs of Internet users, but here again practice and stereotype diverge, and the diary reinvents itself once more for a new generation of diarists. Diaries began to appear online in 1995, typically developing out of the personal home pages of individuals already involved with Internet technologies (Blood, 2002a); such texts featured personal, intimate narratives about the daily lives of the diarists (e.g., Justin Hall's *Links*, one of, if not the, earliest online diary). *Weblogs*, a term coined by blogger Jorn Barger in 1997 (Blood, 2002b), similarly attracted "power users" of the Web (Blood, 2002a). These pages, which featured daily lists of annotated links to other sites, acted as filters for the burgeoning content of the World Wide Web, directing readers to material the blogger found particularly worthwhile. Although some overlap existed between the forms—a diarist could point to links he recommended, and the blogger's annotation might be quite personal—they were distinct in this initial phase. The early Internet life writers were, therefore, users who typically had technological know-how and, perhaps more importantly, Web savvy, some sense of the World Wide Web as a place for community as well as communication.

With the introduction in 1999 of Pitas and Blogger, two free blog-building and publishing tools that did not require knowledge of HTML, the numbers of both diaries and blogs mushroomed (Blood, 2002b). Rebecca Blood (2002a) noted that "[o]nce literally *anyone* could make a weblog, literally anyone did" (x). As these sites and forms became less specialized, blogs and Internet diaries also became less discrete, with more bloggers incorporating journal components, such as narratives of their daily lives, into their traditional "[l]ink-plus-commentary" formula (Powazek, 2002:3); these journal blogs shifted shifted focus from "the Web-at-large" to the personal lives of the writers (Blood, 2002b:10). With the huge influx of new writers into the blogging and online diary communities, both forms evolved to suit their users' needs. Text supported by hyperlinks (pointing either within or outside the site), regular updates, and dated entries—the "basics of blogging" (Miller & Shepherd, 2004:9)—are, increasingly, the hallmarks of so-called "Internet diaries" as well. Collapsing these related forms into one, the latest generation of writers joins personal narratives with commentary about the Web and its contents.

In a series of analyses, the researchers of BROG (Blog Research on Genre Project) have tested hypotheses about blogs, taking note of form (e.g., "filter-style" vs. "personal journal") and features, including the presence of hyperlinks, photographs, and other non-text media, and comments pages; their studies have also examined bloggers' genders and ages. Performing longitudinal content analysis (Herring et al., 2006), content analysis (Herring et al., 2005), and multivariate analyses of stylistic features (Herring & Paolillo, 2006) on samples of randomly selected English-language, primarily text-based weblogs (excluding photo, video, and audio blogs), these authors sought to investigate through empirical methods the conventional ideas and expectations about weblogs, this "popular new mode of computer-mediated communication (CMC)" (Herring & Paolillo, 2006). Because filter-style blogs, the kind championed as the "pure" blog and as an emerging genre quite distinct from diaries (e.g., Miller & Shepherd, 2004; Rak, 2005), receive the most popular attention, the authors expected most of the 2.1 million self-identified blogs to be in this mode (Herring et al., 2005). However, even though their studies excluded online journal sites such as LiveJournal and DiaryLand (in part because they promoted "diaries" and not "blogs"), and sampled "only sites that could be described uncontroversially as weblogs" (Herring et al., 2005), Herring et al. (2005) found that "the blogs in our sample are overwhelmingly of the personal journal type (70.4%), in which the authors report on their lives and inner thoughts and feelings." Moreover, they posited that this focus on the world of the blogger instead of the Web accounts for these texts having fewer links and comments than expected. A recent study of 358 randomly selected Polish weblogs (Trammell et al., 2006) similarly observed that "[t]he blogs analyzed rarely contained hyperlinks in the posts" and that 92.2% of their sample "was categorized as being like a diary or a journal."

The random samples analyzed in these studies move research past generalized observations about blogs that the media have, in their attention to a few examples, turned into dogmatic beliefs about practices that are, in reality, quite varied and less journalistic than may be popularly believed. As Su et al. (2005) noted, "despite the fame high-profile blogs have gained, the majority of blogs are *not* political in nature." Turning from "A-list" blogs, the ones regularly cited by the media and that frequent Technorati's Top 100, list to "ordinary blogs" (Herring et al., 2006), we can see what the majority of bloggers are doing with this form or, more accurately, these forms. In the results of these studies, and in my own research, I read a pattern of Internet life writers reembracing the diary in its most recognized form; tellingly, the pioneering software site Blogger's definition of the blog puts it in the terms that would be most familiar to new users: "A blog is a personal diary" (blogger.com). Originally seduced by the promise of the new embodied by the "blog"—a form free of print culture baggage—online writers who seek to produce personal, daily narratives have, in their Internet texts, turned to an online form that is closest to a traditional print culture genre. Although they might adopt the new name—identifying their sites as blogs and themselves as bloggers—these writers keep producing texts that are recognizable, in form, content, and style, as diary entries (see Bloom, 1996; Bunkers & Huff, 1996, McNeill, 2003; Podnieks, 2000). Their narratives follow the generic convention of print diaries by starting *in medias res*, with the most recent entry appearing as the diary site is visited; they plunge the reader into the immediate events of a life, "what happened to me today," rather than providing any background—beginning at the beginning—as a retrospective account of a life would do. Like print diaries, online diaries focus on the quotidian and the personal; they foreground the diarist's experiences and emotions, are written in the first person, and typically adopt a casual tone. These stylistic features are key elements of diary prose, the "basics of diaries," to adapt Miller and Shepherd's phrase, and users' upholding of them online underscores their generic import.

But although the "New Diarists" continue to write diaries that are quite traditional in many aspects, they have kept the new bells and whistles that the Web allows; indeed, without such features as photographs, comments pages, and links to the writer's e-mail, tags, and lists of linked "friends" or other blogs, diarists have difficulty attracting and maintaining readership—a concern not shared by the majority of print diarists. These features, carry-overs from uniquely digital forms such as the personal home page (Dillon & Gushrowski, 2000), constitute generic cherry-picking on the part of bloggers (or, more accurately in most cases, blogging software designers), who add the digital devices to the diary's repertoire without fundamentally altering the genre: they take what works for their texts (links or contact information or comments, for instance) and keep producing narratives consistent with offline diaries. As Kevin Crowston and Marie Williams (2000) argued, the "[u]se of links and other Web features does not necessarily change the genre of the page" (208). As is seen here, however, they do change how diarists reproduce the genre and maintain the appearance of the "authentic" diary—the personal, chatty, intimate narrative of daily life that the diarist writes for his or her own pleasure, not for (hundreds if not thousands of) readers' perusal.

Intimate *and* Public: Stylistic Features of Online Diaries

Online diarists can uphold diary style while using Internet devices to improve their texts' accessibility and therefore enrich readers' comprehension. Specifically, like their print counterparts, Internet diaries rely on *presupposition* (a term used here, as in Levinson, 1983, in the pragmatic sense, to indicate statements that assert rather than introduce information). Diaries, both online and off, are rife with presupposing expressions, what Levinson called presupposition-triggers, such as proper names, definite reference, deixis (utterances that situate and anchor language, for example, in space ["over there"] and time ["today"]), and anaphorics (words that refer to another word, typically an antecedent). These expressions, signaling information that the writer assumes the addressee shares, provide key signposts for interpreting common ground and pinpointing audience design. Taking as common ground or "shared or noncontroversial knowledge" (Grundy, 2000:10) what they assert, presupposing expressions draw upon the speaker's assumption that her addressees share her worldview and experiences. Thus when Web-diarist Jane Pinckard (2001–2005) wrote, "so i [*sic*] didn't have tickets to the show last night, but we were headed there anyway. if it didn't work out, we could always just have a drink with jee before she left for her trip," she assumes readers know the meaning of "the show," "jee," and "her trip," articulating in consequence a very particular imagined reader.

Studying audience design, as based on presupposition and other factors, in diaries is a rewarding but unexpected arena of research, given the contemporary notion that "true" diaries, as private texts not meant for other eyes, won't have readers, and therefore diarists aren't designing for them. However, these texts, like any other, are addressed (Bakhtin, 1986), and therefore accommodate the range of recipients typically accounted for by audience design theorists (e.g., Bell, 1977; Clark, 1992; Giles & Powesland, 1975). Because of Internet diarists' fidelity to the diary genre's style and content, principles of diary audience design apply online as well (however, as I discuss later, with some innovations made possible by the change in medium). Building on and significantly developing the observations of Goffman (1981), Bell and Clark each outline the participants in conversation, both the addressed and the nonaddressed (Goffman, 1981:9): addressees (the intended recipient), auditors ("present in the group but not directly addressed," Bell, 1997:246), and overhearers, who can be eavesdroppers or bystanders (Clark & Carlson, 1992). The speaker/writer can choose to adopt either a disclosing or nondisclosing attitude toward nonaddressees; the choice of attitude will determine the roles of overhearers as either bystanders or eavesdroppers (Clark & Carlson, 1992). A disclosing speaker (or writer) appears to address one audience with whom they share specific information but simultaneously includes other audiences who may not have this knowledge in common. In text-based communication, parenthetical asides are ways that a writer can discreetly disclose information, allowing an outsider into the "inner circle." Common ground becomes the means for establishing and policing community boundaries, and, in most cases, the speaker/writer has the power to let readers in, or keep them out, based on what information can be presumed to be shared.

In the diary, presuppositions indicate the intimacy of address the diarist adopts in a particular entry, and show—even in a supposedly "private" text—how the writer can project multiple, simultaneous audiences for his or her narrative. Assuming information as common, the diarist fosters a sense of intimacy in the text: he or she assumes that readers (imagined or real) are "in the know," and are therefore addressed by the entry, as we see in Pinckard's writing. At the same time, presupposed knowledge that a reader does not, in fact, share can be a sign of a "private" conversation that readers are, to an extent, "overhearing"—they are not addressed. In contemporary print and digital diaries, this sense of overhearing is important to upholding reader expectations of the private text: If a diary has no other reader than the diarist, the diarist would ostensibly never need to explain anything; she knows what she means. As I have argued elsewhere (McNeill, 2003), even in online diaries, where readers are not only welcome but actively pursued, diarists continue this style of address.

Both on- and offline diarists may address particular kinds of readers specifically, but will design some or all of their entries to be understood by a wider range of participants, even overhearers, striving for maximum accessibility depending on the function that a particular entry serves. As Clark and Carlson (1992) noted, the speaker has a responsibility to design her utterances to be understood by all participants. In this aspect, online diarists, walking the fine generic line by writing diaries that look traditional but that are, quite evidently, very public, have the advantage over print writers. With tools such as links, Internet diarists can maintain the appearance of intimacy, while encouraging new readers—the overhearers—to become addressees by learning the information held as common, by using the archives, for instance, to reconstruct the context. Using hypertext, diarists can create links between relevant entries, allowing readers to put events in context and divine the entire story without the diarist having to explain. For example, Wil Wheaton (2007:http://wilwheaton.typepad.com/) peppers his diary entries with links, connecting offsite to the personal web pages of friends he spent time with or to events he has attended, and internally to previous entries. These links create a complex web of cross-references that lets the reader follow the diarist's experiences without feeling as if they are missing background or crucial connections. Similarly, pioneer Internet diarist Steve Schalchlin of *Living in the Bonus Round* writes a long entry that reflects back ten years to his first online diary entries. Recounting key incidents of that period of his life, he recalls, "Our opening night was one of the greatest nights of my life" (July 31, 2006). The link on "opening night" takes the reader back to the original July 1996 entry, allowing newer readers to understand exactly what Schalchlin means without him having to retell the story, a narrative move that might bore or alienate his longtime readers who may remember the story.

Through these links to other sites and within the diary, both new and old readers will grasp the import of what the diarists describe as it unfolds. Readers of

Web diaries do not need to keep track of references that have significant associations with the current entry but that may have been made days, even years, earlier. Indeed, the long spans between referent and antecedent that have been hallmarks of the traditional print diary may have been made obsolete by hypertext and searchable site archives. The spans will, of course, still exist, but the presence of the linked text announces that a connection is being made and shows the reader the original reference. Similarly, presupposition continues to work as a form of efficiency, but on the Web writers can be accommodating as well as efficient. Links act as the new diary shorthand; they serve the same functions as initials or presupposing expressions do in manuscript diaries, saving diarists from writing out full names, for example, or continuously explaining references, but are much more accessible (and indeed more informative) for a wider number of readers.

Whereas print diarists who publish their texts for a general readership would have to intervene in the entry more obtrusively, through parenthetical asides or footnotes, Web diarists can maintain the illusion of continuity: Wheaton and Schalchlin can carry on their intimate utterances, designed for audiences in the inner circle, on the main page of the diary and supply necessary background information obliquely through links. Even in a public venue, diarists do not typically advertise that they are adopting a disclosing attitude towards overhearers—doing so would compromise the "authenticity" of the reading experience by reminding readers of the diary's public setting. Links allow diarists to preserve that authenticity. "Authentic" diary features—the assumption of shared knowledge, the nondisclosing attitude—remain intact, on the surface, but can be overwritten or redeployed through Web technologies. In the culture of mediated voyeurism/exhibitionism that permeates the Internet, and which carries over to Internet diaries from their print culture association with reading others' "private" (secret) writing, this subtly accessible audience design allows readers to maintain that sense of reading the "real" thing, the thrill of peeking into online lives.

This careful attention to readers' needs and generic expectations—acknowledging the audience's desire for the private and personal, casual and daily diary narrative—demonstrates that bloggers very conscientiously work towards upholding the diary genre rather than inventing a new form to replace it. Perhaps the "blog," in the original news filter format that users so vehemently defended as distinct from the diary, has not been able, in the long term, to meet the needs of the growing number of Internet life writers. Genre theories tell us that, faced with a new or uncertain situation, rhetors will turn to familiar forms and, when possible, adapt them to the novel environment. Because the majority of users have recreated the print diary in an online setting, the early promise of novelty that the blog held has perhaps not been realized, and may never be. But it shapes the production and reception of the diary in this cultural moment, as readers and writers still have the sense of doing something a bit different with the diary when they keep them online instead of on paper, breaking the generic taboo of writing a diary for live, responsive readers.

Conclusion

The Internet diary, then, is a genre overwritten with current and traditional expectations and needs of its users. Returning to my opening questions, I conclude that although we should not view this form as it is practiced by the majority of producers as a new genre, we should also think carefully about casting it as the logical evolution of the diary. Certainly, the mode has taken on new, sometimes multimodal features, and potentially new—or at least different—uses, particularly as it reaches out to wider, responsive audiences. But whether such innovations cause permanent change, enough to fork the genetic path so that print and online diaries diverge, with the potential consequence of one branch weakening and falling out of use, remains to be seen. Given bloggers' return to producing quite standard diary narratives instead of "blogs," I wonder if we might more usefully consider the Web diary as one of the many instances of specific-use diaries that co-exist quite happily, such as diaries dedicated to travel or dreams. The version of the diary that the Web situation creates—what I've called Diary 2.0—might well be not quite an evolution, but perhaps a *re*-evolution, as users turn again to the original-with-a-difference.

Note

1. See, for example, Neale Talbot's (2002) memorable anti-diary rant in *We've Got Blog.*

References

Abkhazi, P. (2002). *A curious cage: A Shanghai journal 1941–1945* (S. W. Jackman, Ed.). Victoria, BC: Sono Nis.

Bakhtin, M. M. (1986). The problem of speech genres (V. W. McGee, Trans.). In C. Emerson & M. Holquist (Eds.), *Speech genres and other late essays* (pp. 60–102). Austin: University of Texas Press.

Bell, A. (1997). Language style as audience design. In N. Coupland & A. Jaworski (Eds.), *Sociolinguistics: A reader and coursebook* (pp. 240–269). London: MacMillan.

Blood, R. (2002a). Introduction. In J. Rodzvilla (Ed.), *We've got blog: How weblogs are changing our culture* (pp. ix–xiii). Cambridge, MA: Perseus.

Blood, R. (2002b). Weblogs: A history and perspective. In J. Rodzvilla (Ed.), *We've got blog: How weblogs are changing our culture* (pp. 7–16). Cambridge, MA: Perseus.

Bloom, L. Z. (1996). "I write for myself and strangers": Private diaries as public documents. In S. Bunkers & C. Huff (Eds.), *Inscribing the daily: Critical essays on women's diaries* (pp. 23–37). Amherst: University of Massachusetts Press.

Bunkers, S., & Huff, C. (Eds.). (1996). *Inscribing the daily: Critical essays on women's diaries* (pp. 23–37). Amherst: University of Massachusetts Press.

Clark, H. (1992). *Arenas of language use.* Chicago: University of Chicago Press.

Clark, H., & Carlson, T. B. (1992). Hearers and speech acts. In H. Clark (Ed.), *Arenas of language use* (pp. 205–247). Chicago: University of Chicago Press.

Crowston, K., & Williams, M. (2000). Reproduced and emergent genres of communication on the World Wide Web. *The Information Society, 16*(3), 201–216.

Culley, M. (Ed.). (1985). *A day at a time: The diary literature of American women from 1764 to the present.* New York: Feminist Press.

Dillon, A., & Gushrowski, B. A. (2000). Genres and the web: Is the personal home page the first uniquely digital genre? *Journal of the American Society for Information Science, 51*(2), 202–205.

Gannett, C. (1992). *Gender and the journal: Diaries and academic discourse.* Albany: State University of New York Press.

Giles, H., & Powesland, P. F. (1975). *Speech style and social evaluation.* London: Academic Press.

Goffman, E. (1981). *Forms of talk.* Philadelphia: University of Pennsylvania Press.

Grundy, P. (2000). *Doing pragmatics.* New York: Oxford University Press.

Herring, S., & Paolillo, J. C. (2006). Gender and genre variation in Weblogs. *Journal of Sociolinguistics, 10*(4), 439–459. Preprint retrieved 6 August 2006, from http://ella.slis.indiana.edu/~herring/jslx.pdf.

Herring, S. C., Scheidt, L. A., Bonus, S., & Wright, E. (2005). Weblogs as a bridging genre. *Information, Technology, and People, 18*(2), 142–171. Preprint retrieved August 9, 2006, from http://www.blogninja.com/it&p.final.pdf.

Herring, S. C., Scheidt, L. A., Kouper, I., & Wright, E. (2006). A longitudinal content analysis of Weblogs: 2003–2004. In M. Tremayne (Ed.), *Blogging, citizenship, and the future of media.* London: Routledge. Preprint retrieved August 7, 2006, from http://ella.slis.indiana.edu/~herring/tremayne.pdf.

Jamieson, K. M. H. (1973). Generic constraints and the rhetorical situation. *Philosophy and Rhetoric, 6*(3), 162–170.

Levinson, S. (1983). *Pragmatics.* Cambridge: Cambridge University Press.

McNeill, L. (2003). Teaching an old genre new tricks: The diary on the Internet. *Biography, 26*(1), 24–47.

Miller, C. R., & Shepherd, D. (2004). Blogging as social action: A genre analysis of the Weblog. In L. J. Gurak, S. Antonijevic, L. Johnson, C. Ratliff, & J. Reyman (Eds.), *Into the blogosphere: Rhetoric, community, and culture of Weblogs.* Retrieved August 8, 2006, from http: blog.lib.umn.edu/blogosphere.

Nardi, B., Schiano, D., & Gumbrecht, M. (2004). Blogging as social activity, or, would you let 900 million people read your diary? *Proceedings of Computer Supported Cooperative Work.* Retrieved August 15, 2006, from http://home.comcast.net/%7Ediane.schiano/CSCW04.Blog.pdf.

Pinckard, J. (2001–2005). *Umami Tsunami.* Retrieved November 30, 2006, from http://www.umamitsunami.com/.

Podnieks, E. (2000). *Daily modernism: The literary diaries of Virginia Woolf, Antonia White, Elizabeth Smart, Anais Nin.* Montreal & Kingston: McGill-Queen's University Press.

Powazek, D. M. (2002). What the hell is a Weblog and why won't they leave me alone? In J. Rodzvilla (Ed.), *We've got blog: How Weblogs are changing our culture* (pp. 3–6). Cambridge, MA: Perseus.

Rak, J. (2005). The digital queer: Weblogs and Internet identity. *Biography, 28*(1), 166–182.

Schalchlin, S. (1997–2007). *Living in the bonus round* (Internet diary). Retrieved August 15, 2006, from http://www.bonusround.com.

Su, N. M., Wang, Y., & Mark, G. (2005). Politics as usual in the blogosphere. *Proceedings of the 4th International Workshop on Social Intelligence Design (SID 2005).* Retrieved December 4, 2006, from http://www.isr.uci.edu/~normsu/papers/Su-Politics-SID05.pdf.

Talbot, N. (2002). Put the keyboard down and back away from the Weblog. In J. Rodzvilla (Ed.), *We've got blog: How Weblogs are changing our culture* (pp. 47–56). Cambridge, MA: Perseus.

Trammell, K. D., Tarkowski, A., & Sapp, A. M. (2006). Republic of blog: Examining Polish bloggers through content analysis. *Journal of Computer-Mediated Communication, 11*(3). Retrieved November 30, 2006, from http://jcmc.indiana.edu/vol11/issue3/trammel.html.

Ulrich, L. T. (1990). *A Midwife's tale: The life of Martha Ballard, based on her diary, 1785–1812.* New York: Vintage.

Vaughan, E. (1985). *The ordeal of Elizabeth Vaughan* (C. M. Petillo, Ed.). Athens: University of Georgia Press.

Wheaton, W. (2004–2007). *WWDN: In exile* (Internet diary). Retrieved August 22, 2006, from http://wilwheaton.typepad.com/.

Chapter 13

Wikis, Wikipedia, and Collaborative Technology

New Challenges and Emergent Language Styles

ANTONELLA ELIA

University of Naples "Federico II"

Abstract

In the last decades, hypertext and hypermedia have started to redefine the concept of authorship, intellectual property, textuality, and the relationship between author and readers. In the mid-1990s, wikis, as a new phenomenon, caught the attention of online communities re-proposing the authorship problem under a new facet. Wikis, as asynchronous Web-based collaborative hypertext authoring systems, represent a new way to edit online documents and conduct discussions. At the time of this writing, the highest-profile wiki project on the Web is Wikipedia (http://www.wikipedia.org), which aims at creating a free hyperlinked multilingual encyclopedia. The features of Wikipedia computer-mediated textuality are explored in this chapter, as well as its discourse community and the emerging WikiSpeak, a new variety of the Netspeak jargon.

Introduction: What Is a Wiki?

The word *wiki* is the abbreviation of "wiki wiki" from the native Hawaiian language, where it is commonly used as an adjective to denote something "quick" or "fast." Wiki is sometimes (ironically) interpreted as the backronym[1] for "What I know Is," which "describes" the limits of knowledge contribution, storage, and exchange. Although a precise definition of "wiki" does not exist (Lamb, 2004), the general consensus is that it is *a Web site that allows users to add and edit content particularly suited for constructive collaborative authoring.*

A wiki enables authors to write collectively in a very simple markup language (called wiki syntax) using a web browser. Some recent variants use a different method as they allow WYSIWYG editing that translates graphically entered formatting instructions into the corresponding HTML tags. In essence, wiki is a simplification of the process of creating HTML pages combined with a system that records each individual change that occurs over time and at any time. In wikipages, the edit button gives the user the opportunity to change the content of the page displayed. Each page also has a "history" function that allows the author (or editor) to review the changes and, if necessary, to revert to an older version. A wiki system also provides various tools that allow the user to easily monitor the constantly changing state of the wiki (Watchlist in Wikipedia) and discuss the issues with regard to wiki content (Discussion pages[2] in Wikipedia).

The original wiki concept is credited to programmer Ward Cunningham (Leuf & Cunningham, 2001) who, in 1994, created the first wiki, the WikiWikiWeb, and installed it on the Web in 1995. In an interview, Cunningham affirmed that he chose wiki-wiki as an alliterative substitute for quick, thereby avoiding naming "this stuff" quick-web. Cunningham's prototype has inspired many successive variants and implementations (Mattison, 2003) as different wiki systems have been set up, some public, some private, with several different uses.

Computer-Mediated Textuality and Hypertext Theories

Vannevar Bush with "Memex" (1945), Ted Nelson with "Project Xanadu" (founded in 1960), and the writers Raymond Queneau (Cent mille milliards de poèmes, 1961), and Julio Cortazar (Cortazar, 1963) were some of the most eminent promoters who revolutionized the traditional concept of linear writing with their hypertextual projects and pioneering experiments.

Nevertheless, in the 1970s, poststructuralist thinkers such as Roland Barthes, Jacques Derrida, Gilles Deleuze and Félix Guattari, Umberto Eco, and Michel Foucault, definitely redefined the long-accepted assumptions about texts, authors, and readers, influencing hypertext theories of authorship and pointing toward innovative theoretical perspectives.

The effects that electronic-writing technologies have on authorship remain an important issue in hypertext theories. In the early 1990s, scholars, such Jay Bolter and George Landow, agreed that the author's role had changed and would have continued to change as textuality migrated from the page to the screen, but they disagreed on the specifics of how print and hypertext-based authorship differ and whether this migration from the page to the screen constitutes a radical break from the age of print. They began to dispute the idea that electronic features, such as hypertextuality, transferred a large degree of textual control from authors to readers, thus blurring the distinction between the role of the author and that of the reader, as the hypertextual reading experience seemed to be a more empowering and creative experience than reading a printed book.

In *Hypertext 2.0*, Landow (1997), described hypertexts as "text composed as lexias (blocks of words moving, or static images, or sounds) linked electronically by multiple paths in an open ended web" (154). He turned to the recent theoretical approaches described in the works of Roland Barthes who conceived of textuality as an open, plural text characterised by multiple paths in which the reader is liberated from the control of the "Author-God." Barthes (Landow, 1997) evoked a notion of a text that has become a:

> galaxy of signifiers, not a structure of signifieds; it has no beginning; it is reversible; we gain access to it by several entrances, none of which can be authoritatively declared to be the main one; the codes it mobilizes extend as far as the eye can reach, they are indeterminable . . . the systems of meaning can take over this absolutely plural text, but their number is never closed, based as it is on the infinity of language. (3)

Landow, elaborating his hypertextual theory, recovered the Barthesian idea reconfiguring the text, the author, and the writing process as the multisequentiality of hypertext permits readers to choose their own paths through a set of possibilities dissolving the fixity of the traditional triangle (author–text–reader).

Wikis: Theoretical Background

The concept of freedom from the control of the *Author-God* evoked by Barthes ("the death of the author") is more pronounced on a wiki than in any other hypertextual environment, because in an open wiki, the reader becomes a "participant author," who is not only allowed, but even spurred to interrupt the process, rewriting, changing, erasing, and modifying the original text or creating new topics. Wiki technology mediates the gap between author and reader because the two actors assume completely interchangeable roles in this new open e-nvironment.[3] WikiText is a "never-ending story" because it is revisable. Furthermore, it is atemporal as nodes continually change through the collaborative writing process, creating a permanently evolving network of intertwined topics reversible and accessible by several entrances, none of which can be authoritatively declared to be the main one. Other inspiring models of wiki rhetoric can find their origin in the ideas expressed by Mikhail Bakhtin and his close circle of contemporaries, who are known for their work on the concept of Heteroglossia and Dialogism. One of the key concepts inherited by the dialogic philosophy is the embryonic concept that all entities (including dialogism itself) are never complete, fixed, or closed off (Stewart, 2003).

For Bakhtin, meaning is in a continuous state of becoming. It is not closed and unchangeable for all of time. Meaning is found in dialogic utterances set in a social situation. Computer-mediated textuality expressed on wikis recognizes a social dialogical and evolving dimension to knowledge in a way very close to the Russian philosopher's ideas. Computer-mediated textuality needs to recognize that any categorization is highly dialogized, contingent, and fluid, and when the phenomenon is portrayed it is just a glimpse of an occurrence that is in continuous change (Stewart, 2003).

For these reasons, computer-mediated texts, especially wikis, have to be defined by a new term as *dialogic genre.* This definition recognizes the importance of both Bakhtinian dialogism and the Barthesian ideal of open textuality free from the control of the single author in favor of a collective and socially reconfigured author. In the Web2.0 generation,[4] wikis as new form of hypertext, are actually blurring the classical distinctions between author and reader, and between producer and consumer of online text or "content" (Graddol, 2004). Although it is already understood that reading involves the production of meaning, wikis allow multiple reader-authors to register different interpretation and analysis directly within the text, and to participate in a new and dynamic collaborative process of fluid and multiple construction of meaning. As Braga and Busnardo (2004) suggested, these developments will offer an entirely new challenge to communicators, greater than simply the navigation of nonlinear texts. Readers will increasingly face multiply authored texts that exist in a condition of constant textual evolution (of form and content), a situation that radically challenges our existing notions of how knowledge is produced, disseminated, and accessed online in the new millennium.

DocumentMode

Wikis offer two different writing modes. The first one is known as DocumentMode.[5] It corresponds to the encyclopedic entries, defined in Wikipedia as "articles." DocumentMode is the conventional mode of editing, respectful of the formal expository writing rules of the encyclopedic genre. DocumentMode articles are coherent and self-contained; they reflect the result of the last update. DocumentMode articles are community property: They may have multiple and changing authors and they are updated to reflect the community consensus. They demonstrate that knowledge is collective and that ideas, not the writers, are the main focus. The style expressed in this "writing space" is expository, extensive, and monological. Texts are detached, accurate, rigid, and refined with a high degree of stylistic formality. Articles are impersonal because they are written in the third person. They are unsigned, highly informative, objective, and respectful of conventions (Elia, 2006). Wikipedia's contributors strictly observe the Wikipedia Manual of Style[6] with the first basic nonnegotiable rule being: "the Neutral Point of View."[7] Articles written in DocumentMode are stylistically less innovative and original than ThreadMode documents (discussed in the next section). Their style is very formal, in that they never make use of first- and second-personal pronouns, acronyms, jargon expressions, or neologisms. The use of jargon is rigorously forbidden; nevertheless if its use is unavoidable, the "banned" expressions must be hyperlinked to pages explaining accurately their meaning so that everyone can understand them.

ThreadMode

The second wiki writing mode, which is the main objective of this chapter, has been defined as ThreadMode (Morgan, 2006). Contributors carry out discussions by posting signed messages in the discussion pages connected to the main article. A ThreadMode page looks like conversational graffiti. It is a written conversation among interested parties, as can be seen in this extract from a page in Meatball Wiki:[8]

> I think that hyperlinks should be red, not blue.—FredFlinstone
> Why do you think that, Fred?—BarneyRubble
> Because, it would make them stand out more.—ff

ThreadMode[9] is stylistically different from the bland, formal, and neutral style expressed in the DocumentMode. A threaded conversation shows many different points of view and lends energy to them, developing a significant sense of community. It has been stated:

> This Wiki seems kinda dead to me. Is it because there are no ThreadMode discussions, which is a symptom of lack of controversy? There's not much fun in reading through a dictionary.—anon.[10]

It is easier to write in ThreadMode than in DocumentMode because contributors do not have to respect any convention or compulsory style defined in any authoritative manual of style. They have only to write, post, and optionally sign freely their paragraphs (anonymous contributions are also very frequent on wikis). The style is fluid and not conventional. It seems to work very well when individual contributions are concise, with a definite objective to a single subtopic. Longer segments that touch on several related topics can lead to a multithreaded mode, in which people respond to multiple parts of a contribution, creating multiple threads that all progress simultaneously and sometimes erupt into a *ThreadMess* or a *ForestFire*.[11]

ThreadMode has a highly personal perspective as posts are written in the first person and are usually signed. Because signatures are often redundant and conversational exchanges chaotic, threads are much more "noisy" and "fuzzy"[12] than in DocumentMode. The style in ThreadMode is dialogical, flexible, and direct. The rhetoric is free and informal. It presents multiple interacting positions and evolves without a predictive structure because the discussion development is changeable and impulsive. ThreadMode style is exploratory, explicit, involving (Tannen, 1989), and rich in new terminology, although less informative than DocumentMode texts. ThreadMode discussions are the evident proof that knowledge is the result of constructivist collaboration and not a lonely production.

Refactoring

The term *refactoring*[13] has its origins in computer programming. On wiki Web sites, refactoring refers to the process of rewriting, reorganizing, and shortening texts, while preserving content. It is not always an easy operation to perform, as its goals are to improve readability while preserving meaning and removing superfluous content without altering the information presented. The process of refactoring implies that the full intentions of the original authors must be preserved at all times. Refactoring discussion pages can be necessary when there is an accumulation of previous unclear or irrelevant posts. Such accumulation may discourage the involvement of potential contributors. Refactoring promotes productive discussions by improving clarity and accessibility. When discussants have reached a consensus, someone will elaborate the new information and suggestions provided. The conclusive remarks of the discussants will be incorporated, modifying or creating the related DocumentMode page. In this process, the informal style of the discussion page will be elaborated into a formal and impersonal style (Morgan, 2006) performing a radical stylistic code switching. Through the refactoring process[14] superfluous and redundant information or poorly formatted or outdated content is removed or stylistically improved.

Wikipedia and Collaborative Technology

The term *Wikipedia* can be applied to three entities: an encyclopedia (the actual body of work created); a project (the effort to make that encyclopedia); and a community (the group of people working on the project). Wikipedia is the widest wiki community. It incorporates Berners-Lee's original concept for the Web. Inventing the Web he took an élitarian communication system and turned it into a mass medium and a powerful force for social change. More than anyone else, he wanted and has fought to keep the Web open, nonproprietary, and free (Berners-Lee, 1999). This is the cultural purpose that Wikipedia also pursues as writing and editing is a collective and collaborative process in this online community.

Wikipedia is the result of the evolution of Nupedia, a project aiming to create an authoritative encyclopaedia inspired and morally supported by Richard Stallman's GNU[15] project. Wikipedia was started in the beginning of 2001 as an attempt to create something similar, a free encyclopedia, that would ultimately be able to compete with the *Encyclopedia Britannica* (although it was developed via a very different and much more open process; see Kohl, Liebert, and Metten, chap. 7, this volume). Vannevar Bush (1890-1974) introduced the concept of what he called the *Memex* in the 1930s. He set out his thoughts in the essay "As We Think" in the *Atlantic Monthly* in July 1945. In this article he predicted that "Wholly new forms of encyclopedias will appear, ready made with a mesh of associative trails running through them." It seems, on reading this quotation, that he had prophetically foreseen the birth of Wikipedia in the new millennium. Nowadays, Wikipedia contains 253 language editions with approximately 7.8 million articles (as of July 2007). The English-language version includes 1.9 million articles. Despite being criticized and attacked, it has been a very successful project, as it has steadily risen in popularity. According to Schiff (2006), Wikipedia is now the 17th most popular site on the Internet, generating more traffic daily than MSNBC.com and the online versions of *The Times* and *The Wall Street Journal* combined. The number of visitors has been doubling every 4 months; the site receives as many as 14,000 hits per second.

Following the open source peer-review maxim, formulated by Eric Raymond (1999) as "given enough eyeballs, all bugs are shallow" (which he defined Linus' law), this allows the encyclopedic project to grow not only in number of articles, but also in terms of the articles' depth and quality of information, which should improve over time through the collective input of knowledgeable readers. Because the review and improvement process is public and ongoing, there is no difference between beta and release versions of the information. Texts continuously change. Peer-review becomes peer-editing, resulting in what Larry Sanger, one of the original project leaders, hailed as the "most promiscuous form of publishing." A reader who sees a mistake or omission in an article can immediately correct it or add the missing information. The quality of the information provided in Wikipedia is difficult to judge. Casual searching reveals that although some articles are very good (both content and form), many others are not. Of course, this is not surprising given the fact that the project is still very young and always in progress. Encyclopedic articles are freely available to anyone as texts are published under the GNU Free Document license (Stalder, 2002).

The different wiki writing modes previously examined, DocumentMode, ThreadMode, and the process of refactoring, are obviously included in Wikipedia.

Before writing encyclopedic articles on the DocumentMode pages, Wikipedians are involved in meta-discussions connected to the quality (content/style) of the articles on the ThreadMode discussion pages. ThreadMode discussions are very important because through them the quality of articles is improved and the collaborative process of co-writing is assured. Wikipedians refactor discussions and articles in order to increase fruition and readability of specific texts. A significant example of how beneficial refactoring can be is represented in Wikipedia by the discussion on the Mother Teresa page.[16]

What is meant by collaborative information architecture in Wikipedia is the process of collectively organizing information and allowing efficient, pleasurable, and stimulating access. Information architecture concerns writing, editing, and page lay out. The role of the information creator in Wikipedia has been expanded from the authoring of text and media to the interplay of ideas.

Wikipedians are allowed to refactor: moving, folding, merging, branching, or re-routing the contributions of others and organizing them in different ways from different perspectives. But Wikipedia is not a virtual dreamland: Occasionally freedom of expression degenerates into disputes and edit wars,[17] which are solved through the intervention of an arbitration committee (ArbCom).[18]

Information architecture is a difficult field in which to collaborate. For example, most contributors have a distinctive intuitive cataloguing method that has to be encoded into rigid standards. Different opinions on how to present information can be easily identified and resolved, whereas divergences on how to *catalogue information have to* find more elaborate solutions. Some difficult tasks associated with information architecture on Wikipedia are for example, creating navigation node pages, creating and assigning categories, writing and maintaining documentation, deleting pages that do not fit into the site, moving pages to "more appropriate places," combining pages, and so on. This is the innovative aspect that is emerging on wikis and Wikipedia, as contributors can edit not only a page, but an entire *site.*

Wilkipedia is "folksonomy"[19] based. Folksonomy is an Internet-based information retrieval methodology consisting of content pages and links collaboratively generated and categorized. Folksonomies are generated by people who may have spent a great deal of time interacting with the content they tag. A folksonomy is conceptually different from traditional taxonomy in that the authors of the labeling system are often the main users and originators of the content to which the labels are applied. The process of folksonomy is intended to make a body of information increasingly easier to search, discover, and navigate over time. In contrast to professionally developed taxonomies, folksonomies are inherently open-ended and can therefore respond quickly to changes and innovations in the way users categorize Wikipedia content. Perhaps the greatest benefit of folksonomy is its relevance in the information retrieval, that is, the capacity of its tags to describe the "aboutness" of a resource. The folksonomic categories are simultaneously personal, social, and (to some degree) systematic, although in an imperfect and provisional way. Folksonomies therefore convey information on multiple levels, including information about the people who create them, and therefore they invite human engagement.

A Unique Online Community

Each discourse community has its own rules on what can be said and how it can be said. Discourse of communities denotes a group of people with certain things in common: a public goal, a body of specialized knowledge, a specialized lexicon, and a set of beliefs about how knowledge is generated. Members also share an understanding of how to communicate with each other and with the larger community. To become a member of a discourse community, contributors must master its theoretical concepts, as well as its language and conventions; this means accepting also its beliefs and values. John Swales (1990) presented six defining features related to a "discourse community":

1. It has a broadly agreed set of common public goals.
2. It has mechanisms of intercommunication among its members.
3. It uses its participatory mechanisms primarily to provide information and feedback.
4. It utilizes and hence possesses one or more genres in the communicative furtherance of its aims.
5. In addition to owning genres, it has acquired some specific lexicon.
6. It has a threshold level of members with a suitable degree of relevant content and discoursal expertise. (21–29)

According to Patricia Bizzell (1992), producing text within a discourse community cannot take place unless writers can define their goals in terms of the community's interpretive conventions. In other words, texts cannot be simply produced. They must fit the standards of the discourse community to which they are appealing. For this reason, being a member of a specific discourse community requires more than just learning its lingo. It requires understanding concepts and expectations set up within that specific community and acting by precise behavioral norms.

The language used in discourse communities has been defined by Gregory (1967) as "diatype." This term describes a type of language variation that is determined by its social use and purpose. According to Halliday (1985), diatype is usually analyzed in terms of "field" (the subject matter or setting), "tenor" (the participants and their relationships), and "mode" (the channel of communication: spoken, written, or mediated). Online discourse communities are spaces where people interact with one another by means of written discourse. Generally speaking, these can include synchronous and asynchronous computer-mediated communication (CMC) tools such as forum, chat, multiuser online games, mailing lists, blogs, wikis, or any other CMC tool. These online environments are primarily text-based, but may contain also multimedia elements such as images, animation, or emoticons. Examples include Internet relay chat (IRC), MSN Messenger, and ICQ.[20]

Wikipedia, as a virtual community, shares all its distinctive references with traditional communities. Nonetheless, it is not a community in the real-world sense because Wikipedians are bound together only by electronic interactions in a virtual space. Their activity has a very specific and pragmatic focus: to create a worldwide encyclopedia. The encyclopedic project has a strong "community" character as there is a shared feeling of belonging. The community actively promotes the development of the project by creating new pages or refactoring old pages, adding or correcting information, and discussing the nature of the content with other users. It is possible to define the Wikipedia community as unique: There are no other communities outside of the Wiki world that combine the attributes cited here.

Wikis and Wikipedia organize their communities around a written project. They are based on writing and not only on reading culture. Online cultures, obviously, organize themselves around their community discourse and their topics. Hence, how topics are named and clustered on the wiki indicate what its culture is about, what its values are, and what its members think about the world. For instance, on the entry page of the Meatball Wiki (an interwiki community), it has been specified that it deals with online cultures, especially how people online come together naturally in groups. The Portland Pattern Repository "WikiWikiPage" explains that it is a Web site written by its users, where anyone can change any page or create new pages. By contrast, the Wikipedia homepage welcomes its visitors with the following slogan: "Welcome to Wikipedia, the free encyclopedia that anyone can edit." Inside its Web site, the Community Portal has a high impact on the reader in terms of content, form, and function. It can be considered a supreme synthesis of the free and collaborative background philosophy. It is not coincidental that the highest frequency keywords recorded in this page are: *help* (22 occurrences), *you* (22 occurrences), *article* (18 occurrences), *collaboration* (8 occurrences), *free* (7 occurrences).

Forums vs. Discussion Pages

The core of the democracy is represented in Wikipedia by ThreadMode discussion pages. They can be associated, to a certain extent, to online forums, even if discussion pages usually tend to be meta-discussions on the topic or on the validity of the content, rather than on the content itself. A comparison between forums and ThreadMode discussion pages is outlined here.

The process of linguistic accommodation is common to both Internet forums and Wikipedia discussion pages. David Crystal (2001) noted a linguistic adjustment in forums observing the dynamics in WELL Community:[21]

> The members accommodate to each other. Although they come from many different backgrounds, and write in many different styles, their contributions progressively develop a shared linguistic character. Everyone comes to use certain types of grammatical construction, jargon, or abbreviations. (147)

This phenomenon has also been observed in Wikipedia discussion pages where contributors communicate, exchange opinions, and discuss technical and editing operations through what has been defined in this chapter as "WikiSpeak,"a shared and specialized high-context language.

On forums, users may post chronological messages, individually answering a discussion thread. Text, images, or audio files may be provided, in most cases, in attachments. The meaning of the forum speech act is aggregative, accumulating ideas and phrases. Quoting is a method used to create meaning through aggregation. Ong (1977) said that speech acts are redundant in oral discourse; it is necessary to repeat previous messages in most of the forums both to remind the reader and to inform new readers of what has been said in the discussion. Thus, the online quoting process acts as a repetition in oral speech (10–13). This phenomenon has not been detected in discussion pages where the different architectural structure allows contributors to immediately add their post in a new paragraph, avoiding to quote the contribution to which they are answering. The last post will be the first one. In case of long discussions, a dynamic Table of Content is added by contributors at the beginning of each page. Its aim is to allow a quick retrieval of the desired topical discussion. Unlike forums, files may never be attached in discussion pages.

Similar to discussion pages, the range of topics discussed on forums is very extensive. As a Web site, running forums may have more than one "room," each dedicated to a different topic. Discussion pages and forums usually provide anyone with the authority to start a new discussion (known as a thread), or to reply to an existing thread. But, whereas discussion pages allow every visitor to post comments in reply, the number of contributors who can post in forums is usually limited and, consequently, so is the range of viewpoints and beliefs expressed. Both in forums and discussion pages, participants' posts are public for many years, especially when an archive is provided. For example, Google Groups includes Usenet articles dating back to 1981. Nevertheless, although posts are permanently stored in discussion pages, sooner or later forum messages will be deleted.

The architecture of the two environments, as already mentioned, is different. In most cases forum interventions are organized in separate messages posted in topical rooms; they are chronologically ordered and only a brief summary, as well as author, date, and time of the post can be viewed. The link has to be clicked to expand the content. By contrast, posts by Wikipedia contributors are published in a unique and long scrollable discussion page. It is a sort of backstage agora, a public and interactive open space whose topic of discussion is related to the main encyclopedic article. Discussion pages are accessible by everyone and, unlike forums, they allow people to edit the messages of others, whereas in the latter only moderators may be authorized to do so in order to monitor and control content. In this way, moderators can control spam and add details and comments. The presence of a moderator and his or her behavior can shape tendencies, address issues, and set the tone of discussions according to a more formal or informal style. David Crystal (2001), referring to forums, stated that "the more specialized is the topic the more likely the content will be focused and several groups use moderator to ensure that the conversation doesn't diverge from the subject too much (go off-topic)" (132).

Official moderators are not provided on discussion pages. This role, as already mentioned, has been substituted by a democratic mechanism of consensus, mutual control, and peer-to-peer review. Nevertheless, one of the roles of the Wikipedia community, as a super-parties entity, is to resolve conflicts among members, if and when they happen, through the ArbCom. The ArbCom deals with only the most serious disputes and cases of rule-breaking. It imposes binding solutions to Wikipedian disputes. It is the last step in the dispute resolution process; it is the final resort to be turned to when all else has failed.

Every online community has a netiquette that defines in a broad and informal way customs and social accepted behaviors for network users. Respecting rules and patterns of behavior is normally assured in the Wikipedia community by the "Wikiquette." It is the community netiquette[22] (see Millar, chap. 11, this volume), the guidelines on how to deal and work with other Wikipedians. Wikiquette rules are more explicit, meticulous, and compulsory than precepts of general forums. This happens for different reasons: First, there is no ubiquitous moderator in discussion pages, and second because Wikiquette has to manage a complex process: the writing of a co-authored encyclopedia. There are some basic rules that have to be respected in the community (e.g., assume good faith, avoid deletions whenever possible, edit, discuss, be polite, sign and date posts to discussion pages [but not articles], work out agreement, respond to questions, etc.). Wikipedia contributors come from many different countries and cultures. They have different points of view and backgrounds. Treating others with sensitivity and respect is the key to collaborating effectively in building an encyclopedia.

WikiSpeak: The Language of the Wikipedia Community

Since the beginning, each communication technology has reshaped our process of socialization and acculturation, simultaneously changing our discursive horizons. Presently, this process has become more complex as the Internet is collapsing in a multitude of new technologies that are emerging very rapidly. The remaining part of this chapter focuses on the nature of Wikipedia's discourse.

As forums and blogs, and as a computer-mediated discourse community, Wikipedia has developed its peculiar wired style that is direct evidence of the evolution of a specialized online discourse connected to the use of a specific web tool. Similarly to other e-environments, Wikipedia's contributors also interact in their backstage community by using a hybrid of spoken and written language. Moreover, they have developed an original lexicon and made fresh linguistic adaptations to suit new online circumstances. This language variety, defined here as WikiSpeak, is an unofficial and high-context subgenre that can be considered as a new expression of netspeak (also known as Netlingo or Net Jargon).

Through computer-mediated discourse analysis (CMDA) it is possible to study micro-level linguistic phenomena, such as online word formation processes, sentence structure, lexical choice, and language code-switching from DocumentMode (encyclopedic entries) to ThreadMode (discussion pages) and vice versa. WikiSpeak is used in the different synchronous and asynchronous CMC channels of the community, such as discussion pages, mailing lists, IRC channels, instant messaging, and personal user pages. Only the channel of the discussion pages has been analyzed in this chapter in order to identify the linguistic features of the WikiSpeak. WikiSpeak's informal register undoubtedly shares both the peculiarities of the face-to-face (FTF) spoken language and the spoken–written language used in Internet forums. Evidence of this has been detected through a quantitative analysis made on a sample of 50 discussion pages randomly selected. Additionally, a contrastive analysis has revealed a clearly visible code switching from the ThreadMode discussion pages to the related DocumentMode encyclopedic articles. Following Biber's (1988) multidimensional analysis methodology, a high occurrence of linguistic features typical of telephone and FTF conversations (Biber, 1988) has been recorded in discussion pages. Specifically, a high frequency of contractions (e.g., I'm, he's, it's, isn't), interjections (question and exclamation marks), negations (don't, no, not), of "be" as main verb and of modal verbs (e.g., can, could, may, might) has been noticed. Different types of deixis, such as personal pronouns (I, you), locative and temporal adverbials (e.g., here, now) as well as a high frequency of proximal and distal deictics (demonstratives: this, that) have also been found.

A comparison of discussion pages versus related encyclopedic article pages has revealed a more informal style conveyed by both a shorter average word length (4.9 vs. 5.3 letters) and a shorter average sentence length (11.7 vs. 22.5 words). Additionally, also the specific average lexical density (38.4 vs. 39.9) of discussion pages was revealed to be lower. The frequency of nominalizations (3.5 vs. 4.7), passives (0.7 vs. 1.5), prepositions (7.4 vs. 10.8), articles (7.4 vs. 9.5), nouns (7.6 vs. 9.2) and conjunctions (7.4 vs. 10.8) has been measured. The lower frequency of the above mentioned linguistic features typical of the formal expository style of the encyclopedic articles is further evidence of the informality of the WikiSpeak conveyed in the discussion pages. One of the main WikiSpeak peculiarities lies in the new lexicon used. A large number of new words (*wikilogisms*[23]) have emerged while Wikipedians talk about technical operations and activities connected to their authorial and collaborative writing work in the ThreadMode discussion pages. Wikipedians have coined new words for their new world, to say new things that do not yet have adequate references. As new knowledge necessarily creates new terms and concepts, rejecting neologisms means to reject cultural evolution. Neologisms are normally strictly connected to the terminological paradigm actually dominating a field of knowledge. Discourse communities have the freedom to break the conventional linguistic code, creating and altering the language in use. A new discourse community is barely conceivable without the use of neologisms, new words, or new interpretations of old words to describe and explain reality in new ways. During the last century, various linguists have developed taxonomies for classifying the different types of word formation. WikiSpeak distinctiveness certainly lies in its lexicon where many word formation processes take place, including several ludic innovations. John Algeo (1999, cited in Shortis, 2001) developed a classification system for word formation based on four factors:

1. Does the word have an etymon based on earlier words?
2. Does the word omit any part of an etymon?
3. Does the word combine two etyma?
4. Are any of the etyma from another language?

He used six groupings: composites (prefixes, suffixes, compounding), shortenings (acronyms, initialisms, clipping, backformation), blending, shifting (functional or semantic shifts), loans, and new creations. He showed the varying percentages for types of word formation in a set of samples taken from different linguistic domains. Algeo (Fig. 13.1) demonstrated the high frequency of composites in his analysis (especially compounds), over other types of word formation. Frequency of shifts and shortenings is also high, but less significant, occurrences of blends and loan words are low and the presence of new creations is irrelevant (Shortis, 2001).

The volume of new words coming into WikiSpeak provides an opportunity to see Algeo's classification in action. To explore the occurrences of the most common new word-formation processes in WikiSpeak, the Wiki glossary available for the newbies in the Wikipedia community pages has been annotated using Algeo's word-formation categories (Table 13.1).

The corpus has been analyzed through the WordSmith Tool software. Table 13.1 reports the annotation system used to analyze the word-formation processes in the WikiSpeak. As a departure from Algeo's analysis, the findings shown in Fig. 13.1B reveal a high frequency of shortenings in discussion pages, particularly initialisms and clippings (very few acronyms have been identified). A slightly lower frequency of composites (most of them compounds) is followed by semantic shifts. The low rate of blends, loans, and new creations is also found in Algeo's analysis (Fig. 13.1A).

In Fig. 13.1A and 1B, the main discordant data are related to the higher number of shortenings in WikiSpeak (initialisms and clippings), whereas the number of composites proves to be very similar. The higher number of wiki shortenings (listed below) is probably due to the need for quick typing and to the use of specific lexicon related to daily technical and editing operations already familiar

Table 13.1. Annotation System Used for Classifying Wikipedian Glossary

Composites	Prefixes	<pr>
	Suffixes	<su>
	Compoundings	<co>
Shortenings	Clippings	<cl>
	Double clippings	<2cl>
	Acronyms	<ac>
	Initialisms	<in>
	BackFormations	<bf>
Blendings		<bl>
Shifts	Functional shifts	<fs>
	Semantic shifts	<ss>
Loans	Other languages	<lo>
New creations	Wikilogisms	<nc>

to the members of the community. With reference to Algeo's classification, the corpus analysis has revealed that a recurrent word-formation process in Wikispeak is the use of affixation (especially prefixes) which alters the original meaning of words (Elia, in press).

Typical Wikipedian prefixes include *de-* (*desysop, dewikify*), *un-* (*unencyclopedic, unwiki*), *sub-* (*subpage, substub*), *trans-* (*transwiki, transclusion*), or the suffix *-bot* (*rambot, vandalbot*), and so on.

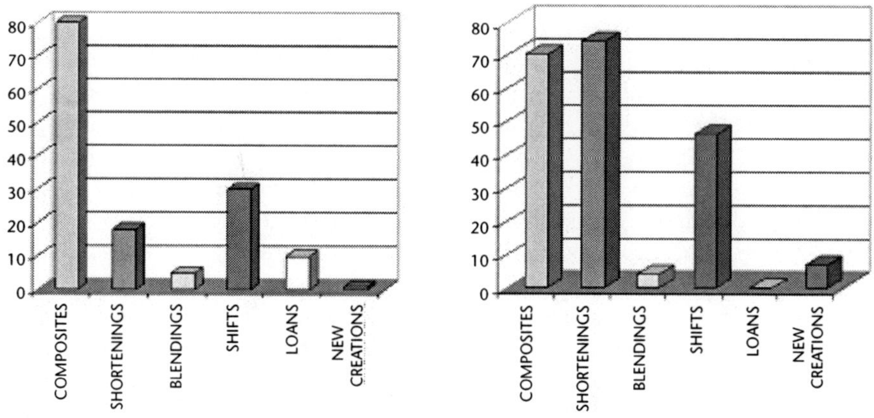

Fig. 13.1(A) Algeo's word formation classification (1990). (B) Word formation classes in Wikipedia.

A popular method of creating wikilogisms is compounding, that is the combination of two existing words to make new words (e.g., *namespace, mediawiki, rollback, infobox, quickpoll*). Some elements turn up repeatedly, for example, wiki (*wikipage, wikibooks, wikilink, wikistress,* etc.). Veterans avoid the excessive use of *wiki*-compounding as it is considered "unencyclopedic" and cliché. However, it is tolerated when it refers to technical terms (e.g., *wikilinks*), when it is an existing subject (such as *Wikimedia Foundation*), or when it is useful in communicating wiki-specific ideas (e.g., *WikiCrime, WikiGnome or WikiHoliday*). When encyclopedic articles are titled with wikilogisms they are immediately elected for deletion as any type of jargon cannot be used in encyclopedic articles.

Because WikiSpeak is a written form of CMC, it makes extensive use of initialisms; on the other hand, the use of acronyms is not very widespread in the community.[24] Some of the most common and recurrent initialisms met in the discussion pages are: *NPOV* (Neutral Point Of View), *COTW* (Collaboration Of The Week), *IFD* (Image For Deletion), *RC* (Recent Changes), *RfM* (Request For Mediation), *VP* (Village Pump), *AOTW* (Acronyms Of The Week). Wiki initialisms are not restricted to words, but can imply also sentences, for example, *WDYS* (what did you say?), *CIO* (check it out), *CID* (consider it done), *RTM* (read the manual), and so on. Some of them are like puzzles as the sound value of the letter, or numeral, acts as a syllable of a word, for example, *B4N* (bye for now), *CYL* (see you later), *U R* (you are), *2L8* (too late), *2g4u* (too good for you), and the like. Basically, it is recommended that a moderate use of TLAs (three-letter acronyms) and initialisms be made in discussions pages, otherwise the message could be incomprehensible. A Wikipedian affirms in a community page[25]: **W**hat **T**he **F**uck? **O**h **M**y **G**od! **T**oo **M**any **D**amn **T**hree **L**etter **A**cronyms. **AR**RR**G**GGHHH!!!! ~!@!~1`2.

Typical of WikiSpeak is the use of clippings, for example, *admin*[26] (administrator), *disambig* (disambiguation), *contribs* (contributions), *diff* (differences) *nom* (nomination), *dupe* (duplicate article), and so on. An original association of the above-mentioned word-formation processes can be noticed, such as the matching of a double clipping to make new compounds, for example, *ArbCom* (Arbitration Committee), *Copyviol* (Copyright violation), *medcom* (Mediation Committee), *Dicdef* (Dictionary definition), *SysOp* (System operator), and the like.

WikiSpeak also makes a large use of blends, for example, *wikipediholic* (wikipedia + alcoholic), *wikiquette* (wiki + netiquette), *wiktionary* (wiki + dictionary), and semantic shifts, for example, *orphan* (a page with no links from other pages), *stub* (a short article), *sandbox* (a page that users may edit however they want), *village pump* (a set of pages set to discuss the technical issues and operations of Wikipedia), and so on (Elia, in press).

Distinctive graphology is also an important feature of WikiSpeak. All orthographic features have been affected. For example, the status of capitalization varies greatly. Similarly to forums, chats, and blogs, there is also a strong tendency to use lowercase (e.g., *i want*) in the discussion pages. The "lower-case default mentality" has a long tradition in cyberspace and causes a distinctive function of any use of capitalization: within Internet the capitals are therefore a specially marked form of communication. Messages wholly in capitals are considered to be "shouting" and usually avoided. Although asterisk and spacing have the same function (e.g., it's VERY important, it's *very* important, it's _very_ important, it's v e r y important), words in capitals add extra emphasis because of their intertextual relation to the comic culture, which is very influential on all artifacts of cyberculture (Wyss, 2000).

A distinctive feature of wiki graphology lies in the way two capitals are used: one initial, one medial. This phenomenon is called *BiCaps* (bicapitalization) or *CamelCase,* and is widespread in discussion pages (e.g., *MediaWiki, WikiProject,* etc.). It is the practice of writing compound words or phrases where the words are joined without spaces, and each word is capitalized within the compound. The name comes from the uppercase "bumps" in the middle of the compound word, suggesting the humps of a camel. CamelCase is a very interesting example of how a programming language influences the wired style. It was originally used in hackers' communities as a word-joiner alternative to the underscore-based style and later in the original wiki markup language to create hypertextual links before the invention of [[_]] double square brackets. Nowadays it has become fashionable in marketing for names of products and companies. Outside these contexts, however, BiCaps are rarely used in formal written English, and most style guides recommend against it.

Spelling practice is also a WikiSpeak distinctiveness. Wikipedians, together with bloggers, chatters, and forum contributors, make use of nonstandard spelling that reflects pronunciation (e.g., *yep, nope, yay, sokay*) or conveys emotions. In this case, a varying number of vowels and consonants (noooooooo, *yayyyyyy*) are used. The use of repeated interjections (*Yes!!!!, WHAT????*), dots (.......) and commas (,,,,,,,) is also very diffused in discussion pages. Nevertheless, punctuation sometimes tends to be minimalist or completely absent in discussion pages. A lot depends on the user's personality: Some Wikipedians make excessive use of it, some are scrupulous about maintaining traditional punctuation, while others do not use it at all. An increased use of symbols, not normally part of the traditional punctuation system, such as the hash (#), hyphens (--) or asterisks (***) has been observed. The replacement of plural –s by –z (e.g., downloadz, filez, gamez) has been found in Wikipedia discussion pages. This original spelling practice has been probably inherited from the crackers' subculture that systematically put at the end of words "z" for "s" to denote an illegal or cracking connection (e.g., *codez, passwordz, MP3z, sitez, FTPz*) or from Black youth culture, which extensively uses it in Hip-Hop lyrics. Spelling and grammatical errors have also been found in discussion pages. They are not assumed to reflect a lack of education, but are simply regarded as typing inaccuracy, the result of a hurried communication in an unconventional writing space.

The limitation of Netspeak, the lack of facial expressions, gestures, body posture, and distance conventions, led to the introduction of smileys (Rezabeck & Cochenoer, 1995). Wikipedians also express their feelings and emotions, making an extensive use of graphic symbols to reflect them. They prefer to use basic smileys (e.g., **:-)** pleasure, humor; **:-(** sadness, dissatisfaction, **; -)** winking, **; -(** crying, etc). However, midget smileys are also very spread (basic smileys without the nose, e.g., **:)** **:(**). By contrast, joke smileys (e.g., **:*)** user is drunk; **:-@** user is screaming; **:-[** , user is a vampire, etc.) are rarely found while smiley icons (e.g.: , etc.) widespread in forums, chats, and blogs, are completely absent in the discussion pages (Elia, in press), as the wiki syntax does not allow an immediate and friendly inclusion of picture files.

All the linguistic variations just mentioned have been found disseminated mainly in the discussion pages and also in other channels used by Wikipedians to communicate in the network of their community pages (see footnote 6). Their occurrence has never been detected in DocumentMode encyclopedic articles. Nonstandard spelling, used without sanction in discussion pages are heavily eschewed in traditional writing and absolutely forbidden in DocumentMode encyclopedic articles.

An Analysis of a Discussion Page: Klodzko

In order to give an empirical demonstration, an analysis of a discussion page (April 18, 2005) related to the article Klodzko[27] is presented here. An analysis of both the Klodzko discussion page and the contributors' user pages have revealed linguistic habits, identity, nationality, and cultural background of the four Wikipedians involved, as well as date and time of the contributions posted (Table 13.2). The nationality of the first two Wikipedians (Halibuttis and Piotr) is Polish, whereas the other two (Nichalp and Nixie) are Indian and Australian. Their different origins and cultural background,[27] whose level is quite high (journalist, doctoral students, engineer), demonstrate how heterogeneous and culturally rich the global collaborative writing process is.

The ThreadMode discussion page contains 496 words organized in 37 lines through six different turn-takings. The length of the interaction was 4 days (from the April 15 to April 18, 2005). The atmosphere and the tone of the discussion was highly positive, revealing the open-mindedness of the contributors, for instance:

> *input from others is particularly important* (Line 1), *I appreciate any comment and/or corrections* (Line 2)

Politeness was manifested through expressions that communicate agreement and through the use of the conditional tense, which conveyed consideration and kindness, for instance:

Table 13.2. Participants in Klodzko Discussion Page

Participant	Origin	Profession	Day	Time (UTC)*
1. HALIBUTT	Poland Klodzko	Journalist with a degree in Spanish studies	April 15, 2005 April 15, 2005	09:30 12:57
2. PIOTR	Poland Krakow	PhD student in computer science and economics	April 15, 2005	10:06
3. NICHALP	India Bombay	Electronics engineer from the University of Bombay	April 17, 2005	20:40
4. NIXIE	Australia Canberra	PhD student at the National Australian University	April 18, 2005	00:31

> *good history, but*... (Line 4), *Agreed*! (Line 11), *yes, my mistake* (Line 13), *I'd expect* (Line 20), *The history section should be shortened* (Line 24);

The Wikipedians'collaborative and helping attitude was expressed through comments such as:

> *See wikipedia references*... (Line 7), *try to avoid*... (Line 7), *I recommend using*... (Line 8).

Occurrences of the new lexicon used in WikiSpeak are found in the use of initialisms, acronyms, clippings and wikilogisms, e.g.:

> *FAs* (Featured articles—Lines 8/17), (or also FArticles—Line 16), *AFAICT* (As Far As I Can Tell—Line 17), *FAC* (Featured Article Candidates—Line 26), *Pics* (Pictures—Line 17), *Dividers club* (Line 17), *UTC* (Coordinated Universal Time), *30k* (thousand = kilobyte →K—Line 20).

It is interesting to note that the plural form has been preserved, even if the nouns have been shortened, as in the case of the acronym FAs and the clipped word

Pics. To convey emotions, a basic smiley (Line 13) and two midget smileys have been used in this discussion page: e.g., :-((dissatisfaction—Line 13), :> (sarcastic remark—Line 20), :) (joking statement—Line 22).

The Klodzko discussion page has been analysed through the software tool Wordsmith. Of 506 tokens, 253 types have been found, thus the lexical density of the linguistic interaction is very high (50%) although slightly lower of the related encyclopedic article (53%). Examples of the informal style in use have been found in the four occurrences of verbal contractions (Fig. 13.2) and in the three occurrences of synthetic negations typical of oral speech (Biber, 1988). The 14 occurrences of the deictic personal pronoun "I" (Fig. 13.3) and the five occurrences of "you" (Fig. 13.4) testify to the high subjectivity and addressivity of the discussion page.

rtant commerce centre but	*that's*	about all; Transport (h
ation chart is a good start,	*I'd*	consider turning it into a
odzko :>). With economy etc.	*I'd*	expect it to double in s
hey have that Klodzko	*doesn't*	:Johannesburg, Marshall, Texa
. As pointed above, there	*isn't*	much to say on the city. Sinc
ht? Wikipedia:References	*doesn't*	say much. Yes, my m
r 2005 (UTC) Agreed.	*I'm*	starting right away You

Fig. 13.2. Occurrences of contractions in Klodzko discussion page.

ake a great featured article.	*I*	appreciate any comments and/or
lation chart is a good start,	*I*	'd consider turning it into
lodzko :>). With economy etc.	*I*	'd expect it to double in
t to double in size, but	*I*	doubt it would became a target
es, my mistake. Will do (	*I*	guess you mean the source for
pr 2005 (UTC) Agreed.	*I*	'm starting right away You
in cases of sources and such	*I*	recommend using Wikipedia:Foot
rticles, the basic difference	*I*	see is the length (which is a
C) These are some things that	*I*	think could be included to mak
accurate and fairly complete.	*I*	think it would make a great f
s is particularly important.	*I*	think it is factually accurate
e the better. Say, Halibutt -	*I*	wonder - why Klodzko? :)
all town of 30k inhabitants,	*I*	would put you on to Kalimpong,
ikipedia:Peer review/Klodzko	*I*	wrote virtually all the text m

Fig. 13.3. Occurrences of the first-person pronoun "I" in Klodzko discussion page.

statistical information that	*you*	can get your hands on that
here any local festivals that	*you*	could add here? --nixie 00:31
dd some more pictures here if	*you*	have some, and are there an
istake. Will do (I guess	*you*	mean the source for the pop ta
I'm starting right away	*you*	mean Wikipedia:Cite sources, r

Fig. 13.4. Occurrences of the second-person pronoun "you" in Klodzko discussion page.

The direct style is confirmed in the five occurrences of question marks (Fig. 13.5) while the explicit mention of authors (e.g., Halibutm Piotrus, Nixie) (Fig. 13.6), in addition to date and time of the published posts, represent a clear contextual reference.

By contrast, in the Klodzko encyclopedic article (1,571 tokens/562 types), no occurrence of first- and second-person pronouns (I/You) has been noted, but only the third-person pronoun "it" (14 occurrences) typical of academic discourse (Biber, 1988) and of formal expository style (Fig. 13.7).

Wikipedia:Cite sources, right	?	Wikipedia:References doesn't s
ource for the pop table, right	?)	Yes. As to other
s. Any other major differences	?	--Halibutt 12:57, Apr 15, 20
-I wonder - why Klodzko	?	:) --Piotr Konieczny aka Proko
tivals that you could add here	?	--nixie 00:31, 18 Apr 2005 (

Fig. 13.5. Occurrences of question mark (?) in Klodzko discussion page.

corrections.	***Halibutt 09:34, Apr 15, 2005 (UTC)***	Good history
konsul	***Piotrus*** Talk ***10:06, 15 Apr 2005 (UTC)***	Agreed. I
ferences?	***--Halibutt 12:57, Apr 15, 2005 (UTC)***	We will wo
other major differences?	**Piotrus** Talk ***09:48, 16, Apr 2005 (UTC)***	The history secti
lk Å contributions)=	***20:40, Apr 17, 2005 (UTC)***	These are some
add here?	***--nixie 00:31, 18 Apr 2005 (UTC)***	

Fig. 13.6. Occurrences of author, date, and time of posts in Klodzko discussion page.

For its historical monuments	it	is sometimes referred to as L
ern Poland, in Lower Silesia.	It	is situated in the centre of t
shores of Nysa Klodzka river.	It	is the capital of Klodzko Coun
re was damaged by landslides.	It	turned out that throughout th
times of Friedrich the Great	it	was one of the biggest fort
anged into a prison. At first	it	was administered by the Abwehr
detention in whole Silesia.	It	was also used as a POW camp fo
ged hands several times, but	it	was finally conquered by Pruss
ieged several times. Although	it	was never captured, the city
dukes retained their powers.	It	was not until 16th century wh
1807 after a long siege, but	it	was returned to Prussia short
province of Silesia. In 1290	it	was sold to Dukes of Swidnica
ukes of Swidnica and in 1301	it	was sold to Dukes of Ziebice.

Fig. 13.7. Occurrences of the third-person pronoun "it" in Klodzko encyclopedic article.

Grammar and spelling errors have been noted in the Klodzko discussion page. There are five typing errors: *aboout* (Line 4), *disatvantage* (Line 16), *referecnes* (Line 6), *the the* (Line 19), *ecomomy* (Line 20). A grammar error has also been found (e.g., *would became* [Line 20]). Mistakes, as already mentioned, are not taken seriously in discussion writing space as here content over form and the communicative function are considered more significant (unlike from DocumentMode articles).

The analysis of this specific discussion page has highlighted that the informal spoken–written style expressed in ThreadMode discussion pages is very different from the formal style expressed in Document Mode articles. Nonetheless, it is not very dissimilar to that which is expressed in Internet forums and in spoken conversation. A peculiar difference resides in the specific lexicon used in WikiSpeak, as it is connected to the new editing process offered by this new Web2.0 collaborative publishing tool.

Conclusion

This chapter explored the influence that hypertextuality has had in redefining the concept of authorship. Wiki has been defined as a new dialogic digital genre as it has recognized the importance of both Bakhtinian dialogism and the Barthesian ideal of open textuality free from the control of the single author in favour of a collective reconfigured author. In the Web2.0, Wikipedia the most outstanding example of the wiki world, has blurred the classical distinction between author and reader, allowing multiple reader-authors to participate in a dynamic and collaborative process of the construction of meaning.

The style of ThreadMode discussion pages has been specifically analyzed in this chapter. This is the writing space where Wikipedians carry out meta-discussions related to the specific encyclopedic articles written in Document Mode pages. The code-switching is highly noticeable here as ThreadMode discussions pages are stylistically different from DocumentMode encyclopedic articles. The ThreadMode style is unconventional and unpredictable and visibly shows the different points of view and personalities of contributors. Posts are signed and written in the first person, using a dialogical and informal style which is exploratory, flexible, and involving (Tannen, 1989), although less informative than DocumentMode articles.

A comparison to forums have offered the opportunity to highlight differences and similarities with wiki discussion pages. Peculiarities of the linguistic style conveyed in Wikipedia discussion pages have been specifically outlined. It has been defined as "WikiSpeak," the variety of the NetSpeak jargon used by Wikipedians while they are engaged in their editing and technical operations. The word-formation process of the new lexicon used has been specifically analyzed. The investigation of a specific discussion page (related to Klodzko article) has brought to light the collaborative atmosphere, the informal register used, different graphology and spelling practice, and last but not least, the new wiki lexicon used (see Fig. 13.8).

Klodzko

I wrote virtually all the text myself, thus input from others is particularly important. I think it is factually accurate and fairly complete. I think it would make a great featured article. I appreciate any comments and/or corrections. Halibutt 09:34, Apr 15, 2005 (UTC)

Good history, but has several problems. 1) While history is extensive, there is almost nothing aboout the city today - its economy, municipal government, etc. Tourist attraction are a good start, but could surely use expantion. 2) no references. If external links were used as referecnes, format them accordingly - see Wikipedia:References. 3) try to avoid external links in the text, in cases of sources and such I recommend using Wikipedia:Footnotes. Do take a look at current cities FAs and see what do they have that Klodzko doesn't: Johannesburg, Marshall, Texas, Marshall, Texas, Sarajevo, Seattle, Washington. --Piotr Konieczny aka Prokonsul Piotrus Talk 10:06, 15 Apr 2005 (UTC)

Agreed. I'm starting right away
You mean Wikipedia:Cite sources, right? Wikipedia:References doesn't say much.

Yes, my mistake :-(

Will do (I guess you mean the source for the pop table, right?)

Yes.

As to other FArticles, the basic difference I see is the length (which is a big disatvantage, since AFAICT the FA is often visited by the *dividers* club) and the number of pics. Any other major differences?

--Halibutt 12:57, Apr 15, 2005 (UTC)

We will worry abour length when we have to. The article is not too short (for History of Klodzko :>). With economy etc. I'd expect it to double in size, but I doubt it would became a target for dividers even then. Pics, of course, will always be useful - the more the better. Say, Halibutt - I wonder - why Klodzko? :) --Piotr Konieczny aka Prokonsul Piotrus Talk 09:48, 16 Apr 2005 (UTC)

The history section should be shortened and detail moved to a main article. As pointed above, there isn't much to say on the city. Since this is a small town of 30k inhabitants, I would put you on to Kalimpong, a town which is currently on FAC for some ideas. More references are also needed. =Nichalp (talk · contributions)= 20:40, Apr 17, 2005 (UTC)

These are some things that I think could be included to make the article more about the city now:

Administration (government), what the the important political issued in the town;
Ecomomy, the article says its an important commerce centre but that's about all;
Transport (highways and railways passing through here), perhaps a more detailed map of the location with respect to highways and other local towns;
Demographics, the historical population chart is a good start, I'd consider turning it into a graph, add any other statistical information that you can get your hands on that helps describe the town;
Tourist attractions, add some more pictures here if you have some, and are there any local festivals that you could add here?

Fig. 13.8. Excerpt from Klodzko discussion page.

Before concluding, a new project, which has recently come out in the news, deserves to be mentioned. Its name is the "Citizendium Project."[29] It is a "citizens" compendium of everything launched by Larry Sanger, co-founder with Jimmy Wales of Wikipedia. It has been described as a progressive fork of Wikipedia. It will start as a mirror of the Wikipedia site and will allow anyone to contribute changes to articles, merging public participation with gentle expert guidance. The final aim of the Citizendium is to become the flagship of a new set of responsibly managed free knowledge projects.

What will this new anti-populist project mean? How much and how will the quality and reliability of the information provided progress? To what extent will the formal style of encyclopedic articles be improved and how will discussion pages and WikiSpeak differ in the future? Will a new cultured community discourse come to life as more scholars, professionals, educators, and experts contribute and edit topics? It will indeed be interesting to observe these linguistic phenomena in future research.

Notes

1. *Backronym* (a blend for back and acronym) is a reverse acronym, that is to say a type of acronym that begins as an ordinary word and is later interpreted as an acronym.

2. *Discussion pages*, also defined talk pages, are community pages (see footnote 5) used to discuss encyclopedic articles. Every Wikipedia article has an associated discussion page

3. E-nvironment is a personal adaptation for electronic environment

4. *Second web generation*, also defined as Web2.0, is an expression coined by O'Reilly Media in 2004. It refers to a second generation of Internet-based services—such as social networking sites (e.g., wikis and blogs), communication tools (e.g., VoIP) and folksonomies that let people collaborate and share information online in previously unavailable ways.

5. DocumentMode (http://www.usemod.com/cgi-bin/mb.pl?DocumentMode)

6. Manual of Style (http://en.wikipedia.org/wiki/Wikipedia:Manual_of_Style)

7. Neutral Point of View (http://en.wikipedia.org/wiki/Wikipedia:Neutral_point_of_view)

8. Meatball Wiki (http://www.usemod.com/cgi-bin/mb.pl) is a meta-wiki dedicated to online communities. Its original goal was to focus on collaborative hypermedia, but current topics range from intellectual property to cyberpunk. Because of its role as "a community about communities," it has become the launching point for various other wiki-based projects.

9. *ThreadMode* (http://www.usemod.com/cgi-bin/mb.pl?ThreadMode)

10. *ThreadMode* (http://www.usemod.com/cgi-bin/mb.pl?ThreadMode)

11. A forest fire is a flame war that spreads uncontrollably into the talk pages of unrelated articles. A forest fire becomes progressively more difficult for any user to keep track of. On Wikipedia, thanks to clear guidelines for article content and a formal dispute-resolution process, this is not a serious problem as on other wikis.

12. By "noise" is meant the quality of lacking any predictable order or plan, whereas "fuzzy" refers to a text that can result in being confused and incoherent.

13. Refactoring (http://en.wikipedia.org/wiki/Refactor)

14. Refactoring talk pages (http://en.wikipedia.org/wiki/Wikipedia:Refactoring_talk_pages)

15. GNU, which stands for Gnu's Not Unix, is an organization founded by Richard Stallman in 1983 with the goal of creating a free operating system, meaning that users are allowed to copy, modify, and redistribute it. The GNU project is now carried out under the patronage of the Free Software Foundation (FSF)

16. Mother Teresa (http://en.wikipedia.org/wiki/Talk:Mother_Teresa)

17. Edit war (http://en.wikipedia.org/wiki/Wikipedia:Edit_wars)

18. The function of the Arbitration Committee is to solve the most serious disputes and the cases of rule-breaking. It is the last step in the dispute resolution process when all else has failed. According to the Arbitration Committee, discussion between users and mediation should be attempted first.

19. The term *folksonomy* is generally attributed to Thomas Vander Wal. It specifically refers to the tagging systems created within Internet communities. It is a combination of the words folks and taxonomy. The term literally means *people's classification management.*

20. ICQ is an instant messaging computer program, which was developed by the company Mirabilis. The first version of the program was released in November 1996 and it soon became the first Internet-wide instant messaging service. The name ICQ is an initialism of "I seek you."

21. WELL (http://www.well.com) stands for Whole Earth Electronic Link. It is a community that was founded in 1985. It had more than 250 groups by mid-2000, referred to as conferences.

22. *Netiquette* is a neologism, it is the contraction of network etiquette. It expresses the conventions of politeness recognized on Usenet, in mailing lists, and other electronic forums. The rules of netiquette are slightly different for different online media.

23. A wikilogism is a form of neologism originating on a wiki project page (http://en.wikipedia.org/wiki/Wikipedia:Wikilogism).

24. Acronyms and initialisms are abbreviated words. There are different interpretations of the two terms. Acronym is a pronounceable word (scuba, Nato radar), usually written in lower case letters and governed by phonological rules, they tend to have a vowel in the middle of consonant clusters. On the other hand, initialisms are free from constraints, are usually written in capital letters, sometimes with a full stop between letters and each individual letter is pronounced (XML,TLA, BBC).

25. Wotta (http://en.wikipedia.org/wiki/WP:WOTTA)

26. The use of the shortening admin for administrator already occurred in office and computer tech jargon

27. Klodzko (http://en.wikipedia.org/wiki/Klodzko)

28. Personal information has been inferred by personal user pages.

29. Citizendium Project (http://citizendium.org)

References

Algeo, J. (1999). *Cambridge history of English language.* Cambridge: Cambridge University Press.

Berners-Lee, T. (1999). *Weaving the Web. The original design and ultimate destiny of the World Wide Web.* San Francisco: HarperCollins.

Biber, D. (1988). *Variation across speech and writing.* Cambridge: Cambridge University Press.

Bizzell, P. (1992). *Academic discourse and critical consciousness.* Pittsburgh: University of Pittsburgh Press.

Bolter, J. D. (1991). *Writing space: Computers, hypertext, and the remediation of print.* Hillsdale, NJ: Erlbaum.

Braga, D. B., & Busnardo, J. (2004). Digital literacy for autonomous learning: Designer problems and learning choices. In I. Snyder & C. Beavis (Eds.), *Doing literacy online: Teaching. learning, and playing in an electronic world* (pp. 45–68). Cresskill, NJ: Hampton Press.

Bush V. (1945). As we may think. *The Atlantic Monthly, 176*(1), 303–308.

Cortazar, Julio. (1963). *Rayuela* [Hopscotch]. Buenos Aires: Editorial Sudamericana.

Crystal, D. (2001). *Language and the Internet.* Cambridge: Cambridge University Press.

December, J. (1993). *Characteristics of oral culture in discourse on the net.* Proceedings of the 12th annual Penn State Conference on Rhetoric and Composition, University Park, Pennsylvania.

Elia, A. (2006). *An analysis of Wikipedia digital writing.* In proceedings of 11th EACL Conference 2006 Trento. Conference of the Association for Computational Linguistics.

Elia, A. (in press). Current trends on the web: Wikipedia and online encyclopedias. In *Rivista di Anglistica*, n. 9. University of Naples 'L'Orientale.'

Graddol, D. (2004). The future of language. *Science, 303*, 1329–1331.

Gregory, M. (1967). Aspects of varieties differentiation. *Journal of Linguistics, 3*, 177–197.

Halliday, M. A. (1985). *Spoken and written language.* Oxford: Oxford University Press.

Lamb, B. (2004). Wide open spaces: Wikis, ready or not. *EDUCAUSE Review, 39*(5), 36–48.

Landow, G. P. (1997). *Hypertext 2.0: The convergence of contemporary critical theory and technology.* Baltimore & London: The John Hopkins University Press.

Leuf, B. & Cunningham, W. (2001). *The Wiki way: Quick collaboration on the web.* New York: Addison-Wesley.

Mattison, D. (2003). *QuickiWiki, SWiki, TWiki, ZWiki and the plone wars: Wiki as a PIM and collaborative content tool* (Vol. 11). Retrieved June 9, 2006, from http://www.infotoday.com/searcher/apr03/mattison.shtml.

Meatball Wiki Web site. http://www.usemod.com/cgi-bin/mb.pl. Retrieved May 18, 2006.

Morgan M. C. (2006). *Blogs and wikis.* Retrieved May 23, 2006, from http://biro.bemidjistate.edu/~morgan/wiki/wiki.ph.

Nelson, T. (1981). *Literary machines.* Available from author. P.O. Box 128, Swarthmore, PA 19081.

Ong, W. (1977). *Interfaces of the word: Studies in the evolution of consciousness and culture.* Ithaca, NY: Cornell University Press.

Portland Pattern Repository Web site. http://c2.com/ppr/. Retrieved May 25, 2006.

Queneau, R. (1961). Cent mille milliards de poèms [Hundred thousand billion poems]. Paris: Gallimard.

Raymond, E. S. (1999). *The cathedral & the bazaar.* Cambridge, MA: O'Reilly.

Rezabeck, L., & Cochenoer, J. (1994). Emoticons: Visual cues for computer-mediated communication. In *Imagery and visual literacy: Selected readings from the Annual Conference of the International Visual Literacy Association.* October 12–16, Tempe, Arizona. Edited by D. G. Beauchamp, R. A. Braden, & R. E. Griffin. Blacksburg, VA: IVLA at Virginia Tech University.

Schiff, S. (2006, July 31) Know it all: Can Wikipedia conquer expertise? *The New Yorker*, pp. 36–43.

Shortis, T. (2001). *The language of ICT.* London: Routledge

Stewart, G. (2003). *Becoming involved: Dialogism and materiality in the development of E-rhetorics.* Retrieved May 29, 2006, from http://www.gavinstewart.net/.

Swales, J. M. (1990). *Genre analysis: English in academic and research settings.* Cambridge: Cambridge University Press.

Tannen, D. (1989). Talking voices: Repetition, dialogue, and imagery in conversational discourse. In *Studies in Interactional Sociolinguistics* (Vol. 6). Cambridge: Cambridge University Press.

Wikipedia Web site. Retrieved June 16, 2005, from http://www.wikipedia.org.

Wyss, E. L. (2000). *Iconicity in the digital world,* Amsterdam & Philadelphia: John Benjamins.

Chapter 14

The Evolution of Web-Site Genres

EVA-MARIA JAKOBS

RWTH Aachen University

Abstract

What happens when new media begin to dominate our communicative practices? Where do their content and forms come from? How can we describe these? This chapter discusses these and other questions with reference to Web-site genres. After a brief review of hypertext and Web-site history, a model for the analysis and description of Web-site genres is presented. This model makes a distinction between Web-site patterns and Web-site genres. Web-site patterns are viewed as a relationship between a problem to be solved, the social purpose of solution, the situational context, and the participants (action type) on the one hand, and the problem solution means (Web-site genre). The Web-site genre is described by prototypical properties such as thematic hierarchies; patterns of structuring, formulation, and visualization; verbal acts; extension; and interactive elements. The discussion also touches on the role of culture and looks at trends such as scenario-based Web-site genres.

The penetration of new media into our everyday lives evokes the question as to what precisely is new about them. As the history of new media shows, a qualitative leap in the use of written text in electronic media only emerges when media-specific forms of presentation are developed, allowing them to be differentiated from traditional forms. This is true of hypertext as a nonlinear form of presentation, which—as is the claim of this chapter—demands both the production and reception of verbal phrases and statements to be adapted to it.

Given the extensive research on hypertext, it is not possible to even begin to provide a comprehensive review. This chapter should be understood as one approach among many, one that has been influenced by the discussion in the German literature on the topic. This approach views hypertext as a principle of presentation that over time has emerged from the experimental stage and has led to a broad range of conventional pragmatic patterns, most of which have not been the focus of linguistic research. On the other hand, these patterns represent a considerable proportion of everyday communicative strategies. We use them in dealing with the Internet: in looking for information, doing one's banking, ordering books from Amazon, looking up items in Wikipedia, reading online newspapers, booking vacations or putting resources for students online.

Text linguistics is only now beginning to take notice of these forms of communicative practice. There have been few attempts to classify hypertext patterns or provide models for their systematic description. The same is true for users: The knowledge that does exist of hypertext characteristics is vague, unsystematic, and often rudimentary. As is often the case, the use of media precedes its theoretical and practical reflection.

This chapter seeks first to examine the entities hypertext, Web site and Web-site pattern with respect to their historical emergence. Afterwards it addresses the question of describing Web-site patterns.

Hypertext

The Term *Hypertext*

At present the term *hypertext* refers both in theory and in practice to very different phenomena. In the most general sense, hypertext is simply a specific form of presentation characterized by the distribution of content to nodes or modules electronically connected via links. The result is a network (see Fig. 14.1).

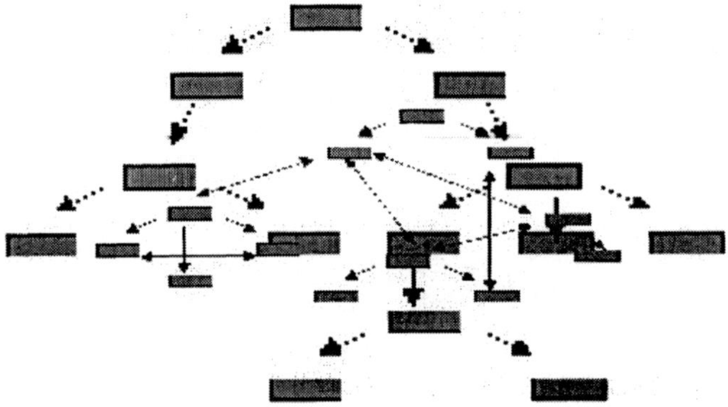

Figure 14.1. Hypertext-like structure.

The size and content of the network can vary considerably, giving rise to one of the many problems associated with defining hypertext. What exactly does the term *hypertext* refer to? How big or small is a hypertext? Hypertextually organized entities can range from extremely large networks (the World Wide Web) to far-reaching networks such as an interactive grammar system to very small entities like the restaurant menu represented at a company's intranet. The content of a node can also be presented in a variety of ways (as text, graphics, animation, video, audio). The choice of content is dependent on the author's intentions—hypertext nodes can, for example, consist solely of technical drawings.

From a linguistic perspective, the interesting hypertext uses are those with a clearly recognizable communicative function. Hypertext will, therefore, be defined as an on- or offline accessible, functionally determined subnet of modules with a certain communicative purpose, organized by an overarching thematic idea (see Jakobs, 2003; Storrer, 2000). The two categories of function and theme provide a contextual framework for understanding parts of the network such as individual modules. This definition excludes collections of electronic artifacts like the World Wide Web. One of the most familiar examples of online accessible hypertext uses is the Web site.

The History of Hypertext and Web Sites

Like all forms of representation, hypertext has a history: The principle of shaping content in modular form and connecting it by way of cross-references has its roots in ideas and solutions that are relatively old. In print, the idea of a nonlinear linking of content is realized by footnotes and references to other texts; solutions like these can be traced at least as far back as the Middle Ages. However, the implementation of this idea as a holistic principle of presentation is relatively recent, its practical realization becoming possible only through the development of new technologies for the production, storage, distribution, and reception of information.

Vannevar Bush is often considered to be the originator of this modern form. A professor at MIT at the age of 29 and later advisor to President Roosevelt, Bush was concerned with the problem of dealing with vast quantities of information. His starting point is what we now call "information overload." In 1939, he recognized that there would soon be more printed information than any person would be able to manage. For Bush the problem was less the quantity of information than the lack or inadequacy of organizational and accessibility mechanisms. He thus developed the idea of a system that allowed users to have access to information directly from their desktop. In 1945 he described his ideas in the famous article "As We May Think" in the magazine *Atlantic Monthly*. The system called Memex (memory extender) is often seen as the forerunner of today's hypertext but was itself never built. A camera attached to the user's forehead was to record everything that came within the user's field of vision (handwritten notes, printed texts, pictures, etc.). This data was then to be stored in microfilm (microfiche) and made accessible via a kind of optical screen. The ordering of the data was to be provided by an associative indexing that permitted the user to mimic human memory by laying down associative trails through the data that could be forwarded to other users. This access was to be individualized: Users were to have the possibility of altering texts, inserting commentaries, or producing links between documents.

Bush's visionary idea was not technically feasible at the time. This became possible only with a new generation of technology. One of the most important technical pioneers is Douglas C. Engelbart, who between 1962 and 1975 developed the AUGMENT (later called NLS) system at the Stanford Research Institute. One fundamental feature of AUGMENT was that users could build links among documents and share each other's links. The hypertext-like system supported cooperative work in an electronic environment. The system was intended to allow distributed publishing and editing, electronic document delivery, and teleconferencing, among other features. AUGMENT used technical tools now standard on the personal computer such as the mouse, multiple windows, and a graphic interface.

Ted Nelson is considered one of the most important hypertext pioneers. In 1965 he coined the terms *hypertext* and *hypermedia*. Nelson sought to develop the idea of a universal knowledge-management and information-production system, which he called XANADU. XANADU was based on the idea of replacing the physical presence of information at the workplace by logical links to spatially distributed forms of memory storage accessible to the user when needed. This information could be verbal, visual, or auditory. The aim was to transform the entire conventional publication system into a global hypertext network that makes all published information available to everyone and enable users to freely recombine any and all documents and add their own content.

The first functioning hypertext system with links originated in the late 1960s and was named HES (Hypertext Editing System). In 1978, A. Lippmen developed the first hypermedia system, ASPEN-MOVIE-MAP, which simulated a trip through the city of Aspen. The development of the program "Guide" in 1986 and later "Hypercard" enabled the transition from research to practical applications. Widespread use was made possible by increasingly easier-to-use software and possibilities afforded by the Internet.

Most hypertexts produced in the 1990s resulted from text conversion (Storrer, 1999). Only a few hypertexts were designed and realized as hypertexts *per se*. In general, this phase can be seen as a period of upheaval: On the one hand some new forms were being derived from older, already existing forms; on the other hand it was a time of experimentation with new solutions (style, graphics, etc.). Gradually new forms of presentation were developed that exploited the potential of electronic media for communicative purposes. One feature of this development is the extensive use of the new forms of presentation, which among other things, led to an explosive growth in the number of Web sites. The desire to be part of the process appears to have been more important than the quality of the product. A good example for this period in media history is the development of the online newspaper (see Jakobs, 1998).

The beginning of the 21st century witnessed a consolidation phase. Norms of presentation became standardized among the user community and a wide range of pragmatic patterns became available both online and offline. Well-defined Web genres have emerged; we see predictable characteristics in Web portals, university Web sites, the Web sites of political parties, and so on. A feature of this period of genre development is an increased interest in quality and quality assurance procedures replacing the interest in experimenting with forms (Jakobs, 2003).

Web Sites and Web-Site Patterns

"The" Web Site

The term *Web site* designates a communication form. It is a collective term for a broad range of hypertext implementations produced according to various pragmatic patterns. According to this definition there is no such thing as "the" Web site, rather there is a variety of communicative artifacts that may be analyzed and assessed as implementations (or tokens) of Web site patterns (or types). Examples of Web site types are online municipal sites, interactive job exchanges and auction houses, personal web pages (Schütz et al., 2005), and online insurance sites. The confusing assortment of expressions used to describe such types testifies to a gradual process of differentiation—often there are no generally established terms for individual pragmatic patterns. Despite this, Web sites continue to address a broad spectrum of tasks, interests, and needs in private, public, and professional contexts.

Many Web sites and hypertexts are in turn composed of "building blocks" or subunits (Jakobs & Lehnen, 2005). In this chapter at least two types of subunits or module types are distinguished: functional and topic-related

FUNCTIONAL MODULES. There are three types of functional modules.

- *System-related modules* have the function of supporting the user's orientation and navigation in the Web site as a whole (e.g., units like home page, site overview, search engine. See Fig. 14.2).
- *Task-related modules* allow the user to complete noverbal actions (e.g., to transfer money or to download a text. See Fig. 14.3).
- *Interaction-related modules* enable verbal interaction with other persons (e.g., via contact page, e-mail, guest book. See Fig. 14.4).

TOPIC-RELATED MODULES. These modules present the content or topics of the hypertext or Web site in question (see Fig. 14.5).

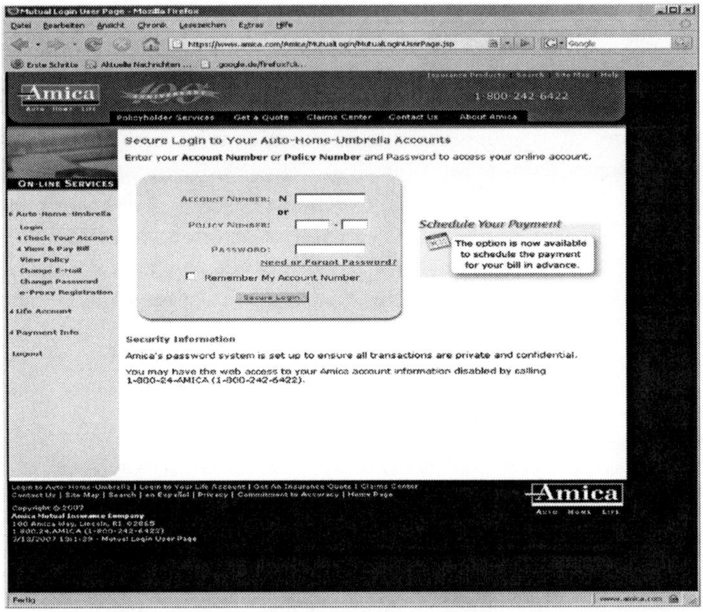

Figure 14.2. Examples for a system-related module (LogIn) (online insurance site, http://www.amica.com/Amica/MutualLogin/MutualLoginUserPage.jsp).

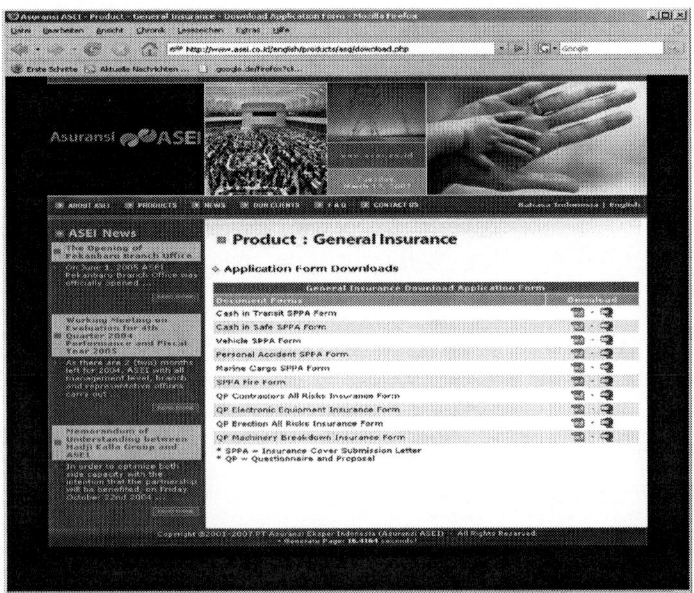

Figure 14.3. Task-related module (main function: download) (online insurance site, http://www.asei.co.id/english/products/asg/download.php).

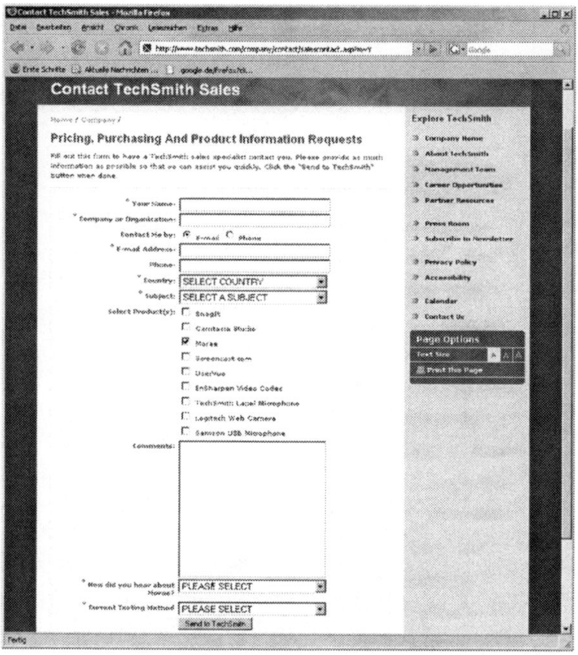

Figure 14.4. Interaction-related module with guestbook (college Web site, http://www.techsmith.com/company/contact/salescontact.asp).

Figure 14.5. Topic-related module (topic: drive) (company Web site, http://www.porsche.com/uk/aboutporsche/porschephilosophy/whatwedo/sportscar/).

Some of these subunits represent integrated tokens of e-paper genres, such as company reports that can be downloaded as a whole, whereas other subunits represent integrated tokens of hypertext genres (e.g., an interactive job exchange as part of an online newspaper). The overarching structure serves as a contextualising and organising framework, as may be seen in the case of an interactive marketplace.

A Web-site pattern such as an interactive marketplace is scenario-based. Its production and reception presupposes what in psychology is described as "scenario knowledge." We expect that certain areas of our lives, like a visit to a restaurant (generally), proceed according to a kind of script. Our knowledge of scenarios like buying and bidding comprises the knowledge of social situations and their components, knowledge of the expected participants, their roles, forms of behaviour, social relationships to one another, and typical actions (see Bucher & Jäckel, 2002). Scenario-based Web-site genres draw on the participant's verbal knowledge of such scenarios, which, by providing patterns and metaphors for action (e.g., the shopping basket metaphor) thus assist users in negotiating (or navigating) their way through the Web site. Without this kind of knowledge certain hypertext implementations like that of Ebay, a virtual auction house, would not function at all.

Another interesting aspect of pattern description is the distinction between closed and open hypertexts. Closed hypertexts, such as an integrated interactive help system, contain a fixed number of modules and are rather static, whereas open hypertexts are "texts in motion" (Storrer, 2000), open to modification and alteration. This characteristic is worth noting because it is precisely the lack of relative completeness that permits the distinction between text and hypertext.

Another typical feature of hypertext and Web-site entities is multimodality. This is also a feature of print but in the case of hypertext the possibilities of the combination and interplay of codes (animation, film, sound, etc.) are much more numerous and complex.

Website Patterns

How can we describe the pragmatic patterns that guide the production and reception of hypertext implementations? Do we require new analytic models or can we modify and adapt existing ones? In the following the latter course has been chosen, utilizing a model (Jakobs, 2003) based on a functional-pragmatic approach by Sandig (1997). According to Sandig, text patterns are part of our communicative knowledge. As members of a discourse community we know that we can solve certain problems under certain circumstances with communicative means, in this case with a text genre. The systematic relationship between a standard problem (e.g., how to cook a meal) and its contextual embodiment (I'm at home in Aachen) on the one hand (action type) and the problem-solving means (to consult a cookbook with recipes) on the other hand is called a text pattern. A text pattern is thus understood as a cognitive schema that delivers pragmatic knowledge for the production and reception of communicative artifacts. This approach assumes texts to be situation-bound cultural artifacts that perform specific tasks, whose success depends to a great extent on their ability to prove their worth in particular situations.

The action type encompasses the social function or purpose of the pattern (e.g., to give cooking instructions), the situational context, the choice of presentational medium (e.g., a book) and code (e.g., written language), as well as the participants (e.g., author and reader). The problem-solving means text genre (e.g., recipe) has prototypical attributes, some obligatory, some optional, of varying importance and realized to varying degrees. These attributes can be described as a typical verbal act hierarchy (with an embedded topic structure), patterns of sequencing, formulating and visualization, as well as a typical extension. They include communicative options and alternatives, to be employed according to specific goals and in specific contexts.

In order to describe Web-site patterns, the fundamentals of Sandig's approach can be adapted with some modifications. The modifications are less concerned with the categories of acting type than with the modeling of the problem-solution means Web-site genre (see Table 14.1). They deal with the kind and number of descriptive categories as well as the distinction of different descriptive levels. These modifications can here be only briefly examined using the example of an online insurance site.

Table 14.1. Categories of Web-site Pattern

Action Type	Problem Solution Means Web-Site Genre With Prototypical Features
Social purpose	Topic patterns
Situational context • Domain • Culture • Period	Structuring patterns
Media	Action repertoire
Choice of medium and code	Formulation patterns
Participants • Roles • Relationship	Visualization patterns
	Extension Interactive elements

Adapted from Sandig (1997)

Action Type

Purpose: Functional-pragmatic approaches of text genre classification mostly focus on the category of purpose. Web-site genres are created for a vast range of specific purposes. Farkas and Farkas (2002) grouped these purposes into a number of broad categories:

1. education,
2. entertainment,
3. providing news, public information, and specialized information,
4. E-commerce: promotion/selling/support,
5. Web portals,
6. persuasion,
7. building and sustaining community, and
8. personal and artistic expression.

Especially in the case of overlapping functions (infotainment, edutainment) hypertexts and Web sites are difficult to categorize. Another approach is described by Storrer (2000), who said the combination of two categories—purpose and institution—allows us to distinguish hypertexts of different complexity as communicative entities.

Sandig (1997) assumed verbal patterns to be determined by their social purpose, which comprises both the solution to standard problems (the social meaning of the pattern) and the ways in which the problems are solved (e.g., a Web-site pattern). These are in turn the result of the choice of a particular action means—here, the Web-site pattern. An example may suffice. Germans are said to avoid risks; in contrast to Americans, they seek to counter risk-filled situations in life by taking insurance policies. They may do so either by visiting an insurance broker or company in person or by using online insurance Web sites. The purpose of an online insurance site is to inform the user about different products, to provide advice, to initiate contract procedures, and to support damage claims. The user seeks certainty and cost and time savings; the company seeks to minimize expenses and maximize profits by acquiring new clients. In this way, online insurance Web sites can complement or offer sensible alternatives to standard solutions.

The Situational Context. A second essential component concerns the framing of the problem solution by dimensions like domain, time, and culture. Each domain (e.g., public vs. business) has its own norms, values, and conventions. They guide the producer and their design decisions as well as the expectations and behavior of the recipients. On the other hand, Web-site design is influenced by time-related aspects. Historically, hypertexts represent a recent form of communication that has gradually emerged on the basis of existing patterns and presentation conventions. By examining their development we may observe styles and trends that have changed in the course of time, and hence the criteria for their evaluation. Blinking elements and a flashy presentation dominated the Web sites of the 1990s, whereas recently the trend has been to less colorful, unadorned designs.

The cultural framing of communicative problem solutions is a further crucial aspect, yet has received comparatively little attention. As is often the case, most Web-site research originates in the West and analyses Western hypertexts and Web sites. Closer examination reveals that the cultural context can have an immense influence on Web-site design. Studies by Bucher (2004) show, for example, that in many cases a one-to-one-translation will fail. The reception and evaluation of a communicative artifact is deeply influenced by its underlying philosophy. For example, Western design and reception are oriented to style principles of various schools or movements (Bauhaus' "form follows function," constructivism, the De Stijl movement), principles that had notable impact on newspaper design. In contrast, Chinese Web design criteria are oriented to the "principle of fullness," an integral part of Chinese folk culture.[1] This would explain why Chinese portals seem to be designed according to the maxim "as much as possible all at once" and hence appear to Western eyes as cluttered and confused.

Choice of Medium and Code. The choice of medium determines to a certain extent the choice of code and thus the form of presentation. The choice of Web-site genres as a means of problem solving necessarily entails using the Internet and production and reception media like the personal computer. These in turn enable a broad spectrum of presentation codes and modes, as evidenced by the trend to multimodality.

The Participants: Hypertexts involve a variety of participants with different social roles and intentions. An online insurance site or a virtual department store generally has an "author" or "owner" (a company, an institution) represented by several interest groups (an executive board, marketing and/or public relations departments), and an addressee, which can in turn encompass several subgroups. Beyond the immediate participants, the nature of the media involved determines the need for other participants: The "author" (the insurance company) commissions professionals to produce the insurance Web site, who assume different tasks and roles (as writer, editor, graphic designer, programmer, etc.) in the design process. This plurality of authors often leaves its traces on the final product.

A second potential conflict concerns the relation between the individual motives, interests, and goals of participants on the one hand, and/or the socially agreed upon purpose of the problem solution on the other hand. Whereas customers seek quick answers to their questions (finding information, being provided with options, settling their claims), insurance companies seek to present themselves and their products in the best possible light. The promotional style in which they do so is often not appreciated, particularly on the Internet.

Summary. With respect to Web-site pattern description, the most important action-type category is purpose. It allows us to recognize and delimit Web sites as functionally and thematically determined entities. Other categories like the given situational context (domain, time, and culture) and medium and code contextualize the use of design repertoires in the sense of possibilities and restrictions. How the participants deal with possibilities and restrictions depends on their roles, intentions, and capabilities.

Web-Site Genres as Problem-Solving Means

In this approach, the Web-site genre is conceptualized as a problem-solution means. Web-site genres are collections of content and design patterns with prototypical features. Producers of Web sites know that they have a certain repertoire of and leeway in the design possibilities at their disposal. Web-site users expect typically solutions and orient their behavior and value judgments to familiar patterns. The patterns guide the selection, organization, and structuring of content (topic hierarchies, structuring patterns), its verbalization and visualization, the extension of parts of the system and the system as a whole, as well as the integration and design of interactive elements.

Topic Patterns. The characterization of hypertext as a functionally and thematically determined communicative entity entails placing a relatively strong emphasis on the categories *content* and *topic*. Many Web-site genres have typically more than one topic: They are "poly-thematic." The relationship among the topics is organized by an overarching thematic "roof." Another typical feature of many Web-site genres is the topic hierarchy. The top of a hierarchy is represented by a single module that provides access to the hierarchy. Terms for modules with this function are *home page* and *main page*.

Structuring Patterns. In contrast to Sandig's approach, sequencing patterns in this approach are seen as a subcategory of the category structuring pattern. With respect to hypertext characteristics, two levels of description may be distinguished:

1. the order of nodes as parts of the whole system, and
2. the order of elements on the level of a single node or module.

Each level can be analyzed with respect to tree degrees of extension: (a) global (the structure of the whole entity), (b) middle, (c) local (the position of one element, see Fig. 14.6).

The description and analysis of first-level structures yet has received comparatively little attention. Farkas and Farkas (2002) distinguished four global patterns: linear structure, multipath structure, Web-like structure, and matrix structure. Another approach may be to distinguish mono-sequential, poly-sequential, and nonsequential hypertexts. Nonsequential Web-site genres do without paths. The network structure demands the use of search engines and interactive overviews to permit access to content (e.g., hyperbolic trees). Web-site genres often combine

Level \ Degree of extension	Global	Middle	Local
The order of nodes as part of the Web site	Example: poly-sequential Web site	Example: thematic subhierarchy of nodes	Example: relation between direct linked nodes
The order of content on a node/module	Example: Beginning–Middle–End Pattern	Example: Paragraph Structure	Example: Text-Image Sequence

Fig. 14.6. Structuring patterns—levels and depth of description.

patterns. Thus, an online insurance site may present its products poly-sequentially on the basis of thematic hierarchies and their derived paths.

In the case of sequential or poly-sequential Web-site genres the organizing principles can be:

1. Role-related: Insurances make a difference between "potential" customer and clients with structural consequences. "Potential" customers can only examine the "open" or public space of an online insurance site; clients have the permission to use the content behind the "iron wall" of the LogIn-border (see Fig. 14.2).
2. Topic- and/or task-related: To follow our example of an online insurance site, some topics are treated "publicly" (e.g., product presentations or contract conclusions), whereas other topics are dealt in the LogIn-space (claim settlements).
3. Scenario-based: The content is organized according to well-defined processes as part of a scenario (e.g., following the steps of a contract conclusion).
4. Behavior-related (e.g., individualized order of content in an online newspaper).
5. According to the user's local position (e.g., the order of hotel lists as part of interactive travel guides used by mobile phones with General Packet Radio Services).

Several structuring principles (see 4 and 5) are new and only possible in electronic environments.

For the description and analyses of second-level structures, a broad range of linguistic approaches and categories is found in the literature. There is a lack of research concerning structuring patterns for FAQs, Unified Resource Locations (URLs), form and download pages, or for position-specific elements (e.g., position of the telephone number on a node or module).

ACTION REPERTOIRE. Characteristic features of hypertext-like modularity, navigation via links and other navigation means, and the requirement of an electronic environment demand specific types of verbal and nonverbal acts that must be considered when describing Web-site genres. They extend the already complex inventory of verbal acts associated with text genres. One example may serve to illustrate this.

> *Combination of speech act and (hyper)link function* (See Fig. 14.7): The utterance *The Cricket World Cup is the premier international championship of men's One-day International cricket* is a speech act (a declarative). On the other hand it contains elements (*Cricket World Cup*, *One-day International cricket*) with a double function: As a part of the utterance the element is a meaning-constitutive part of the text; as a marked element it is an embedded link. As a link the element has for its part a double function: (a) it allows the user to navigate

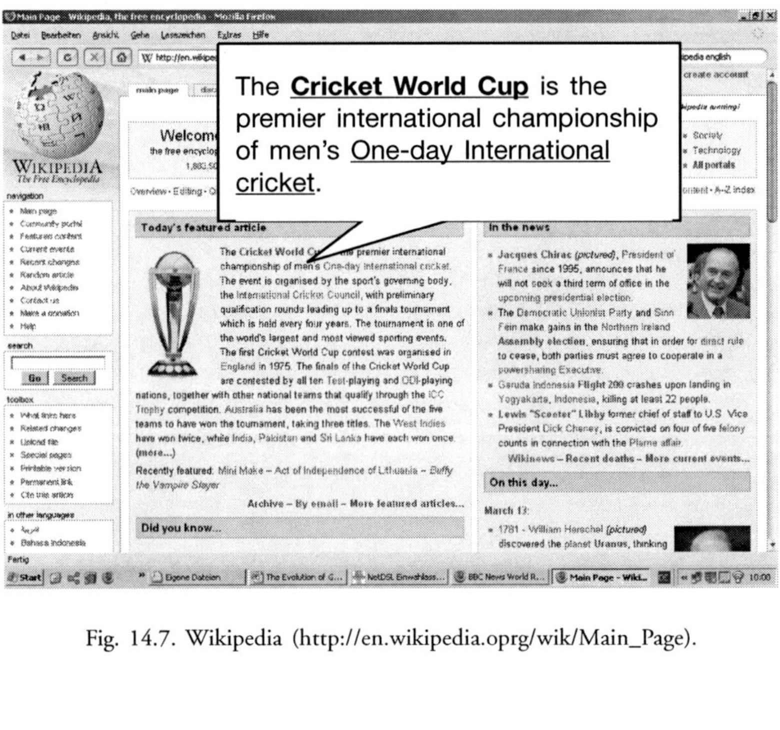

Fig. 14.7. Wikipedia (http://en.wikipedia.oprg/wik/Main_Page).

to another part of the text or module (internal link) or to another module of the same or another hypertext or Web-site system (external link);[2] (b) at the same time it labels the topic or function of the link target or destination (e.g., a module with the topic *Cricket World Cup).*

FORMULATION PATTERNS. Formulation patterns encompass decisions at the level of word selection, grammar, and style. Media-specific characteristics concern hypertext-specific lexis and formulation patterns for Web-site components (e.g., the URL) or for links and other navigation aids (site maps, navigation bars, etc.). The patterns for naming links vary according to the function of the link:

- Topic-related linkage: Naming of the topic (*Cricket World Cup*), term with cataphoric function with or without ellipses (*More . . .*).
- Navigation-related linkage: Imperative plus deictic expession (*Click here.*), naming of a link destination (*Home*) or a direction (*back*). Formulation patterns for links typically combine verbal and visual elements, for example a verbal expression plus an underline, italics or color as highlighting. The combination of Icon (e.g., ) and link label (e.g., *home*) implemented as roll-over elements is another example of double-coding by verbal and visual means.

The formulation level clearly shows the influence of a medial context. The general "scroll ban" imposes brevity and conciseness, the danger of clicking further and of losing the user requires short meaningful titles, the navigation and orientation problem demands precise instructions.

VISUALISATION PATTERNS. A great deal of research is needed in the area of visualization patterns and the diversity of usable code (e.g., for the presentation of the node structure or content). Including moving images, audio material, braille, or sign language requires a wide range of descriptive instruments that in some cases do not exist, as with the interaction of codes (text–image, text–moving image, text–animation, spoken language–text–animation–music, etc.). One solution may be to adapt approaches to text–image relationships in printed texts (e.g., Kress & van Leeuwen, 1996; Sandig, 2000) or film analyses. However, this will only make sense if the associated models and theories of text processing can be transferred to hypertexts and Web sites as well. To some extent, new approaches are needed.

EXTENSION. The category extension is closely connected to the distinction of first- and second-level structures. At the first level—the number and order of nodes—the category is not very useful; theoretically hypertexts can be arbitrarily large. However, at the second level—the content of a single node/module—the average scope or extension of what we can see on screen is a very important question. Features of technological devices (size and quality of the screen, "keyhole" reception) are limiting factors. Users, especially older users, prefer not to scroll—sometimes they do not even recognize that a page is scrollable (see Becker, 2004; Urban, 2007)

Interactive Elements. A constitutive feature of hypertexts is their interactivity, which must be taken into account in an adequate fashion when describing hypertext genres. In this approach there is a clear distinction between the following

- *Interactivity in the sense of the verbal interaction of individuals.* Interactive elements of this type allow the user to communicate with other people (e.g., chat rooms, forums).
- *Interactivity in the sense of computer responses to human input (e.g., links, payment forms, search engines, etc.)* (see Bucher, 2004). Interactive elements possess affinities to specific formulation, structuring, and visualization patterns.

The potential for human interaction and/or human–machine interactivity in Web-site genres varies according to the quantity and quality of the communication means employed.

Summary

This review of the evolution of Web site pattern is intended to contribute to our understanding of the changes affecting communication in our everyday lives. It points to certain inadequacies in current research, such as the description of code combinations (like text–image) and their effects, culture-related phenomena of Web communication and evaluation criteria (Jakobs, 2005; Jakobs & Lehnen, 2006). Research into Web site patterns and genres and their communicative function in discourse communities can give us a deeper insight into current and future communication practises and media developments.

Notes

1. In contrast to the "aesthetic of emptiness" that derives from Zen Buddhism and that is common among upper classes.
2. External links connect hypertext systems and hence generate one of the many forms of intertextuality.

References

Becker, S. A. (2004). A study of web usability for older adults seeking online health resources. *ACM Transactions on Computer-Human Interaction (TOCHI), 11*(4), 387–406.

Bucher, H.-J. (2004). Online-Aktivität—ein hybrider Begriff für eine hybride Kommunikationsform. Begriffliche Klärungen und empirische Rezeptionsbefunde [Online-activity—a hybrid term for a hybrid form of communication. Conceptual clearances and empiric reception results]. In C. Bieber & C. Leggewie (Eds.), *Interaktivität. Ein transdisziplinärer Schlüsselbegriff* [Interactivity: A transdisciplinary key concept] (pp. 132–167). Frankfurt am Main: Campus.

Bucher, H.-J., & Jäckel, M. (2002). E-Business-Plattformen im Usability-Test [E-business-platforms in usability testing]. In G. Roters, O. Turecek, & W. Klingler (Eds.), *Content im Internet. Trends und Perspektiven* [Content in the internet: Trends and prospects] (pp. 67–82). Berlin: Vistas.

Farkas, D. K., & Farkas, J. B. (2002). *Principles of web design*. New York: Longman.

Jakobs, E.-M. (1998). Online-Zeitungen: Potentiale und Prozesse [Online-newspapers: Potentials and processes]. In H. Strohner, L. Sichelschmidt, & M. Hielscher (Eds.), *Medium Sprache* [Medium language] (pp. 91–108). Frankfurt am Main: Lang.

Jakobs, E.-M. (2003). Hypertextsorten [Hypertext genres]. *Zeitschrift für germanistische Linguistik. Themenheft Deutsche Sprache im Internet und in den neuen Medien, 31*, 232–273.

Jakobs, E.-M. (2005). Bewertungsperspektiven auf Websites [Perspectives on website evaluation]. In A. Schütz, S. Habscheid, W. Holly, J. Krems, & G. G. Voß (Eds.), *Neue Medien im Alltag. Befunde aus den Bereichen Arbeit, Lernen und Freizeit* [New media in everyday life: Results from the fields work, learning and recreation] (pp. 71–86). Lengerich: Pabst.

Jakobs, E.-M., & Lehnen, K. (2005). Hypertext—Klassifikation und Evaluation [Hypertext—classification and evaluation]. In T. Siever, P. Schlobinski, & J. Runkehl (Eds.), *Websprache.net. Sprache und Kommunikation im Internet* [Weblanguage.net. Language and communication in the internet] (pp. 159–184). Berlin & New York: de Gruyter.

Jakobs, E.-M., & Lehnen, K. (2006). Linguistische Aspekte der Website-Evaluation. Sprachwissenschaftliche Anforderungen und Verfahren im Nutzertest [Linguistic aspects of website evaluation: Linguistic requirements and methods in the user test]. In J. K. Androutsopoulos, J. Runkehl, P. Schlobinski, & T. Siever (Eds.), *Neuere Entwicklungen in der linguistischen Internetforschung* [New developments in linguistic internet research] (pp. 263–282). Hildesheim: Georg Olms Verlag.

Kress, G., & van Leeuwen, T. (1996). *Reading images. The grammar of visual design*. London & New York: Routledge.

Sandig, B. (1997). Formulieren und Textmuster. Am Beispiel von Wissenschaftstexten [Formulation and text patterns: On the example of scientific texts]. In E.-M. Jakobs & D. Knorr (Eds.), *Schreiben in den Wissenschaften* [Writing in the sciences] (pp. 25–44). Frankfurt am Main: Lang.

Sandig, B. (2000). Text als prototypisches Konzept [Text as a prototypic concept]. In M. Mangasser-Wahl (Ed.), *Prototypentheorie in der Linguistik* [Prototype theory in linguistics] (pp. 93–112). Tübingen: Stauffenburg.

Schütz, A., Marcus, B., Machalik, F., & Renner, K.-H. (2005). Selbstdarstellung im Internet. Die Analyse privater Homepages [Self-projection in the internet: An analysis of private homepages]. In A. Schütz, S. Habscheid, W. Holly, J. Krems, & G. G. Voß (Eds.), *Neue Medien im Alltag. Befunde aus den Bereichen Arbeit, Lernen und Freizeit*

[New media in everyday life: Results from the fields work, learning and recreation] (pp. 257–274). Lengerich: Pabst.

Storrer, A. (1999). Was ist eigentlich eine Homepage? Neue formen der wissensorganisation im World Wide Web [What actually is a homepage? New forms of organizing knowledge in the World Wide Web]. *Sprachreport*, *1*, 2–8.

Storrer, A. (2000). Was ist "hyper" am Hypertext? [What is "hyper" on hypertext?] In W. Kallmeyer (Ed.), *Sprache und neue Medien. Jahrbuch des Instituts für deutsche Sprache* [Language and new media: Annuals of the Institute for German Language] (pp. 222–249). Berlin & New York: de Gruyter.

Urban, I. (2007). *Altersspezifische Webangebote? Kommunikative Usability am Beispiel der Website des Aachener Verkehrsverbund.* Unpublished master's thesis, RWTH Aachen.

About the Contributors

Alexander T. Bergs is professor of English Linguistics at the University of Osnabrück, Germany. His areas of expertise include historical sociolinguistics, Construction Grammar, varieties of English, and 19th century English. (Further information: http://www.anglistik.uni-osnabrueck.de/bergs.)

Antonella Elia is a Ph. D. student in English for Special Purposes at the University of Naples Federico II. Her expertise lies in questions connecting languages, humanities, and computing. (Further information: http://www.dipstat.unina.it/dottorato/dottorandi_schede/elia.htm.)

Jürg Häusermann is professor of media analysis and media production at the Faculty of Modern Languages of the University of Tübingen, Germany. He has trained television, radio, and print journalists and worked as a freelance journalist. (For further information: http://homepages.uni-tuebingen.de/haeusermann/.)

Eva-Maria Jakobs is professor for text linguistics and technical communication at the Institute of Linguistics and Communication Studies. She also leads the Institute for Professional Communication and Business Media (both at RWTH Aachen University, Germany) as director. Her key activities lie in the fields of text linguistics, new media, business and technical communication, and text production. (For further information: http://www.tl.rwth-aachen.de/team/leitung/index.html.)

Rodney H. Jones is associate professor in English and Communication at City University of Hong Kong. His research interests include computer mediated communication, discourse and sexuality, literacy, and multimodal discourse analysis. (For further information: http://personal.cityu.edu.hk/~enrodney/.)

Christian Kohl, M.A. studied computational linguistics, English, and German Linguistics at the University of Trier. He is currently the head of the electronic publishing division of de Gruyter and is also working on his PhD thesis at the University of Koblenz, Landau, titled "Knowledge construction on the Internet."

His research interests are electronic publishing, semantic web and topic maps, and collaborative writing.

Wolf-Andreas Liebert, Ph.D., is professor of German linguistics at the University of Koblenz-Landau, Campus Koblenz. His research interests focus on social discourse, metaphors, and knowledge construction. He is engaged in open content encyclopedias since 2001 (http://erste.oekonux-konferenz.de/dokumentation/texte/liebert.html.).

Martin Luginbühl is Research Fellow at the Linguistic Department of the German Studies Department at the University of Zurich. His publications are in the field of conversation analysis, media linguistics, and text linguistics. (For further information: http://www.ds.unizh.ch/mluginbu//)

Thomas Metten, M.A., is head of the press office of the University of Koblenz-Landau, Campus Koblenz, research associate for the research-project *Ompetencies in Science-Transfer* at the Institute for German Language and Literature and also PhD student. His research interests are the construction of knowledge, new media, aesthetic experience, und the theory of images.

Laurie McNeill teaches in the Department of English at the University of British Columbia, and specializes in auto/biography studies. She is the editor of "Reconsidering Genre," a special issue of *Life Writing*, and has published on subjects including obituaries, memorial responses to trauma, and online life writing.

Sharon Millar is Associate Professor at the Institute of Language and Communication, University of Southern Denmark. Her research interests include normative mechanisms in language, the development and nature of metalinguistic awareness and the discursive construction of ethnic, national and supranational identities. (For further information: http://www1.sdu.dk/Hum/english/people/sharon_millar.shtml.)

Anabel Quan-Haase is Assistant Professor of Information and Media Studies and Sociology at the University of Western Ontario. Her research investigates how information and communication technologies transform organizations, communities, and educational settings. (For further information: http://publish.uwo.ca/~aquanhaa/.)

Charley Rowe is Honorary Assistant Professor in the School of English at the University of Hong Kong. Her research interests include sociolinguistics; computer-mediated communication, particularly e-mail interaction; and dialectology. (For further information: http://www.hku.hk/english/staff/crowe/index.htm.)

Stefania Spina is researcher in Linguistics at the University for Foreigners Perugia, Italy, where she teaches *Linguistic Computing*. Her main research interest is Corpus Linguistics. (For further information: http://culturitalia.uibk.ac.at/s_spina/personale/personale.htm.)

Eva L. Wyss is currently International Visiting Professor for Gender Studies at the Rheinland-Pfalz Universities of Mainz, Trier and Koblenz-Landau. She also serves as COO of the Zürcher Hochschulinstitut für Schulpädagogik und Fachdidaktik (Institute of Education, Zürich). Her research interests include intimate discourse, media linguistics, and written communication. (For further information: www.evawyss.ch.)

Author Index

Subject Index

Lightning Source UK Ltd.
Milton Keynes UK
UKOW051531231111

182567UK00001B/99/P

9 781572 739307